BEING and TIME

SUNY SERIES IN CONTEMPORARY CONTINENTAL PHILOSOPHY

Dennis J. Schmidt, editor

BEING and TIME

MARTIN HEIDEGGER

Translated by
Joan Stambaugh

Revised and with a Foreword by
Dennis J. Schmidt

STATE UNIVERSITY OF NEW YORK PRESS
Albany

Originally published by Max Niemeyer Verlag.
Tübingen, © 1953.

Published by
STATE UNIVERSITY OF NEW YORK PRESS, ALBANY

© 2010 State University of New York

For information, contact
State University of New York Press, Albany, NY
www.sunypress.edu

Production, Laurie Searl
Marketing, Fran Keneston

Library of Congress Cataloging-in-Publication Data

Heidegger, Martin, 1889–1976.
[Sein und Zeit. English]
Being and time : a revised edition of the Stambaugh translation /
Martin Heidegger ; translated by Joan Stambaugh ; revised and
with a foreword by Dennis J. Schmidt.
p. cm. — (SUNY series in contemporary continental philosophy)
Includes bibliographical references and index.
ISBN 978-1-4384-3275-5 (hardcover : alk. paper)
ISBN 978-1-4384-3276-2 (pbk. : alk. paper)
1. Ontology. 2. Space and time. I. Stambaugh, Joan, 1932–
II. Schmidt, Dennis J. III. Title.
B3279.H48S43
2010
111—dc22   2010015994

10 9 8 7 6 5 4 3 2

Dedicated to

EDMUND HUSSERL

in admiration and friendship

Todtnauberg in Baden, Black Forest
8 April 1926

CONTENTS

PART ONE

The Interpretation of Dasein in Terms of
Temporality and the Explication of Time as the
Transcendental Horizon of the Question of Being

DIVISION ONE

The Preparatory Fundamental Analysis of Dasein

Chapter One

The Exposition of the Task of a Preparatory Analysis of Dasein

Chapter Two

Being-in-the-World in General as the
Fundamental Constitution of Dasein

Chapter Three

The Worldliness of the World

Chapter Four

Being-in-the-World as Being-with and Being a Self: The "They"

Chapter Five

Being-in as Such

Chapter Six

Care as the Being of Dasein

DIVISION TWO

Dasein and Temporality

Chapter Five

Temporality and Historicity

Chapter Six

*Temporality and Within-Timeness as the
Origin of the Vulgar Concept of Time*

FOREWORD

Published in the Spring of 1927, *Being and Time* was immediately recognized as an original and groundbreaking philosophical work. Reviewers compared it to an "electric shock" and a "lightning strike," and there was praise for the "philosophical brilliance" and "genius" of its young author, Martin Heidegger (he was only thirty-seven years old). *Being and Time*, and Heidegger, quickly became the focus of debates and controversy, as well as an inspiration for new impulses in thinking. Indeed, the publication of *Being and Time* was an intellectual event of such consequence that it seems right to describe it with a comment Goethe made in another context: "from here and today a new epoch of world history sets forth."

Prior to the publication of *Being and Time* Heidegger had achieved some fame on the basis of his lecture courses at the University of Marburg. The courses were challenging and stimulating, and it is no accident that many of his students during these years would become original thinkers in their own right. Hannah Arendt later spoke of "the rumor of a hidden king" circulating among university students in Germany. Hans-Georg Gadamer described Heidegger's classes as "an elemental event" in which "the boldness and radicality of [Heidegger's] questioning took one's breath away." Among students at least, it was clear that during his years at Marburg (1923–1928) Heidegger was laying the groundwork for a genuine philosophical revolution. But until the publication of *Being and Time*, that revolution remained only a rumor, since Heidegger had not published anything for a decade.

Despite his renown as a teacher, this absence of publications placed enormous pressures upon Heidegger to finally bring into print the ideas that he had been developing in his lecture courses. The reason for this is clear: in 1925 Heidegger was passed over by the Ministry

of Education for a promotion to Paul Natorp's chair in philosophy
at Marburg due to his lack of publications. Faced with this rejection,
Heidegger worked intensely over the next year to finish the project
of *Being and Time* until he was finally able to present "a virtually
complete manuscript" to his teacher, Edmund Husserl, on Husserl's
sixty-seventh birthday (April 8, 1926). Later that same year, Heidegger
was again nominated for Natorp's chair, and this time he submitted
the galley proofs of *Being and Time* in support of his nomination. The
Ministry of Education's response was to reject Heidegger's nomination
yet again, returning the proofs of *Being and Time* with the comment
"insufficient." After *Being and Time* was published a few months later,
its reception made it abundantly clear that this genuinely new philosophical voice and viewpoint was destined to have a profound impact
upon philosophy, and that Heidegger had indeed opened up a new
path for thinking. In 1928 Husserl retired from teaching at the University of Freiburg and Heidegger was offered Husserl's chair; although
Heidegger eventually received an offer from Marburg, he accepted the
chance to move to Freiburg as Husserl's successor instead.

One consequence of this pressure to publish was that Heidegger
decided to publish *Being and Time* in installments rather than wait until
the entire text was finished as he had outlined it. The conception of
the fundamental ontological project undertaken in *Being and Time* was,
however, fully articulated from the outset: in §8 of the Introduction,
Heidegger outlines the plan of *Being and Time* as divided into two
parts, each with three divisions. The first installment of the text that
Heidegger published in 1927 consisted only of the first two divisions
of the first part; two-thirds of the planned text—the last division of the
first part and the entire second part of the projected text—were still to
be written. Initially, Heidegger planned on completing the project of
Being and Time as he had originally outlined it, but by 1929 or 1930, he
had abandoned that plan. The text that we now have, and that stands
as the complete text of *Being and Time*, is thus the "incomplete" version
that was published in the spring of 1927. Heidegger announces this in
his "Author's Preface to the Seventh German Edition" (1953) when he
writes: "The designation 'First Half,' which previous editions bore, has
been deleted. After a quarter century, the second half could no longer
be added without the first being presented anew. Nonetheless, its path
still remains a necessary one even today if the question of being is to
move our Dasein" (H xxvii).

What this means for one who sets out to read *Being and Time*
is that this book needs to be understood as a torso of its own intentions and the fragment of a larger project. This poses a special interpretive problem. Since *Being and Time* continually rewrites itself, that
is, since it repeatedly revisits earlier analyses in the light of their

own conclusions, one needs to read it with this incompletion always in view: what was projected, even if never completed, was to have been the basis for a reexamination of the text that we do have. Although significant portions of its "missing" sections were eventually written and published elsewhere (the 1927 lecture course, *The Basic Problems of Phenomenology*, is especially significant in this regard), Heidegger never attempted to "complete" *Being and Time*, and he made no clear effort to link the portions that were published elsewhere to the plan of *Being and Time*. When one bears in mind that *Being and Time* remains a fragment, that Heidegger never explicitly continued to work out the plan of that project, and that he never systematically incorporated the work he did there into his later work, one is left with a genuine question: is the project of fundamental ontology laid out in *Being and Time* able to be carried out to completion? Or is it the case that this project cannot be completed according to the conditions that it sets for itself? Heidegger would ask this same question—even if only tacitly—for many years.

Heidegger's most extensive, significant, and explicit self-assessment of *Being and Time* is perhaps found in his *Contributions to Philosophy (of the Event)*. There, as elsewhere, Heidegger's view was that the project of fundamental ontology as it was outlined in *Being and Time* could not, in the end, be completed. Indeed, it is striking that much of the vocabulary and structures that he so painstakingly developed in *Being and Time* would disappear from his work over the next decade. Despite this, Heidegger was both emphatic and unwavering in his insistence that working through the project of *Being and Time*, while coming to understand its insights and its limits, was a decisive and ineluctable step on the path that thinking needs to take in our times. Thus, while Heidegger never resumed this project, he never repudiated it either. On the contrary, he frequently returned to it as he reexamined and rethought his own path of thinking. Despite—or even perhaps because of—the aborted character of his aims in *Being and Time*, this book was a constant touchstone for Heidegger's own path of thinking and it would become a landmark in the history of philosophy that continues to challenge us today.

Being and Time begins by referring to a doubled forgetfulness characterizing our historical present: we have not only forgotten the question of being [Sein] and how to ask it, but this forgetting itself is no longer noticed, it too has been forgotten. This forgetting is not an "error" or a "mistake" that we can simply correct: one does not forget the question of being the way one forgets one's keys, nor is the question of being a question that one can ask in the form that most questions take. Rather, the sources of such a question, as well as the way it is forgotten, are, as Heidegger's analysis shows, rooted in our way of being. Factical life is necessarily bound up in a movement that conceals as well as reveals. The same structures that disclose factical life shroud it in obscurity. As Heidegger put it elsewhere, "Das Leben ist diesig, es nebelt sich immer wieder ein" ["Life is immanent and hazy, it always and again encloses itself in a fog"]. Although our way of being always involves us in the fog of forgetfulness which, for the most part, we do not even notice, our way of being equally involves us in opening up and clearing the question of being. Asking the question of what it means "to be"—and "not to be"—defines and distinguishes each of us is and determines the way we live our lives. Our way of being in the world is shaped fundamentally, and in every respect, by the fact that each of us is always preoccupied with this question, even if it is not thematized as such and even when it is forgotten. Yet, precisely because we have forgotten how to ask this question and because we have become numb to what moves us to pose the question of being, the first and largest task of *Being and Time* is to recover and renew it. In order to do this, Heidegger undertakes a phenomenological analysis of that being [Seiende] who asks such a question, the being who is concerned about its own being [Sein], namely, Dasein. To this end, the published portion of the original plan of *Being and Time* contains the "fundamental preparatory analysis of Dasein" (Part One) and the repetition of this analysis in the light of temporality which the existential analytic had disclosed as the primordial meaning of Dasein's way of being (Part Two). The move from this analysis of the being of the questioner to the question of being itself was ultimately never carried out.

No brief summary of either this analysis or its repetition can ever be adequate to the range and subtlety of *Being and Time*. The sweep of Heidegger's analysis here is stunning: truth, death, anxiety, praxis, others, tools, language, mood, guilt, history, existence, and time are just some of the many topics addressed here. Likewise, the innovative way in which such topics are unfolded and discussed becomes apparent in the language with which Heidegger carries out this analysis. Starting with the word "Dasein"—which is not another name for "consciousness," "subject," or "human being"—Heidegger introduces a vocabulary that challenges the reader. We find neologisms alongside commonplace words that are used in quite uncommon senses. We read, for instance, of "being-in-the-world," "being-with," "worldliness," "thrownness," "attunement," "temporality," "conscience," "facticity," "everydayness," "equipment," "anxiety," "authenticity," "care," "objective presence," "equiprimordiality," and "taking care"—to name just a few of the words that need to be listened to carefully. The language of this text frequently needs to be heard in a different register than the one to which one might be accustomed: meanings are stretched, roots of words are highlighted, small and seemingly insignificant words—"in," for instance—are shown to be complex and in need of careful reflection. Heidegger argued that "it is the business of philosophy to preserve the power of the most elemental words" (H 220). Reading *Being and Time*, one experiences this elemental power of language.

Of course, all of this makes the task of one who translates Heidegger especially difficult. Piled atop the usual struggles of translation, there are special difficulties facing the translator of *Being and Time*: Heidegger's suspicion about the inherited language of philosophizing and his neologisms, which are intended to force the reader to think from a fresh perspective, make the burden of the translator especially challenging. Joan Stambaugh faced these challenges with the insight born of her long connection to Heidegger's work. Thanks to Stambaugh, we have a translation in which one hears something of Heidegger's own voice. Nonetheless, all translation operates in a realm defined by nuance, options, multiple legitimate possibilities, apparent impossibilities, and resonances that complicate every otherwise fine choice. Translation is interpretation. It is also a sort of treason no matter how deep its fidelity. While I have taken advantage of the last fifteen years of commentary on this translation to make modifications to Stambaugh's original translation, there should be no doubt: this is still the Stambaugh translation of *Being and Time*. It is not possible to give an indication of all of the changes made in this revised edition; however, some general remarks might be helpful.

One difference in this new edition is that the German word being translated is frequently identified by being inserted in square brackets. Doing this solves a number of problems such as those arising from the difficulty in distinguishing "being" ["Sein"] from "beings" ["Seiende"]. Whenever there was any chance of confusion about the translation of these words, the German words were inserted. Likewise, it is not uncommon for Heidegger to make connections between ideas by having words echo one another. Typically, these echoes cannot be maintained in an English translation. Thus repetitions in German words like "environment" or "surrounding world" ["**Um**welt"], "dealings" or "association" ["**Um**gang"], "circumspection" ["**Um**sicht"]; "considerateness" ["Rück**sicht**"], "tolerance" ["Nach**sicht**"], "sight" ["**Sicht**"]; "nullity" ["**Nicht**igkeit"], "nothingness" ["**Nicht**s"], "the not" ["**Nicht**heit"] cannot be repeated in English, but the links that unite these, and many other words, can at least be made visible even to the reader who does not read German.

Greek characters are now used for the Greek words. The "Lexicon of Greek Expressions" is available at the end of the book for those who might find a transliteration and translation of many of those words helpful. Heidegger's notes, which were previously published as endnotes, are now footnotes. The marginal comments from Heidegger's own readings of *Being and Time* are included at the bottom of the page (indicated by symbols) along with the footnotes (indicated by numbers). The numbers in the margins indicate the page number that has been used in the German text since the seventh edition.

Most of Stambaugh's translations of key words remain as before. There is, however, one change that needs to be explained and defended, namely, the decision that Dasein would no longer be written as Da-sein. The reason this change needs some comment is simple: as Professor Stambaugh writes in her Translator's Preface "it was Heidegger's express wish that in future translations the word *Da-sein* should be hyphenated throughout *Being and Time*." Among the many considerations that have led me to no longer write Dasein with a hyphen I will note only the following. First, Heidegger expressed his wish in this matter prior to the publication of the final version of *Sein und Zeit* in his *Gesamtausgabe*, so he could have made the same change in the German edition. He did not make a change, but continued to write Dasein without a hyphen. This is a translation of *Sein und Zeit*, not the place for corrections to be made that would move the translation away from the German original. Second, there are a number of places in *Being and Time* where Heidegger does write Da-sein with a hyphen. To write Dasein with a hyphen in every case covers over a distinction that Heidegger himself makes and finds significant. Third, in later texts, most notably his *Beiträge*, Heidegger will write Da-sein with a hyphen, reverting to the non-hyphenated form mostly only to speak about *Being and Time*. There again Heidegger is, in some real sense, distinguishing Da-sein from Dasein. Finally, since many scholars writing in English use the German text when writing about Heidegger, English language scholarship has, by and large, not adopted the decision to translate Dasein as Da-sein. Professor Stambaugh discussed her translation of *Being and Time* with Heidegger many times and rightly followed his wish. But it has become clear over the years that the reasons not to do this are overwhelming. Too many differences are obscured when we do not stay with the practice that Heidegger himself retained in the final edition of *Sein und Zeit*. For the most part, however, the changes in this revised translation consist of modifications of sentences and other changes that, hopefully, will bring Stambaugh's translation even more in line with Heidegger's original text. Of course, if I have introduced new problems in doing this, I am responsible for those. My intention in making revisions was not to produce a different translation, but to present what I hope will prove to be a somewhat better version of the Stambaugh translation of *Being and Time*.

Heidegger himself thanked Joan Stambaugh for her translation of *Being and Time* by presenting her with a copy of the poem that is reproduced on the cover of this book. Joan Stambaugh must be thanked yet again: first, for her translation and the contribution that translation has made to Heidegger studies; second, for giving me permission to make the revisions in this new edition. Translators struggle over every word and the result is hard won. A translator who allows her work to be modified by another is truly generous.

Thanks are also due to Dr. Hermann Heidegger for his support of this translation from the beginning, as well as for this revised edition. Finally, I have benefited greatly from comments and suggestions offered by colleagues and friends—more than I can name here—as I have worked through this revised translation of *Being and Time*.

Dennis J. Schmidt
State College, Pennsylvania
2 April 2010

TRANSLATOR'S PREFACE

There are many reasons that *Being and Time* poses special problems for its translator and for the readers of an English language translation. Three aspects of the text are especially noteworthy and so need to be commented upon here. First, one needs to bear in mind that, in *Being and Time*, Heidegger has introduced a large number of German neologisms. Words, such as *Befindlichkeit*, which would be readily intelligible to a German reader since it has a clear relation to an everyday German phrase [*wie befinden Sie sich?*—"How are you doing?"], appear here as Heidegger's own coinage and so pose a challenge even to a German reader. Typically, Heidegger's neologisms have strong connections to everyday phrases or words, and so exhibit a curious mix of strangeness and familiarity. Second, Heidegger frequently employs quite common vocabulary in uncommon ways. Here the most visible example is his use of the word *Dasein* which, besides having a long history as a philosophic term (it is, for example, one of the categories in Hegel's *Science of Logic*), is a word that belongs to everyday conversation. One of Heidegger's intentions in *Being and Time* is to re-appropriate that word and give it new meaning without completely repudiating its everyday sense. Again, a German reader might find a sort of alchemy of familiarity and strangeness in Heidegger's use of such words. Third, naturally one finds Heidegger using quite traditional philosophical vocabulary in *Being and Time*, but he goes to great lengths to move such words into new and rather untraditional senses. Words such as *Wahrheit* ["truth"] and *Sein* ["being"] are only the most obvious examples of the problem a translator and reader face.

Such are some of the sorts of translation problems that are especially amplified by the nature of this text. There remain of course the usual problems that belong to the project of translating German into

English. But these problems, while demanding a certain vigilance on the part of both the translator and the reader, are not insurmountable. Some three decades ago Macquarrie and Robinson published a translation of *Being and Time* and in so doing were at the forefront of bringing Heidegger's work into English. That translation itself came to shape the way in which Heidegger's work was discussed in English. After three decades of translations of Heidegger's other works by a variety of translators, some of the decisions that were made in that early translation of *Being and Time* might now need to be reconsidered.

The present translation attempts to take into account the insights of the past thirty years of Heidegger scholarship in English. This translation was begun some time ago and has undergone changes over the years as colleagues have offered suggestions. Some individual translations will no doubt still provoke controversy and, because Heidegger's language is so rich and multivalent, would likely do so no matter how they were translated. But it is hoped that this translation will remedy some of the infelicities and errors of the previous translation, and open a productive debate about some of the more original and still puzzling language of this text. To that end, a few very brief remarks about some of the terminological decisions may be in order to help orient the reader to some of the choices made here and to some of the alternative translations that might well be borne in mind by the reader. However, in the end, the translation will need to justify itself in the reading.

It was Heidegger's express wish that in future translations the word *Da-sein* should be hyphenated throughout *Being and Time*, a practice he himself instigated, for example, in chapter 5 of Division One. Thus the reader will be less prone to assume he or she understands it to refer to "existence" (which is the orthodox translation of *Dasein*) and with that translation surreptitiously bring along all sorts of psychological connotations. It was Heidegger's insight that human being is *uncanny*: we do not know who, or what, that is, although, or perhaps precisely because, we *are* it.

As is the case with German nouns in general, the terms *Da-sein* and *Angst*, which remain untranslated, retain capitalization: no other English expressions are capitalized, including the term "being." Capitalizing "being," although it has the dubious merit of treating "being" as something unique, risks implying that it is some kind of Super Thing or transcendent being. But Heidegger's use of the word "being" in no sense refers the word to something like *a* being, especially not a transcendent Being. Heidegger does not want to substantivize this word, yet capitalizing the word in English does tend to imply just that. The later words for being, *Ereignis* ["appropriation," "belonging-together"] and *Das Geviert* ["the Fourfold"] express *relations* that first constitute any possible *relata* or things, and thus confirm this nonsubstantializing intention.

Another peculiarly problematic word is *Verfallen* which does not literally mean "falling," but speaks rather of a "falling prey" to something (the world). In other words, it is a kind of "motion" that is unable to go anywhere. While one might render the substantive (*das Verfallen / die Verfallenheit*) as "ensnarement," I choose to translate it as "entanglement" to avoid the connotation of a trap.

The word *Befindlichkeit*, which I have translated as "attunement," needs some qualification and special comment. Another legitimate translation for this word is "disposition" (here one could refer to the French translation which uses *disposition*). But despite quite compelling reasons to use "disposition," I decided that "attunement" was the better choice if only because it seems better able to avoid suggesting that there are psychological connotations carried in Heidegger's analysis of *Befindlichkeit*.

The translation of the word *Bewandtnis* here as "relevance" also needs a comment. While one might opt for a different translation, namely "situation," it seems that such a choice leaves the meaning of the word too broad and vague, and that it fails to capture the distinction Heidegger makes between *Bewandtnis*, which has primarily to do with things, and *Situation*, which is more applicable to people.

The crucial trio of words *Sorge, Besorgen,* and *Fürsorge* was rendered as "care," "taking care," and "concern," respectively. A shift from the previous translation of *Being and Time* needs to be noted here. Macquarrie and Robinson had used "concern" as a translation for *besorgen*; however, I have chosen to use the word "concern" as a translation for *Fürsorge*, which involves human issues. I have translated *besorgen* as "taking care" (as opposed to the Macquarrie and Robinson translation of "solicitude") because it refers more to errands and matters that one takes care of or settles.

Heidegger sometimes uses the word *Andere* with an article (*die Andere*), but for the most part the word appears without an article. I have followed him in this and so use "others" rather than "the others." The emphasis on the other(s), so strongly expressed in Sartre as a threat to subjectivity, is lacking in Heidegger. After all, we belong to others; we are others too.

Vorlaufen. Anticipation is perhaps too weak. Macquarrie and Robinson's "running forward in thought" seemed a bit awkward. But it may be the better choice.

Nähe was consistently rendered as "nearness," whereas in earlier versions I often used "closeness" as well.

The word *Wiederholung*, which I have translated as "retrieval," could also be translated as "recapitulation" since that word is used in music to refer to what Heidegger seems to intend by *Wiederholung*. In music (specifically in the sonata form) recapitulation refers to the return of the initial theme after the whole development section.

Because of its new place in the piece, that same theme is now heard differently.

One final word, which is a departure from the previous translation, needs to be noted. *Augenblick*, which I have translated as "moment" (as opposed to Macquarrie and Robinson's "moment of vision"), should help pick up the temporal connotations of the word and also shed some of the almost mythical sense of the previous translation. Here the reader might do well to associate the word with the English expression "in the twinkling of an eye" or the French *coup d'oeil*.

The nonnumerical notes at the bottom of the pages are Heidegger's own notes made in the margin on the occasion of subsequent rereading.

I wish to thank Dennis Schmidt for his valuable and helpful suggestions. Thanks are due to Theodore Kisiel for the lexicon. I also wish to thank the Alexander von Humboldt-Stiftung for its generous support enabling me to complete this translation.

Joan Stambaugh
New York, 1995

AUTHOR'S PREFACE TO THE SEVENTH GERMAN EDITION

The text of *Being and Time* first appeared in the spring of 1927 in the *Jahrbuch für Phänomenologie und phänomenologische Forschung*, Vol. VIII, edited by E. Husserl, and it was published simultaneously as a separate volume.

The present reprint, which is the seventh edition, is unchanged with respect to the text, but has been newly revised with regard to quotations and punctuation. The page numbers of this reprint agree with those of the earlier editions except for minor deviations.

The designation "First Half," which previous editions bore, has been deleted. After a quarter century, the second half could no longer be added without the first being presented anew. Nonetheless, its path still remains a necessary one even today, if the question of being is to move our Dasein.

For the elucidation of that question the reader may refer to my *Einführung in die Metaphysik* which is being published by the same press as this text. That work presents the text of a lecture course delivered in the summer semester of 1935.

... δῆλον γὰρ ὡς ὑμεῖς μὲν ταῦτα (τί ποτε βούλεσθε σημαίνειν ὁπόταν   1
ὂν φθέγγησθε) πάλαι γιγνώσκετε, ἡμεῖς δὲ πρὸ τοῦ μὲν ᾠόμεθα, νῦν δ'
ἠπορήκαμεν . . .

[Plato, *Sophist* 244a]

"For manifestly you have long been aware of what you mean when
you use the expression *'being'* [*'seiend'*]. We, however, who used to
think we understood it, have now become perplexed."

Do we in our time have an answer to the question of what we
really mean by the word 'being' ['seiend']? Not at all. So it is fitting
that we should raise anew *the question of the meaning of being* [*Sein*]. But
are we nowadays even perplexed at our inability to understand the
expression 'being' ['Sein']? Not at all. So first of all we must reawaken
an understanding for the meaning of this question. The aim of the following treatise is to work out the question of the meaning of *"being"*
[*"Sein"*] and to do so concretely. The provisional aim is the interpretation of *time* as the possible horizon for any understanding whatsoever
of being.

But the reasons for making this our aim, the investigations which
such a purpose requires, and the path to its achievement, call for some
introductory remarks.

The Exposition of the Question of the Meaning of Being

CHAPTER ONE
The Necessity, Structure, and Priority of the Question of Being

§ 1. *The Necessity of an Explicit Repetition of the Question of Being*

This question has today been forgotten—although our time considers itself progressive in again affirming "metaphysics." All the same we believe that we are spared the exertion of rekindling a γιγαντομαχία περὶ τῆς οὐσίας ["a Battle of Giants concerning Being," Plato, *Sophist* 245e6-246e1]. But the question touched upon here is hardly an arbitrary one. It sustained the avid research of Plato and Aristotle but from then on ceased to be heard *as a thematic question of actual investigation*. What these two thinkers achieved has been preserved in various distorted and "camouflaged" forms down to Hegel's *Logic*. And what then was wrested from phenomena by the highest exertion of thought, albeit in fragments and first beginnings, has long since been trivialized.

Not only that. On the basis of the Greek point of departure for the interpretation of being a dogma has taken shape which not only declares that the question of the meaning of being is superfluous, but even sanctions its neglect. It is said that "being" is the most universal and the emptiest concept. As such it resists every attempt at definition. Nor does this most universal and thus indefinable concept need any definition. Everybody uses it constantly and also already understands what is meant by it. Thus what troubled ancient philosophizing and kept it so by virtue of its obscurity has become obvious, clear as day, such that whoever persists in asking about it is accused of an error of method.

At the beginning of this inquiry the prejudices that constantly instill and repeatedly promote the idea that a questioning of being is not needed cannot be discussed in detail. They are rooted in ancient ontology itself. That ontology can in turn only be interpreted adequately under the guidance of the question of being which has been clarified and answered beforehand. One must proceed with regard to the soil from which the fundamental ontological concepts grew and with reference to the suitable demonstration of the categories and their completeness. We therefore wish to discuss these prejudices only to the extent that the necessity of a repetition of the question of the meaning of being becomes evident. There are three such prejudices.

1. "Being"* is the most "universal" concept: τὸ ὄν ἐστι καθόλου μάλιστα πάντων.[1] Illud quod primo cadit sub apprehensione, est ens, cuius intellectus includitur in omnibus, quaecumque quis apprehendit. "An understanding of being is always already contained in everything we apprehend in beings."[2] But the "universality" of "being" is not that of genus. "Being" does not delimit the highest region of beings so far as they are conceptually articulated according to genus and species: οὔτε τὸ ὄν γένος ["being is not a genus"].[3] The "universality" of being "surpasses" the universality of genus. According to the designation of medieval ontology, "being" is a transcendens. Aristotle himself understood the unity of this transcendental "universal," as opposed to the manifold of the highest generic concepts with material content, as the unity of analogy. Despite his dependence upon Plato's ontological position, Aristotle placed the problem of being on a fundamentally new basis with this discovery. To be sure, he too did not clarify the obscurity of these categorical connections. Medieval ontology discussed this problem in many ways, above all the Thomist and Scotist schools, without gaining fundamental clarity. And when Hegel finally defines "being" as the "indeterminate immediate," and makes this definition the foundation of all the further categorial explications of his Logic, he remains within the perspective of ancient ontology—except that he gives up the problem, raised early on by Aristotle, of the unity of being in contrast to the manifold of "categories" with material content. If one says accordingly that "being" is the most universal concept, that cannot mean that it is the clearest and that it needs no further discussion. The concept of "being" is rather the most obscure of all.

---

* the being [das Seiend], beingness [Seiendheit].

1. Aristotle, Metaphysics, III.4.1001a21.
2. Thomas Aquinas, Summa Theologiae II.1, qu. 94, a. 2.
3. Artistotle, Metaphysics III.3.998b22.

2. The concept of "being" is indefinable. This conclusion was   *4*
drawn from its highest universality.[4] And correctly so—if *definitio fit*
per genus proximum et differentiam specificam [if "definition is achieved
through the proximate genus and the specific difference"]. Indeed,
"being" cannot be understood as a being. *Enti non additur aliqua natura*:
"being" cannot be defined by attributing beings to it. Being cannot
be derived from higher concepts by way of definition and cannot be
represented by lower ones. But does it follow from this that "being"
can no longer constitute a problem? Not at all. We can conclude only
that "being" ["Sein"] is not something like a being [Seiendes].* Thus
the manner of definition of beings which has its justification within
limits—the "definition" of traditional logic which is itself rooted in
ancient ontology—cannot be applied to being. The indefinability of
being does not dispense with the question of its meaning but forces
it upon us.

3. "Being" is the self-evident concept. "Being" is used in all knowing and predicating, in every relation to beings [Seienden] and in every
relation to oneself, and the expression is understandable "without further ado." Everybody understands: "the sky *is* blue," "I *am* happy,"
and similar statements. But this average comprehensibility only demonstrates the incomprehensibility. It shows that an enigma lies *a priori*
in every relation and being toward beings as beings. The fact that we
live already in an understanding of being and that the meaning of
being is at the same time shrouded in darkness proves the fundamental
necessity of retrieving the question of the meaning of "being."

If what is "self-evident" and this alone—"the covert judgments of
common reason" (Kant)—is to become and remain the explicit theme
of our analysis (as "the business of philosophers"), then the appeal to
self-evidence in the realm of basic philosophical concepts, and indeed
with regard to the concept "being," is a dubious procedure.

However, consideration of the prejudices has made it clear at the
same time that not only is the *answer* to the question of being lacking,
but even the question itself is obscure and without direction. Thus to
retrieve the question of being means first of all to work out adequately
the *formulation* of the question.

---

* no! rather: a decision about beyng [Seyn] cannot be made with the help of such
conceptuality.

4. Cf. Pascal, *Pensées et Opuscules*, ed. Brunschvicg (Paris, 1912), p. 169: "One cannot
undertake to define being without falling into this absurdity; for one cannot define a
word without beginning in this way: '*It is* . . .'. This beginning may be expressed or
implied. Thus, in order to define being one must say '*It is* . . .' and hence employ the
word to be defined in its definition."

§ 2. *The Formal Structure of the Question of Being*

The question of the meaning of being must be *formulated*. If it is a—or even *the*—fundamental question, such questioning needs the suitable transparency. Thus we must briefly discuss what belongs to a question in general in order to be able to make clear that the question of being is an *eminent* one.

Every questioning is a seeking. Every seeking takes its lead beforehand from what is sought. Questioning is a knowing search for beings in their thatness and whatness. The knowing search can become an "investigation," as the revealing determination of what the question aims at. As questioning about . . . questioning has *what it asks about* [*Gefragtes*]. All asking about . . . is in some way an inquiring of. . . . Besides what is asked, what is *interrogated* [*Befragtes*] also belongs to questioning. What is questioned is to be defined and conceptualized in the investigating, that is, the specifically theoretical, question. As what is really intended, what is to be *ascertained* [*Erfragtes*] lies in what is questioned; here questioning arrives at its goal. As an attitude adopted by a being [*Seienden*], the questioner, questioning has its own character of being [*Sein*]. Questioning can come about as "just asking around" or as an explicitly formulated question. What is peculiar to the latter is the fact that questioning first becomes lucid in advance with regard to all the above-named constitutive characteristics of the question.

The question to be *formulated* is about the meaning of being. Thus we are confronted with the necessity of explicating the question of being with regard to the structural moments cited.

As a seeking, questioning needs prior guidance from what it seeks. The meaning of being must therefore already be available to us in a certain way. We intimated that we are always already involved in an understanding of being. From this grows the explicit question of the meaning of being and the tendency toward its concept. We do not *know* what "being" means. But already when we ask, "what *is* 'being' ['Sein']?" we stand in an understanding of the "is" without being able to determine conceptually what the "is" means. We do not even know the horizon upon which we are supposed to grasp and pin down the meaning. *This average and vague understanding of being is a fact.*

No matter how much this understanding of being wavers and fades and borders on mere verbal knowledge, this indefiniteness of the understanding of being that is always already available is itself a posi-

tive phenomenon which needs elucidation. However, an investigation of the meaning of being will not wish to provide this at the outset. The interpretation of the average understanding of being attains its necessary guideline only with the developed concept of being. From

the clarity of that concept and the appropriate manner of its explicit
understanding we shall be able to discern what the obscure or not
yet elucidated understanding of being means, namely, what kinds of
obfuscation or hindrance of an explicit elucidation of the meaning of
being are possible and necessary.

Furthermore, the average, vague understanding of being can be
permeated by traditional theories and opinions about being in such a
way that these theories, as the sources of the prevailing understanding,
remain hidden. What is sought in the question of being is not completely
unfamiliar, although it is at first totally ungraspable.

What is *asked about* in the question to be elaborated is being, that
which determines beings as beings, that in terms of which beings have
always been understood no matter how they are discussed. The being
of being "is" itself not a being. The first philosophical step in understanding the problem of being consists in avoiding the μῦθόν τινα
διηγεῖσθαι,[5] in not "telling a story," that is, not determining beings
as beings by tracing them back in their origins to another being—as
if being [Sein] had the character of a possible being [Seienden]. As
what is asked about, being thus requires its own kind of demonstration which is essentially different from the discovery of beings. Hence
what is to be *ascertained*, the meaning of being, will require its own
conceptualization, which again is essentially distinct from the concepts
in which beings receive their determination of meaning.

Insofar as being constitutes what is asked about, and insofar as
being means the being of beings, beings themselves turn out to be
what is *interrogated* in the question of being. Beings are, so to speak,
interrogated with regard to their being. But if they are to exhibit the
characteristics of their being without falsification they must for their
part have become accessible in advance as they are in themselves. The
question of being demands that the right access to beings be gained
and secured in advance with regard to what it interrogates. But we call
many things "existent" ["seiend"],* and in different senses. Everything
we talk about, mean, and are related to is existent [seiend] in one way   7
or another. What and how we ourselves are is also existent. Being
[Sein] is found in thatness and whatness, reality, the objective presence

---

* "seiend" is translated here as "existent." On occasions when it needs to be translated
as "being" this will be noted so that it is not confused with "Sein." The first appearance
of "seiend" is in the "Exergue" (H xxix) where Heidegger uses it to translate the Greek
word "ὄν" in his translation of the passage from Plato's *Sophist*. [TR]

5. Plato, *Sophist* 242c.

of things [Vorhandenheit],* subsistence, validity, existence [Dasein],†
and in the "there is" ["es gibt"]. In *which* being is the meaning of
being to be found;‡ from which being is the disclosure of being to get
its start? Is the starting point arbitrary, or does a certain being have
priority in the elaboration of the question of being? Which is this exem-
plary§ being [Seiende] and in what sense does it have priority?

If the question of being is to be explicitly formulated and
brought to complete clarity concerning itself, then the elaboration of
this question requires, in accord with what has been elucidated up to
now, explication of the ways of regarding being, of understanding and
conceptually grasping its meaning, preparation of the possibility of the
right choice of the exemplary being, and elaboration of the genuine
mode of access to this being [Seienden]. Regarding, understanding and
grasping, choosing, and gaining access to, are constitutive attitudes of
inquiry and are thus themselves modes of being of a particular being,
of *the* being we inquirers ourselves in each case are. Thus to work
out the question of being means to make a being—one who questions—transparent in its being. Asking this question, as a mode of *being*
of a being, is itself essentially determined by what is asked about in

---

* "Vorhandenheit," "Vorhanden," and "Vorhandensein" are difficult words to translate
into English and cannot always be translated in a consistent manner. While "objective
presence" is used in this instance, the notion of something being "vorhanden" does not
necessarily imply that it has the character of being an "object" or something "objective."
Since talk of object entails the notion of a subject as well, this qualification is especially
important because it is always necessary to distinguish Dasein from anything like a
subject. While one can refer to a "Gegenstand" ["object"] as being "vorhanden," not
all that has the character of "Vorhandenheit" has the character of an object. Although
the translation "objective presence" can be problematic, alternative translations such as
"presence" or "being present" that are frequently used here—as well as other alternatives not chosen—have liabilities as well. In other words, there is no clear solution to
the translation of these important words. Nonetheless, depending upon the context, one
can usually find an acceptable translation. Whatever translation is used, the reader is
well advised to keep these caveats in mind and to give special attention to Heidegger's
own account of the meaning of this set of words. In doing this, one should note that the
full understanding of "Vorhandenheit" cannot be developed without careful attention to
the set of words that name another way in which we encounter that which is present
within our world: "Zuhandenheit," "Zuhanden," and "Zuhandensein." To this end, see
esp. H 69ff, where Heidegger shows how "handiness" ["Zuhandenheit"] underlies how
it is that we encounter what is "present" ["Vorhanden"]. [TR]
† neither the usual concept nor any other.
‡ two different questions are aligned here; misleading, above all in relation to the role
of Dasein.
§ Misleading. Dasein is exemplary because it is the co-player [das Bei-spiel] that in its
essence as Da-sein (perduring [wahrend] the truth of being) plays to and with being—
brings it into the play of resonance.

it—being.* This being [Seiende], which we ourselves in each case are and which includes inquiry among the possibilities of its being, we formulate terminologically as *Dasein*. The explicit and lucid formulation of the question of the meaning of being requires a prior suitable explication of a being (Dasein) with regard to its being.†

But does not such an enterprise fall into an obvious circle? To have to determine beings *in their being* beforehand and then on this foundation first pose the question of being—what else is that but going around in circles? In working out the question do we not "presuppose" something that only the answer can provide? Formal objections such as the argument of "circular reasoning," an argument that is always easily raised in the area of investigation of principles, are always sterile when one is weighing concrete ways of investigating. They do not offer anything to the understanding of the issue and they hinder penetration into the field of investigation.

But in fact there is no circle at all in the formulation of our question. Beings can be determined in their being without the explicit concept of the meaning of being having to be already available. If this were not so there could not have been as yet any ontological knowledge. And probably no one would deny the factual existence of such knowledge. It is true that "being" ["Sein"] is "presupposed" in all previous ontology, but not as an available *concept*—not as the sort of thing we are seeking. "Presupposing" being has the character of taking a preliminary look at being in such a way that on the basis of this look beings that are already given are tentatively articulated in their being. This guiding look at being grows out of the average understanding of being in which we are always already involved *and which ultimately*‡ *belongs to the essential constitution of Dasein itself*. Such "presupposing" has nothing to do with positing a principle from which a series of propositions is deduced. A "circle in reasoning" cannot possibly lie in the formulation of the question of the meaning of being, because in answering this question it is not a matter of grounding by deduction, but rather of laying bare and exhibiting the ground.

"Circular reasoning" does not occur in the question of the meaning of being. Rather, there is a notable "relatedness backward or forward" of what is asked about (being) [Sein] to asking as a mode of being of a being. The way what is questioned essentially engages our questioning belongs to the innermost meaning of the question of being. But this only means that the being that has the character of Dasein has

8

---

* Da-sein: as held out into the nothingness of beyng [Seyn], held as relation.
† But the meaning of being is not drawn from this being.
‡ i.e., from the beginning.

a relation to the question of being itself, perhaps even a distinctive one. But have we not thereby demonstrated that a particular being has a priority with respect to being and that the exemplary being that is to function as what is primarily *interrogated* is pregiven?* In what we have discussed up to now neither has the priority of Dasein been demonstrated nor has anything been decided about its possible or even necessary function as the primary being to be interrogated. But certainly something like a priority of Dasein has announced itself.

§ 3. *The Ontological Priority of the Question of Being*

The characterization of the question of being, under the guideline of the formal structure of the question as such, has made it clear that this question is a unique one, such that its elaboration and even its solution require a series of fundamental reflections. However, what is distinctive about the question of being will fully come to light only when that question is sufficiently delineated with regard to its function, intention, and motives.

9   Up to now the necessity of a retrieval of the question was motivated partly by its venerable origin but above all by the lack of a definite answer, even by the lack of any adequate formulation of the question. But one can demand to know what purpose this question should serve. Does it remain solely, or *is* it at all, only a matter of free-floating speculation about the most general generalities—*or is it the most basic [prinzipiellste] and at the same time most concrete question?*

Being is always the being of a being. The totality of beings can, with respect to its various domains, become the field where particular areas of knowledge are exposed and delimited. These areas—for example, history, nature, space, life, human being, language, and so on—can in their turn become thematized as objects of scientific investigations. Scientific research demarcates and first establishes these areas of knowledge in a rough and ready fashion. The elaboration of the area in its fundamental structures is in a way already accomplished by pre-scientific experience and interpretation of the domain of being to which the area of knowledge is itself confined. The resulting "fundamental concepts" comprise the guidelines for the first concrete disclosure of the area. Whether or not the importance of the research always lies in such establishment of concepts, its true progress comes about not so much in collecting results and storing them in "handbooks" as in being forced to ask questions about the basic constitution of each area, these questions being chiefly a reaction to increasing knowledge in each area.

---

* Again as above (H 6–7), an essential simplification and yet correctly thought. Dasein is not an instance of being for the representational abstraction of being; rather, it is the site of the understanding of being.

The real "movement" of the sciences takes place in the revision of these basic concepts, a revision which is more or less radical and lucid with regard to itself. A science's level of development is determined by the extent to which it is *capable* of a crisis in its basic concepts. In these immanent crises of the sciences the relation of positive questioning to the matter in question becomes unstable. Today tendencies to place research on new foundations have cropped up on all sides in the various disciplines.

The discipline which is seemingly the strictest and most securely structured, *mathematics*, has experienced a "crisis in its foundations." The controversy between formalism and intuitionism centers on obtaining and securing primary access to what should be the object of this science. Relativity theory in *physics* grew out of the tendency to expose nature's own coherence as it is "in itself." As a theory of the conditions of access to nature itself it attempts to preserve the immutability of the laws of motion by defining all relativities; it is thus confronted by the question of the structure of its pre-given area of knowledge, that is, by the problem of matter. In *biology* the tendency has awakened to get behind the definitions that mechanism and vitalism have given to organism and life and to define anew the kind of being of living beings as such. In the *historical and humanistic disciplines* the drive toward historical actuality itself has been strengthened by the transmission and portrayal of tradition: the history of literature is to become the history of critical problems. *Theology* is searching for a more original interpretation of human being's being toward God, prescribed by the meaning of faith itself and remaining within it. Theology is slowly beginning to understand again Luther's insight that its system of dogma rests on a "foundation" that does not stem from a questioning in which faith is primary and whose conceptual apparatus is not only insufficient for the range of problems in theology but rather covers them up and distorts them.

Fundamental concepts are determinations in which the area of knowledge underlying all the thematic objects of a science attains an understanding that precedes and guides all positive investigation. Accordingly these concepts first receive their genuine evidence and "grounding" only in a correspondingly preliminary research into the area of knowledge itself. But since each of these areas arises from the domain of beings themselves, this preliminary research that creates the fundamental concepts amounts to nothing else than interpreting these beings in terms of the basic constitution of their being. This kind of investigation must precede the positive sciences—and it *can* do so. The work of Plato and Aristotle is proof of this. Laying the foundations of the sciences in this way is different in principle from "logic" limping along behind, investigating here and there the status of a science in terms of its "method." Such laying of foundations is productive logic in the sense that it leaps ahead, so to speak, into a particular realm of being, discloses it for the first time in its constitutive being, and makes the acquired structures available to the positive sciences as lucid directives for inquiry. Thus, for example, what is philosophically primary is not a theory of concept-formation in historiology, nor the theory of historical knowledge, nor even the theory of history as the object of historiology; what is primary is rather the interpretation of genuinely historical beings with regard to their historicality. Similarly, the positive result of Kant's *Critique of Pure Reason* consists in its approach to working out what belongs to any nature whatsoever, and not in a "theory" of knowledge. His transcendental logic is an *a priori* logic of the realm of being called nature.

But such inquiry—ontology taken in its broadest sense without reference to specific ontological directions and tendencies—itself still needs a guideline. It is true that ontological inquiry is more original than the ontic inquiry of the positive sciences. But it remains naïve and opaque if its investigations into the being of beings leave the meaning of being in general undiscussed. And precisely the ontological task of a genealogy of the different possible ways of being (a genealogy which is not to be construed deductively) requires a preliminary understanding of "what we really mean by this expression 'being' ['Sein']."

The question of being thus aims not only at an a priori condition of the possibility of the sciences, which investigate beings as this or that kind of being and which thus always already move within an understanding of being, but also at the condition of the possibility of the ontologies which precede the ontic sciences and found them. *All ontology, no matter how rich and tightly knit a system of categories it has at its disposal, remains fundamentally blind and perverts its innermost intent if it has not previously clarified the meaning of being sufficiently and grasped this clarification as its fundamental task.*

Ontological research itself, correctly understood, gives the question of being its ontological priority over and above merely resuming an honored tradition and making progress on a problem until now opaque. But this scholarly, scientific priority is not the only one.

§ 4. *The Ontic Priority of the Question of Being*

Science in general can be defined as the totality of fundamentally coherent true propositions. This definition is not complete, nor does it get at the meaning of science. As ways in which human beings behave, sciences have this being's (the human being's) kind of being. We define this being terminologically as *Dasein*. Scientific research is neither the sole nor the most immediate kind of being of this being that is possible. Moreover, Dasein itself is distinctly different from other beings. We must make this distinct difference visible in a preliminary way. Here the discussion must anticipate subsequent analyses which only later will become truly demonstrative.

Dasein is a being that does not simply occur among other beings. Rather it is ontically distinguished by the fact that in its being this being is concerned *about* its very being. Thus it is constitutive of the being of Dasein to have, in its very being, a relation of being to this being. And this in turn means that Dasein understands itself in its being [Sein] in some way and with some explicitness. It is proper to this being that it be disclosed to itself with and through its being. *Understanding of being is itself a determination of being of Dasein [Seinsverständnis ist selbst eine Seinsbestimmtheit des Daseins].** The ontic distinction of Dasein lies in the fact that it *is* ontological.

To be ontological does not yet mean to develop an ontology. Thus if we reserve the term ontology for the explicit, theoretical question of the being of beings, the ontological character of Dasein referred to here is to be designated as pre-ontological. That does not signify simply being [seiend] ontical, but rather being [seiend] in the manner of an understanding of being.

We shall call the† very being [Sein] to which‡ Dasein can relate in one way or another, and somehow always does relate, *existence* [*Existenz*]. And because the essential definition of this being cannot be accomplished by ascribing to it a what that specifies its material content, because its essence lies rather in the fact that in each instance it has to be its being [Sein] as its own, the term Dasein, as a pure expression of being, has been chosen to designate this being [Seienden].

Dasein always understands itself in terms of its existence, in terms of its possibility to be itself or not to be itself. Dasein has either chosen these possibilities itself, stumbled upon them, or in each instance already grown up in them. Existence is decided only by each Dasein itself in the manner of seizing upon or neglecting such possibilities. We come to terms with the question of existence always only through existence itself. We call *this* [*hierbei* führende] kind of understanding of oneself *existentiell* understanding. The question of existence is an ontic "affair" of Dasein. For this question the theoretical transparency of the ontological structure of existence is not necessary. The question of structure aims at the analysis of what constitutes existence.§ We shall call the coherence of these structures *existentiality*. Its analysis does not have the character of an existentiell understanding but rather an *existential* one. The task of an existential analysis of Dasein is prescribed   *13*

---

* But in this case being not only as the being of human being (existence). That becomes clear from the following. Being-in-the-world includes *in itself* the relation of existence to being in the whole: the understanding of being.

† that

‡ as its own

§ Thus not a philosophy of existence [Existenzphilosophie].

with regard to its possibility and necessity in the ontic constitution of Dasein.

But since existence defines Dasein, the ontological analysis of this being always requires in advance a consideration of existentiality. However, we understand existentiality as the constitution of being of the being [Seienden] that exists. But the idea of being in general already lies in the idea of such a constitution of being. And thus the possibility of carrying out the analysis of Dasein depends upon the prior elaboration of the question of the meaning of being in general.

Sciences and disciplines are ways of being of Dasein in which Dasein also relates to beings that it need not itself be. But being in a world belongs essentially to Dasein. Thus the understanding of being that belongs to Dasein just as originally implies the understanding of something like "world" and the understanding of the being of beings accessible within the world. Ontologies which have beings unlike Dasein as their theme are accordingly founded and motivated in the ontic structure of Dasein itself. This structure includes in itself the determination of a pre-ontological understanding of being.

Thus *fundamental ontology*, from which alone all other ontologies can originate, must be sought in the *existential analysis of Dasein*.

Dasein accordingly takes priority in several ways over all other beings. The first priority is an *ontic* one: this being is defined in its being by existence. The second priority is an *ontological* one: on the basis of its determination as existence Dasein is in itself "ontological." But just as originally Dasein possesses—in a manner constitutive of its understanding of existence—an understanding of the being of all beings unlike itself. Dasein therefore has its third priority as the ontic-ontological condition of the possibility of all ontologies. Dasein has proven itself to be that which, before all other beings, is ontologically the primary being to be interrogated.

However, the roots of the existential analysis, for their part, are ultimately *existentiell*; i.e. they are *ontic*. Only when philosophical research and inquiry themselves are grasped in an existentiell way—as a possibility of being of each existing Dasein—does it become possible at all to disclose the existentiality of existence and therewith to get hold of a sufficiently grounded set of ontological problems. But with this the ontic priority of the question of being has also become clear.

14

The ontic-ontological priority of Dasein was already seen early on, without Dasein itself being grasped in its genuine ontological structure or even becoming a problem with such an aim. Aristotle says, ἡ ψυχὴ τὰ ὄντα πώς ἐστιν.[6] The soul (of the human being) is in a certain way a

---

6. *De anima* III.8.431b21; cf. III.5.430a14ff.

being [das Seiende]. The "soul" which constitutes the being of human being discovers in its ways to be—αἴσθησις and νόησις—all beings with regard to their thatness and whatness, that is to say, always also in their being [Sein]. Thomas Aquinas discussed this statement—which refers back to Parmenides' ontological thesis—in a manner characteristic of him. Thomas is engaged in the task of deriving the "transcendentals," the characteristics of being that lie beyond every possible generic determination of a being in its material content, every *modus specialis entis*, and that are necessary attributes of every something, whatever it might be. For him the *verum* too is to be demonstrated as being such a *transcendens*. This is to be accomplished by appealing to a being which, in conformity with its kind of being, is suited to "come together," that is to agree with any being whatsoever. This distinctive being, the *ens, quod natum est convenire cum omni ente* [the being whose nature it is to meet with all other beings], is the soul (*anima*).[7] The priority of "Dasein" over and above all other beings, which emerges here without being ontologically clarified, obviously has nothing in common with a vapid subjectivizing of the totality of beings.

The demonstration of the ontic-ontological distinctiveness of the question of being is grounded in the preliminary indication of the ontic-ontological priority of Dasein. But the analysis of the structure of the question of being as such (§ 2) came up against the distinctive function of this being within the formulation of that very question. Dasein revealed itself to be that being which must first be elaborated in a sufficiently ontological manner if the inquiry is to become transparent. But now it has become evident that the ontological analysis of Dasein in general constitutes fundamental ontology, that Dasein consequently functions as the being that is to be *interrogated* fundamentally in advance with respect to its being.

If the interpretation of the meaning of being is to become a task, Dasein is not only the primary being to be interrogated; in addition to that it is the being that always already in its being is related to *what is sought* in this question. But then the question of being is nothing else than the radicalization of an essential tendency of being that belongs to Dasein itself, namely, of the pre-ontological understanding of being.

15

---

7. *Quaestiones de veritate*, qu. 1, a. 1c; cf. the occasionally stricter exposition, which deviates from what was cited, of a "deduction" of the transcendentals in the brief work *De natura generis.*

CHAPTER TWO

The Double Task in Working Out the Question of Being: The Method of the Investigation and Its Outline

§ 5. *The Ontological Analysis of Dasein as Exposing the Horizon for an Interpretation of the Meaning of Being in General*

In designating the tasks involved in "formulating" the question of being, we showed that not only must we pinpoint the particular being that is to function as the primary being to be interrogated, but also that an explicit appropriation and securing of correct access to this being is required. We discussed which being it is that takes over the major role within the question of being. But how should this being [Seiende], Dasein, become accessible and, so to speak, come into view in an interpretation that is comprehensible?

The ontic-ontological priority that has been demonstrated for Dasein could lead to the mistaken opinion that this being [Seiende] would also have to be what is primarily given ontically-ontologically, not only in the sense that such a being could be grasped "immediately," but also that the prior giveness of its kind of being [Seinsart] would be just as "immediate." True, Dasein is ontically not only what is near or even nearest—we ourselves *are* it, each of us. Nevertheless, or precisely for this reason, it is ontologically what is farthest. True, it belongs to its most proper being to have an understanding of being and to sustain a certain interpretation of it. But this does not at all mean that the most readily available pre-ontological interpretation of its own being could be adopted as an adequate guideline, as though this understanding of being had to arise from a thematically ontological reflection on the most proper constitution of its being. Rather, in accordance with the kind of being belonging to it, Dasein tends to understand its own being [Sein] in terms of *the* being [Seienden] to which it is essentially,

15

16   continually, and most closely related—the "world."* In Dasein itself
and therewith in its own understanding of being, as we shall show,
the way the world is understood is ontologically reflected back upon
the interpretation of Dasein.

The ontic-ontological priority of Dasein is therefore the reason
why the specific constitution of the being of Dasein—understood in the
sense of the "categorial" structure that belongs to it—remains hidden
from it. Dasein is ontically "nearest" to itself, ontologically farthest
away; but pre-ontologically certainly not foreign to itself.

We have merely precursorily indicated that an interpretation of
this being is confronted with peculiar difficulties rooted in the mode
of being of the thematic object and the way it is thematized. They do
not result from some shortcoming of our powers of knowledge or lack
of an appropriate conceptuality—a lack seemingly easy to remedy.

Not only does an understanding of being belong to Dasein, but
this understanding also develops or decays according to the actual
manner of being of Dasein at any given time; for this reason it has
a wealth of interpretations at its disposal. Philosophical psychology,
anthropology, ethics, "politics," poetry, biography, and historiography
pursue in different ways and to varying extents the behavior, faculties, powers, possibilities, and destinies of Dasein. But the question
remains whether these interpretations were carried out in as original
an existential manner as their existentiell originality perhaps merited.
The two do not necessarily go together, but they also do not exclude
one another. Existentiell interpretation can require existential analysis, provided philosophical knowledge is understood in its possibility
and necessity. Only when the fundamental structures of Dasein are
adequately worked out with explicit orientation toward the problem of
being will the previous results of the interpretation of Dasein receive
their existential justification.

Hence the first concern in the question of being must be an analysis of Dasein. But then the problem of gaining and securing the kind of
access that leads to Dasein truly becomes crucial. Expressed negatively,
no arbitrary idea of being and reality, no matter how "self-evident" it
is, may be brought to bear on this being in a dogmatically constructed
way; no "categories" prescribed by such ideas may be forced upon
Dasein without ontological deliberation. The manner of access and
interpretation must instead be chosen in such a way that this being
can show itself to itself on its own terms. And furthermore, this manner should show that being as it is *initially and for the most part*—in
its average *everydayness*. Not arbitrary and accidental structures but
17   essential ones are to be demonstrated in this everydayness, structures

---

* i.e., here in terms of what is objectively present.

that remain determinative in every mode of being of factical Dasein. By looking at the fundamental constitution of the everydayness of Dasein we shall bring out in a preparatory way the being of this being.

The analytic of Dasein thus understood is wholly oriented toward the guiding task of working out the question of being. Its limits are thereby determined. It cannot hope to provide a complete ontology of Dasein, which of course must be supplied if something like a "philosophical" anthropology is to rest on a philosophically adequate basis. With a view to a possible anthropology or its ontological foundation, the following interpretation will provide only a few "parts," although not inessential ones. However, the analysis of Dasein is not only incomplete but at first also *preliminary*. It only brings out the being of this being without interpreting its meaning. Its aim is rather to expose the horizon for the most primordial interpretation of being. Once we have reached that horizon the preparatory analytic of Dasein requires repetition on a higher, genuinely ontological basis.

The meaning of being [Sein] of that being [Seienden] we call Dasein will prove to be *temporality* [*Zeitlichkeit*]. In order to demonstrate this we must repeat our interpretation of those structures of Dasein that shall have been indicated in a preliminary way—this time as modes of temporality. While it is true that with this interpretation of Dasein as temporality the answer to the guiding question about the meaning of being in general is not already given,* the soil from which we may reap it will nevertheless be prepared.

We intimated that a pre-ontological being [Sein] belongs to Dasein as its ontic constitution. Dasein *is* in such a way that, by being [seiend], it understands something like being. Remembering this connection, we must show that *time* is that from which Dasein tacitly understands and interprets something like being at all. Time must be brought to light and genuinely grasped as the horizon of every understanding and interpretation of being. For this to become clear we need an *original explication of time as the horizon of the understanding of being, in terms of temporality as the being of Dasein which understands being*. This task as a whole requires that the concept of time thus gained be distinguished from the common understanding of it. The latter has become explicit in an interpretation of time which reflects the traditional concept that has persisted since Aristotle and beyond Bergson. We must thereby make clear that, and in what way, this concept of time and the common understanding of time in general originate from temporality. In this way the common concept of time receives again its rightful autonomy—contrary to Bergson's thesis that time understood in the common way is really space.

18

---

* καθόλου, καθ' αὐτό.

"Time" has long served as the ontological—or rather ontic—criterion for naïvely distinguishing the different regions of beings. "Temporal" beings (natural processes and historical events) are separated from "atemporal" beings (spatial and numerical relationships). We are accustomed to distinguishing the "timeless" meaning of propositions from the "temporal" course of propositional statements. Further, a "gap" between "temporal" being and "supratemporal" eternal being is found, and the attempt is made to bridge the gap. "Temporal" here means as much as being [seiend] "in time," an obscure enough definition to be sure. The fact remains that time in the sense of "being in time" serves as a criterion for separating the regions of being. How time comes to have this distinctive ontological function, and even with what right precisely something like time serves as such a criterion, and most of all whether in this naïve ontological application of time its genuinely possible ontological relevance is expressed, has neither been asked nor investigated up to now. "Time," especially on the horizon of the common understanding of it, has chanced to acquire this "obvious" ontological function "of itself," as it were, and has held onto it until today.

In contrast we must show, on the basis of the question of the meaning of being which shall have been worked out, *that—and in what way—the central range of problems of all ontology is rooted in the phenomenon of time correctly viewed and correctly explained.*

If being is to be conceived in terms of time, and if the various modes and derivatives of being in their modifications and derivations are in fact to become intelligible through a consideration of time, then being itself—and not only beings that are "in time"—is made visible in its "temporal" ["zeitlich"] character. But then "temporal" can no longer mean only "being in time ["in der Zeit seiend"]." The "atemporal" and the "supratemporal" are also "temporal" with respect to their being; this not only by way of privation when compared to "temporal" beings, which are "in time," but in a *positive* way which, of course, must first be clarified. Because the expression "temporal" belongs to both pre-philosophical and philosophical usage, and because that expression will be used in a different sense in the following investigations, we shall call the original determination of the meaning of being and its characters and modes which devolve from time its *temporal* [*temporale*] determination. The fundamental ontological task of the interpretation of being as such thus includes the elaboration of the *temporality of being* [*Temporalität des Seins*]. In the exposition of the problem of temporality the concrete answer to the question of the meaning of being is first given.

Because being is in each instance comprehensible only in regard to time, the answer to the question of being cannot lie in an isolated and blind proposition. The answer is not grasped by repeating what is stated propositionally, especially when it is transmitted as a free-floating result, so that we merely take notice of a "standpoint" which perhaps deviates from the way the matter has been previously treated. Whether the answer is "novel" is of no importance and remains extrinsic. What is positive about the answer must lie in the fact that it is *old* enough to enable us to learn to comprehend the possibilities prepared by the "ancients." In conformity to its most proper sense, the answer provides a directive for concrete ontological research, that is, a directive to begin its investigative inquiry within the horizon exhibited—and that is all it provides.

If the answer to the question of being thus becomes the guiding directive for research, then it is sufficiently given only if the specific mode of being of previous ontology—the vicissitudes of its questioning, its findings, and its failures—becomes visible as necessary to the very character of Dasein.

§ 6. *The Task of a Destruction of the History of Ontology*

All research—especially when it moves in the sphere of the central question of being—is an ontic possibility of Dasein. The being of Dasein finds its meaning in temporality. But temporality is at the same time the condition of the possibility of historicity as a temporal mode of being of Dasein itself, regardless of whether and how it is a being "in time." As a determination, historicity is prior to what is called history (world-historical occurrences). Historicity means the constitution   20 of being of the "occurrence" of Dasein as such; upon its ground something like "world history," and belonging historically to world history, is possible. In its factical being Dasein always is how and "what" it already was. Whether explicitly or not, it *is* its past. It is its own past not only in such a way that its past, as it were, pushes itself along "behind" it, and that it possesses what is past as a property that is still objectively present and at times has an effect on it. Dasein "is" its past in the manner of *its* being which, roughly expressed, on each occasion "occurs" out of its future. In its manner of existing at any given time, and thus also with the understanding of being that belongs to it, Dasein grows into a customary interpretation of itself and grows up on that interpretation. It understands itself initially in terms of this interpretation ,and, within a certain range, constantly does so. This understanding discloses the possibilities of its being and regulates them. Its own past—and that always means that of its "generation"—does not *follow after* Dasein but rather always already goes ahead of it.

This elemental historicity of Dasein can remain concealed from it. But it can also be discovered in a certain way and be properly cultivated. Dasein can discover, preserve, and explicitly pursue tradi-

tion. The discovery of tradition, and the disclosure of what it "trans-
mits" and how it does this, can be undertaken as a task in its own
right. Dasein thus assumes the mode of being that involves historical
inquiry and research. But the discipline of history—more precisely,
the historicality underlying it—is possible only as the kind of being
belonging to inquiring Dasein, because Dasein is determined by historicity in the ground of its being. If historicity remains concealed from
Dasein, and so long as it does so, the possibility of historical inquiry
and discovery of history is denied it. If the discipline of history is
lacking, that is no evidence *against* the historicity of Dasein; rather
it is evidence for this constitution of being in a deficient mode. Only
because it is "historic" in the first place can an age lack the discipline
of history.

On the other hand, if Dasein has seized upon its inherent possibility not only of making its existence transparent, but also of inquiring
into the meaning of existentiality itself, that is to say, of provisionally
inquiring into the meaning of being in general; and if insight into the
essential historicity of Dasein has opened up in such inquiry, then it is
inevitable that inquiry into being, which was designated with regard
to its ontic-ontological necessity, is itself characterized by historicity.
The elaboration of the question of being must therefore receive its
directive to inquire into its own history from the most proper ontological sense of the inquiry itself, as a historical one; that means to
become historical in a disciplined way in order to come to the positive appropriation of the past, to come into full possession of its most
proper possibilities of inquiry. The question of the meaning of being is
led to understand itself as historical in accordance with its own way
of proceeding, that is, as the provisional explication of Dasein in its
temporality and historicity.

The preparatory interpretation of the fundamental structures of
Dasein with regard to its usual and average way of being—in which
it is also first of all historical—will make the following clear: Dasein
not only has the inclination to be entangled in the world in which it is
and to interpret itself in terms of that world by its reflected light; at the
same time Dasein is also entangled in a tradition which it more or less
explicitly grasps. This tradition deprives Dasein of its own leadership
in questioning and choosing. This is especially true of *that* understanding (and its possible development) which is rooted in the most proper
being of Dasein—the ontological understanding.

The tradition that hereby gains dominance makes what it "trans-
mits" so little accessible that initially and for the most part it covers it
over instead. What has been handed down is handed over to obviousness; it bars access to those original "wellsprings" out of which the
traditional categories and concepts were in part genuinely drawn. The

tradition even makes us forget such a provenance altogether. Indeed, it makes us wholly incapable of even understanding that such a return is necessary. The tradition uproots the historicity of Dasein to such a degree that it only takes an interest in the manifold forms of possible types, directions, and standpoints of philosophizing in the most remote and strangest cultures, and with this interest tries to veil its own groundlessness. Consequently, in spite of all historical interest and zeal for a philologically "objective" interpretation, Dasein no longer understands the most elementary conditions which alone make a positive return to the past possible in the sense of its productive appropriation.

At the outset (§ 1) we showed that the question of the meaning of being was not only unresolved, not only inadequately formulated, but despite all interest in "metaphysics" has even been forgotten. Greek ontology and its history, which through many twists and turns still define the conceptual character of philosophy today, are proof of the fact that Dasein understands itself and being in general in terms of the "world." The ontology that thus arises deteriorates into a tradition, which allows it to sink to the level of the obvious and become mere material for reworking (as it was for Hegel). Greek ontology thus uprooted becomes a fixed body of doctrine in the Middle Ages. But its systematic is not at all a mere joining together of traditional elements into a single structure. Within the limits of its dogmatic adoption of the fundamental Greek conceptions of being, this systematic contains a great deal of unpretentious work which does make advances. In its *scholastic* mold, Greek ontology makes the essential transition via the *Disputationes metaphysicae* of Suarez into the "metaphysics" and transcendental philosophy of the modern period; it still determines the foundations and goals of Hegel's *Logic*. Certain distinctive domains of being become visible in the course of this history and become the primary leitmotives for the subsequent range of problems (Descartes' *ego cogito*, subject, ego, reason, spirit, person), but, corresponding to the thorough neglect of the question of being, they remain unquestioned with respect to being and the structure of their being. But the categorial content of traditional ontology is transferred to these beings with corresponding formalizations and purely negative restrictions, or else dialectic is called upon to help with an ontological interpretation of the substantiality of the subject.

If the question of being is to achieve clarity regarding its own history, a loosening of the sclerotic tradition and a dissolution of the concealments produced by it is necessary. We understand this task as the *destruction* of the traditional content of ancient ontology which is to be carried out along the *guidelines of the question of being*. This destruction is based upon the original experiences in which the first,

22

and subsequently guiding, determinations of being were gained.

This demonstration of the provenance of the fundamental ontological concepts, as the investigation which displays their "birth cer-
tificate," has nothing to do with a pernicious relativizing of ontological
standpoints. The destruction has just as little the *negative* sense of disburdening ourselves of the ontological tradition. On the contrary, it
should stake out the positive possibilities in that tradition, and that
always means to stake out its *limits*. These are factically given with a
specific formulation of the question and the prescribed demarcation of
the possible field of investigation. Destruction does not relate itself in a
negative way to the past: its critique concerns "today" and the dominant way we treat the history of ontology, whether it is conceived as
the history of opinions, ideas, or problems. Destruction does not wish
to bury the past in nullity; it has a *positive* intent. Its negative function
remains tacit and indirect.

The destruction of the history of ontology essentially belongs to
the formulation of the question of being and is possible solely within
such a formulation. Within the scope of this treatise, which has as its
goal a fundamental elaboration of the question of being, the destruction can be carried out only with regard to the fundamentally decisive
stages of this history.

In accord with the positive tendency of this destruction, the question must first be asked whether and to what extent in the course of
the history of ontology in general the interpretation of being has been
thematically connected with the phenomenon of time. We must also
ask whether the problematic of temporality [Temporalität], which necessarily belongs here, was fundamentally worked out or could have
been. Kant is the first and only one who traversed a stretch of the
path toward investigating the dimension of temporality [Temporal-
ität]—or allowed himself to be driven there by the compelling force
of the phenomena themselves. Only when the problem of temporality
[Temporalität] is pinned down can we succeed in casting light on the
obscurity of his doctrine of schematism. Furthermore, in this way we
can also show *why* this area had to remain closed to Kant in its real
dimensions and in its central ontological function. Kant himself knew
that he was venturing forth into an obscure area: "This schematism of
our understanding as regards appearances and their mere form is an
art hidden in the depths of the human soul, the true devices of which
are hardly ever to be divined from Nature and laid uncovered before
our eyes."[1] What it is that Kant shrinks back from here, as it were,
must be brought to light thematically and in principle if the expression
"being" ["Sein"] is to have a demonstrable meaning. Ultimately the

23

---

1. Kant, *Kritik der reinen Vernunft*, B 180–81.

phenomena to be explicated in the following analysis under the rubric of "temporality" ["Temporalität"] are precisely those that determine the *most covert* judgments of "common reason," the analysis of which Kant calls the "business of philosophers."

In pursuing the task of destruction along the guideline of the problem of temporality [Temporalität] the following treatise will attempt to interpret the chapter on the schematism and the Kantian doctrine of time developed there. At the same time we must show why Kant could never gain insight into the problem of temporality [Temporalität]. Two things prevented this insight: first, the neglect of the question of being in general, and second, in conjunction with this, the lack of a thematic ontology of Dasein or, in Kantian terms, the lack of a preliminary ontological analytic of the subjectivity of the subject. Instead, despite all his essential advances, Kant dogmatically adopted Descartes' position. Furthermore, although Kant takes this phenomenon back into the subject his analysis of time remains oriented toward the traditional, vulgar understanding of it. It is this that finally prevented Kant from working out the phenomenon of a "transcendental determination of time" in its own structure and function. As a consequence of this double effect of the tradition, the decisive *connection* between *time* and the "*I think*" remain shrouded in complete obscurity. It did not even become a problem.

By taking over Descartes' ontological position Kant neglects something essential: an ontology of Dasein. In terms of Descartes' innermost tendency this omission is a decisive one. With the "*cogito sum*" Descartes claims to prepare a new and secure foundation for philosophy. But what he leaves undetermined in this "radical" beginning is the manner of being of the *res cogitans*, more precisely *the meaning of being of the "sum."* Working out the tacit ontological foundations of the "*cogito sum*" will constitute the second stage of the destruction of, and the path back into, the history of ontology. The interpretation will demonstrate not only that Descartes had to neglect the question of being altogether, but also why he held the opinion that the absolute "being-certain" of the *cogito* exempted him from the question of the meaning of the being of this being [Seienden].

However, with Descartes it is not just a matter of neglect and thus of a complete ontological indeterminateness of the *res cogitans sive mens sive animus* [the thinking thing, whether it be mind or spirit]. Descartes carries out the fundamental reflections of his *Meditations* by applying medieval ontology to this being [Seiende] which he posits as the *fundamentum inconcussum* [unshakable foundation]. The *res cogitans* is ontologically determined as *ens*, and for medieval ontology the meaning of the being of the *ens* is established in the understanding of it as *ens creatum*. As the *ens infinitum* God is the *ens increatum*. But

24

createdness, in the broadest sense of something having been produced, is an essential structural moment of the ancient concept of being. The ostensibly new beginning of philosophizing betrays the imposition of an ill-fated prejudice. On the basis of this prejudice later times neglect a thematic ontological analysis of "the mind" ["Gemütes"] which would be guided by the question of being; likewise they neglect a critical confrontation with the inherited ancient ontology.

Everyone familiar with the medieval period sees that Descartes is "dependent" upon medieval scholasticism and uses its terminology. But with this "discovery" nothing is gained philosophically as long as it remains obscure to what a profound extent medieval ontology influences the way posterity determines or fails to determine the *res cogitans* ontologically. The full extent of this influence cannot be estimated until the meaning and limits of ancient ontology have been shown by our orientation toward the question of being. In other words, the destruction sees itself assigned the task of interpreting the foundation of ancient ontology in light of the problematic of temporality [Temporalität]. Here it becomes evident that the ancient interpretation of the being of beings is oriented toward the "world" or "nature" in the broadest sense and that it indeed gains its understanding of being from "time." The outward evidence of this—but of course *only* outward—is the determination of the meaning of being as παρουσία or οὐσία, which ontologically and temporally means "presence" ["Anwesenheit"]. Beings are grasped in their being [Sein] as "presence"; that is to say, they are understood with regard to a definite mode of time, the *"present"* [*"Gegenwart"*].

The problem of Greek ontology must, like that of any ontology, take its guideline from Dasein itself. In the ordinary and also the philosophical "definition," Dasein, that is, the being of human being, is delineated as ζῷον λόγον ἔχον, that creature whose being is essentially determined by its ability to speak. λέγειν (cf. § 7b) is the guideline for arriving at the structures of being of the beings we encounter in speech and discussion. That is why the ancient ontology developed by Plato becomes "dialectic." The possibility of a more radical conception of the problem of being grows with the continuing development of the ontological guideline itself, that is, with the "hermeneutics" of the λόγος. "Dialectic," which was a genuine philosophic embarrassment, becomes superfluous. *Thus* Aristotle "no longer has any sympathy" for it, *because* he places it on a more radical foundation and transcends it. λέγειν itself, or νοεῖν—the simple apprehension of something objectively present in its sheer objective presence [Vorhandenheit], which Parmenides already used as a guide for interpreting being—has the temporal structure of a pure "making present" of something. Beings, which show themselves in and for this making present and which are understood as genuine beings, are accordingly interpreted with regard to the present; that is to say, they are conceived as presence (οὐσία).

However, this Greek interpretation of being comes about without any explicit knowledge of the guideline functioning in it, without taking cognizance of or understanding the fundamental ontological function of time, without insight into the ground of the possibility of this function. On the contrary, time itself is taken to be one being among others. The attempt is made to grasp time itself in the structure of its being within the horizon of an understanding of being which is oriented toward time in an inexplicit and naïve way.

Within the framework of the following fundamental elaboration of the question of being a detailed temporal interpretation of the foundations of ancient ontology—especially of its scientifically highest and purest stage, that is, in Aristotle—cannot be offered. Instead, we offer an interpretation of Aristotle's treatise on time,[2] which can be taken as a way of *discerning* the basis and limits of the ancient science of being.

Aristotle's treatise on time is the first detailed interpretation of this phenomenon that has come down to us. It essentially determined all the subsequent interpretations of time, including that of Bergson. From our analysis of Aristotle's concept of time it becomes retrospectively clear that the Kantian interpretation moves within the structures developed by Aristotle. This means that Kant's fundamental ontological orientation—despite all the differences implicit in a new inquiry—remains Greek.

The question of being attains true concreteness only when we carry out the destruction of the ontological tradition. By so doing we can prove the inescapability of the question of the meaning of being and thus demonstrate what it means to talk about a "retrieval" of this question.

In this field where "the matter itself is deeply veiled,"[3] any investigation should avoid overestimating its results. For such inquiry is constantly forced to face the possibility of disclosing a still more original and more universal horizon from which it could draw the answer to the question: what does 'being' mean? We can discuss such possibilities seriously and with a positive result only if the question of being has been reawakened and we have reached the point where we can come to terms with it in a controlled fashion.

---

2. Aristotle, *Physics* IV.10.217b29–14.224a17.
3. Kant, *Kritik der reinen Vernunft*, B 121.

§ 7. *The Phenomenological Method of the Investigation*

With the preliminary characterization of the thematic object of the investigation (the being of beings, or the meaning of being in general) its method too would appear to be already prescribed. The task of ontology is to set in relief the being of beings and to explicate being itself. And the method of ontology remains questionable in the highest degree as long as we wish merely to consult historically transmitted ontologies or similar efforts. Since the term ontology is used in a formally broad sense for this investigation, the approach of clarifying its method by tracing the history of that method is automatically precluded.

In using the term ontology we do not specify any particular philosophical discipline standing in relation to others. It should not at all be our task to satisfy the demands of any established discipline. On the contrary, such a discipline can be developed only from the objective necessity of particular questions and procedures demanded by the "things themselves."

With the guiding question of the meaning of being the investigation arrives at the fundamental question of philosophy in general. The treatment of this question is *phenomenological*. With this term the treatise dictates for itself neither a "standpoint" nor a "direction," because phenomenology is neither of these and can never be as long as it understands itself. The expression "phenomenology" signifies primarily a *concept of method*. It does not characterize the what of the objects of philosophical research in terms of their content, but the *how* of such research. The more genuinely effective a concept of method is and the more comprehensively it determines the fundamental conduct of a science, the more originally is it rooted in confrontation with the things themselves and the farther away it moves from what we call a technical device—of which there are many in the theoretical disciplines.

28 The term "phenomenology" expresses a maxim that can be formulated: "To the things themselves!" It is opposed to all free-floating constructions and accidental findings; it is also opposed to taking over concepts only seemingly demonstrated; and likewise to pseudo-questions which often are spread abroad as "problems" for generations. But one might object that this maxim is, after all, abundantly self-evident and, moreover, an expression of the principle of all scientific knowledge. It is not clear why this commonplace should be explicitly put in the title of our research. In fact, we are dealing with "something self-evident" which we want to get closer to, insofar as that is important for the clarification of the procedure in our treatise. We shall explicate only the preliminary concept of phenomenology.

The expression has two components: phenomenon and logos. Both go back to the Greek terms φαινόμενον and λόγος. Viewed extrinsically, the word phenomenology is formed like the terms theology, biology, sociology, translated as the science of God, of life, of the community. Accordingly, phenomenology would be the *science of phenomena*. The preliminary concept of phenomenology is to be exhibited by characterizing what is meant by the two components, "phenomenon" and "logos," and by establishing the meaning of the word which is the result of their *combination*. The history of the word itself, which originated presumably with the Wolffian school, is not important here.

A. The Concept of Phenomenon

The Greek expression φαινόμενον, from which the term "phenomenon" derives, comes from the verb φαίνεσθαι, meaning "to show itself." Thus φαινόμενον means: what shows itself, the self-showing, the manifest. φαίνεσθαι itself is a *middle voice* construction of φαίνω, to bring into daylight, to place in brightness. φαίνω belongs to the root φα-, like φῶς, light or brightness, that is, that within which something can become manifest, visible in itself. Thus the meaning of the expression *phenomenon* is *established as what shows itself in itself*, what is manifest. The φαινόμενα, "phenomena," are thus the totality of what lies in the light of day or can be brought to light. Sometimes the Greeks simply identified this with τὰ ὄντα (beings). Beings can show themselves from themselves in various ways, depending on the mode of access to them. The possibility even exists that they can show themselves as they are *not* in themselves. In this self-showing beings "look like . . .". Such self-showing [Sichzeigen] we call *seeming [Scheinen]*. And so the expression φαινόμενον, phenomenon, means in Greek: what looks like something, what "seems" [Scheinbar], "semblance" ["Schein"]. φαινόμενον ἀγαθόν means a good that looks like—but "in reality" is not what it gives itself out to be. It is extremely important for a further understanding of the concept of phenomenon to see how what is named in both meanings of φαινόμενον ("phenomenon" as self-showing and "phenomenon" as semblance) are structurally connected. Only because something claims to show itself in accordance with its meaning at all, that is, claims to be a phenomenon, *can* it show itself *as* something it is *not*, or *can* it "only look like . . .". The original meaning (phenomenon: what is manifest [das Offenbare]) already contains and is the basis of φαινόμενον ("semblance"). We attribute to the term "phenomenon" the positive and original meaning of φαινόμενον terminologically, and separate the phenomenon of semblance [Schein] from it as a privative modification. But what *both* terms express has at first nothing at all to do with what is called "appearance" ["Erscheinung"] or even "mere appearance" [bloße Erscheinung].

Thus, one speaks of "appearances of symptoms of illness." What is meant by this are occurrences in the body that show themselves and in this self-showing as such "indicate" something that does *not* show itself. When such occurrences emerge, their self-showing coincides with the presence [Vorhandensein] of disturbances that do not show themselves. Appearance, as the appearance "of something," thus precisely does *not* mean that something shows itself; rather, it means that something which does not show itself announces itself through something that does show itself. Appearing is* a *not showing itself*. But this "not" must by no means be confused with the privative not which determines the structure of semblance. What does *not* show itself, in *the* manner of what appears, can also never seem. All indications, presentations, symptoms, and symbols have this fundamental formal structure of appearing, although they do differ among themselves.

Although "appearing" is never a self-showing in the sense of phenomenon, appearing is possible only *on the basis* of a *self-showing* of something. But this, the self-showing that makes appearing possible, is not appearing itself. Appearing is an *announcing* of itself through something that shows itself. If we then say that with the word "appearance" we are pointing to something in which something appears without itself being an appearance, then the concept of phenomenon is not thereby delimited but *presupposed*. However, this presupposition remains hidden because the expression "to appear" in this definition of "appearance" is used in two senses. That in which something "appears" means that in which something makes itself known, that is, does not show itself; in the expression "without itself being an 'appearance'" appearance means the *self-showing*. But this self-showing essentially belongs to the "wherein" in which something makes itself known. Accordingly, phenomena are *never* appearances, but every appearance is dependent upon phenomena. If we define phenomenon with the help of a concept of "appearance" that is still unclear, then everything is turned upside down, and a "critique" of phenomenology on this basis is surely a remarkable enterprise.

The expression "appearance" itself in turn can have a double meaning. First, *appearing* in the sense of announcing itself as something that does not show itself and, second, in the sense of what does the announcing—that which in its self-showing indicates something that does not show itself. Finally, one can use appearing as the term for the genuine meaning of phenomenon as self-showing. If one designates these three different states of affairs as "appearance" confusion is inevitable.

However, this confusion is considerably increased by the fact that "appearance" can take on still another meaning. If one understands that which does the announcing—that which in its self-showing indicates

30

---

* in this case

the nonmanifest—as what comes to the fore in the nonmanifest itself, and radiates from it in such a way that what is nonmanifest is thought of as what is essentially *never* manifest, if this is so, then appearance is tantamount to production [Hervorbringung] or to what is produced [Hervorgebrachtes]. However, this does not constitute the real being of the producing or productive [Hervorbringenden], but is rather appearance in the sense of "mere appearance." What does the announcing and is brought forward indeed shows itself in such a way that, as the emanation of what it announces, it precisely and continually veils what it is in itself. But then again, this not-showing which veils is not semblance. Kant uses the term "appearance" in this twofold way. On the one hand, appearances are for him the "objects of empirical intuition," that which shows itself in intuition. This self-showing (phenomenon in the genuine, original sense) is, on the other hand, "appearance" as the emanation of something that makes itself known but *conceals* itself in the appearance.

Since a phenomenon is constitutive for "appearance" in the sense of announcing itself through a self-showing, and since this phenomenon can turn into semblance in a privative way, appearance can also turn into mere semblance. Under a certain kind of light someone can look as if he were flushed. The redness that shows itself can be taken as announcing the presence of fever; this in turn would indicate a disturbance in the organism.   *31*

Phenomenon—the self-showing in itself—means a distinctive way something can be encountered. On the other hand, *appearance* means a referential relation in beings themselves such that what does the *referring* (the announcing) can fulfill its possible function only if it shows itself in itself—only it if is a "phenomenon." Both appearance and semblance are themselves founded in the phenomenon, albeit in different ways. The confusing multiplicity of "phenomena" designated by the terms phenomenon, semblance, appearance, and mere appearance, can be unraveled only if the concept of phenomenon is understood from the very beginning as the self-showing in itself.

But if in the way we grasp the concept of phenomenon we leave undetermined which beings are to be addressed as phenomena, and if we leave altogether open whether the self-showing is actually a particular being or a characteristic of the being of beings, then we are dealing solely with the *formal* concept of phenomenon. If by the self-showing we understand those beings that are accessible, for example, through empirical intuition in Kant's sense, then the formal concept of phenomenon can be used legitimately. In this usage phenomenon has the meaning of the *vulgar* concept of phenomenon. But this vulgar concept is not the phenomenological concept of phenomenon. In the horizon of the Kantian problematic what is understood phenomenologically by the term phenomenon (disregarding other differences) can be illustrated when we say that what already shows itself in appearances, prior to and always accompanying what we commonly understand as phenomena (though unthematically), can be brought thematically to self-showing. What thus shows itself in itself ("the forms of intuition") are the phenomena of phenomenology. For, clearly, space and time must be able to show themselves in this way. They must be able to become phenomena if Kant claims to make a valid transcendental statement when he says that space is the *a priori* "wherein" of an ordering.

Now if the phenomenological concept of phenomenon is to be understood at all (regardless of how the self-showing may be more closely determined), we must inevitably presuppose insight into the sense of the formal concept of phenomenon and the legitimate use of phenomenon in its vulgar meaning. However, before determining the preliminary concept of phenomenology we must delimit the meaning of λόγος, in order to make clear in which sense phenomenology can be "a science of" phenomena.

B. The Concept of Logos

32   The concept of λόγος has many meanings in Plato and Aristotle, indeed in such a way that these meanings diverge without a basic meaning positively taking the lead. This is in fact only an illusion which lasts so long as an interpretation is not able to grasp adequately the basic meaning in its primary content. If we say that the basic meaning of λόγος is discourse [Rede], this literal translation becomes valid only when we define what discourse itself means. The later history of the word λόγος, and especially the manifold and arbitrary interpretations of subsequent philosophy, constantly conceal the authentic meaning of discourse— which is manifest enough. λόγος is "translated," and that always means interpreted, as reason, judgment, concept, definition, ground, relation. But how can "discourse" be so susceptible of modification that λόγος means all the things mentioned, and indeed in scholarly usage? Even if λόγος is understood in the sense of a statement, and statement as "judgment," this apparently correct translation can still miss the fundamental meaning—especially if judgment is understood in the sense of some contemporary "theory of judgment." λόγος does not mean judgment, in any case not primarily, if by judgment we understand "connecting two things" or "taking a position" (by endorsing or rejecting).

Rather, λόγος as discourse really means δηλοῦν, to make manifest "what is being talked about" in discourse. Aristotle explicates this function of discourse more precisely as ἀποφαίνεσθαι.[4] λόγος lets some-

---

4. Cf. *De interpretatione*, chaps. 1–6. See further, *Metaphysics* VII.4 and *Nicomachean Ethics* VII.

thing be seen (φαίνεσθαι), namely what is being talked about, and indeed *for* the speaker (who serves as the medium) or for those who speak with each other. Discourse "lets us see," ἀπὸ . . . from itself, what is being talked about. In discourse (ἀπόφανσις), insofar as it is genuine, *what* is said should be derived *from* what is being talked about. In this way spoken communication, in what it says, makes manifest what it is talking about and thus makes it accessible to another. Such is the structure of λόγος as ἀπόφανσις. Not every "discourse" suits *this* mode of making manifest, in the sense of letting something be seen by indicating it. For example, requesting (εὐχή) also makes something manifest, but in a different way.

When fully concrete, discourse (letting something be seen) has the character of speaking or vocalization in words. λόγος is φωνή, indeed   33 φωνὴ μετὰ φαντασίας—vocalization in which something always is sighted.

Only *because* the function of λόγος as ἀπόφανσις lies in letting something be seen by indicating it can λόγος have the structure of σύνθεσις. Here synthesis does not mean to connect and conjoin representations, to manipulate psychical occurrences, which then gives rise to the "problem" of how these connections, as internal, correspond to what is external and physical. The συν here has a purely apophantical meaning: to let something be seen in its *togetherness* with something, to let something be seen *as* something.

Furthermore, because λόγος lets something be seen, it can *therefore* be true or false. But everything depends on staying clear of any concept of truth construed in the sense of "correspondence" or "accordance" ["Übereinstimmung"]. This idea is by no means the primary one in the concept of ἀλήθεια. The "being true" of λόγος as ἀληθεύειν means: to take beings that are being talked *about* in λέγειν as ἀποφαίνεσθαι out of their concealment; to let them be seen as something unconcealed (ἀληθές); to *discover* them. Similarly "being false," ψεύδεσθαι, is tantamount to deceiving in the sense of *covering up*: putting something in front of something else (by way of letting it be seen) and thereby passing it off *as* something it is *not*.

But because "truth" has this meaning, and because λόγος is a specific mode of letting something be seen, λόγος simply may *not* be appealed to as the primary "place" of truth. If one defines truth as what "genuinely" pertains to judgment, which is quite customary today, and if one invokes Aristotle in support of this thesis, such a procedure is without justification and the Greek concept of truth thoroughly misunderstood. In the Greek sense what is "true"—indeed more originally true than the λόγος we have been discussing—is αἴσθησις, the simple sense perception of something. To the extent that an αἴσθησις aims at its ἴδια—that is, the beings which are genuinely accessible only *through* it and *for* it, for example, looking at colors—perception is always true. This means that looking always discovers colors, hearing always discovers tones. What is in the purest and most original sense "true"—that is, what only discovers in such a way that it can never cover up anything—is pure νοεῖν, straightforwardly observant apprehension of the simplest determinations of the being of beings as such. This νοεῖν can never cover up, can never be false; at worst it can be a *nonapprehending*, ἀγνοεῖν, not sufficing for straightforward, appropriate access.

34   What no longer takes the form of a pure letting be seen, but rather in its indicating always has recourse to something else and so always lets something be seen *as* something, acquires with this structure of synthesis the possibility of covering up. However, "truth of judgment" is only the opposite case of this covering up; it is a *multiply-founded* phenomenon of truth. Realism and idealism alike thoroughly miss the meaning of the Greek concept of truth from which alone the possibility of something like a "theory of Ideas" can be understood as philosophical *knowledge*.

And because the function of λόγος lies in letting something be seen straightforwardly, in *letting* beings be *apprehended*, λόγος can mean *reason*. Furthermore, because λόγος is used in the sense not only of λέγειν but also of λεγόμενον (what is pointed to as such), and because the latter is nothing other than the ὑποκείμενον (what always already lies present at the *basis* of all relevant speech and discussion), λόγος qua λεγόμενον means ground, *ratio*. Finally, because λόγος as λεγόμενον can also mean what is addressed, as something that has become visible in its relation to something else in its "relatedness," λόγος acquires the meaning of *relation* and *relationship*.

This interpretation of "apophantic speech" may suffice to clarify the primary function of λόγος.

C. The Preliminary Concept of Phenomenology

When one brings to mind concretely what has been exhibited in the interpretation of "phenomenon" and "logos" one is struck by an inner relation between what is meant by these terms. The expression "phenomenology" can be formulated in Greek as λέγειν τὰ φαινόμενα. But λέγειν means ἀποφαίνεσθαι. Hence phenomenology means: ἀποφαίνεσθαι τὰ φαινόμενα—to let what shows itself be seen from itself, just as it shows itself from itself. That is the formal meaning of the type of research that calls itself "phenomenology." But this expresses nothing other than the maxim formulated above: "To the things themselves!"

Accordingly, the term "phenomenology" differs in meaning from such expressions as "theology" and the like. Such titles designate the objects of the respective disciplines in terms of their content. "Phenomenology" neither designates the object of its researches nor is it a title that describes their content. The word only tells us something about the *how* of the demonstration and treatment of *what* this discipline considers. Science "of" the phenomena means that it grasps its objects in *such* a way that everything about them to be discussed must be directly indicated and directly demonstrated. The basically tautological expression "descriptive phenomenology" has the same sense. Here description does not mean a procedure like that of, say, botanical morphology. The term rather has the sense of a prohibition, insisting that we avoid all nondemonstrative determinations. The character of description itself, the specific sense of the λόγος, can be established only from the "content" ["Sachheit"] of what is "described," that is, of what is to be brought to scientific determinateness in the way phenomena are encountered. The meaning of the formal and vulgar concept of the phenomenon formally justifies our calling every way of indicating beings as they show themselves in themselves "phenomenology."

Now what must be taken into account if the formal concept of phenomenon is to be deformalized to the phenomenological one, and how does this differ from the vulgar concept? What is it that phenomenology is to "let be seen"? What is it that is to be called "phenomenon" in a distinctive sense? What is it that by its very essence becomes the *necessary* theme when we indicate something *explicitly*? Manifestly it is something that does *not* show itself initially and for the most part, something that is *concealed* [*verborgen*] in contrast to what initially and for the most part does show itself. But, at the same time, it is something that essentially belongs to what initially and for the most part shows itself, indeed in such a way that it constitutes its meaning and ground.*

But what remains *concealed* in an exceptional sense, or what falls back and is *covered up* [*Verdeckung*] again, or shows itself only in a "*disguised*" ["*verstellt*"] way, is not this or that being but rather, as we have shown in our foregoing observations, the *being* of beings. It can be covered up to such a degree that it is forgotten and the question about it and its meaning altogether omitted. Thus, phenomenology has taken into its "grasp" thematically as its object that which, in terms of its ownmost content, demands that it become a phenomenon in a distinct sense.

Phenomenology is the way of access to, and the demonstrative manner of determination of, that which is to become the theme of ontology. *Ontology is possible only as phenomenology.* The phenomenological concept of phenomenon, as self-showing, means the being of beings—its meaning, modifications, and derivatives. This self-showing is nothing arbitrary, nor is it something like an appearing. The being of beings can least of all be something "behind which" something else stands, something that "does not appear."

---

* Truth of being.

Essentially, nothing else stands "behind" the phenomena of phenomenology. Nevertheless, what is to become a phenomenon can be concealed. And it is precisely because phenomena are initially and for the most part *not* given that phenomenology is needed. Being covered up is the counterconcept to "phenomenon."

There are various ways phenomena can be covered up. In the first place, a phenomenon can be covered up in the sense that it is still completely *undiscovered*. There is neither knowledge nor lack of knowledge about it. Moreover, a phenomenon can be *submerged* [*verschüttet*]. This means it was once discovered but then got covered up again. This covering up can be total, but more commonly, what was once discovered may still be visible, though only as semblance. However, where there is semblance there is "being." This covering up as "dissimulation" ["Verstellung"] is the most frequent and the most dangerous kind, because here the possibilities of being deceived and misled are especially pernicious. Within a "system" the structures and concepts of being that are available, but concealed with respect to their autochthony, may perhaps claim their rights. On the basis of their integrated structure in a system they present themselves as something "clear" which is in no need of further justification and which therefore can serve as a point of departure for a process of deduction.

The covering up itself, whether it be understood in the sense of concealment, being submerged, or disguised, has in turn a twofold possibility. There are accidental coverings and necessary ones, the latter grounded in the enduring nature of the discovered. It is possible for every phenomenological concept and proposition drawn from genuine origins to degenerate when communicated as a statement. It gets circulated in a vacuous fashion, loses its autochthony, and becomes a freefloating thesis. Even in the concrete work of phenomenology itself there lurks the possibility of a calcification and of the inability to grasp what was originally "grasped." And the difficulty of this research consists precisely in making it self-critical in a positive sense.

The way of encountering being and the structures of being in the mode of phenomenon must first be *wrested* from the objects of phenomenology. Thus the *point of departure* of the analysis, the *access* to the phenomenon, and *passage through* the prevalent coverings must secure their own method. The idea of an "originary" and "intuitive" grasp and explication of phenomena must be opposed to the naïveté of an accidental, "immediate" and unreflective "beholding."

On the basis of the preliminary concept of phenomenology just delimited, the terms *"phenomenal"* and *"phenomenological"* can now be given fixed meanings. What is given and is explicable in the way we encounter the phenomenon is called "phenomenal." In this sense we speak of phenomenal structures. Everything that belongs to the manner

37

of indication and explication, and that constitutes the conceptuality this research requires, is called "phenomenological."

Because phenomenon in the phenomenological understanding is always just what constitutes being, and furthermore because being is always the being of beings, we must first of all bring beings themselves forward in the right way if we are to have any prospect of exposing being. These beings must likewise show themselves in the way of access that genuinely belongs to them. Thus the vulgar concept of phenomenon becomes phenomenologically relevant. The preliminary task of "phenomenologically" securing that being which is to serve as our example, as the point of departure for the analysis proper, is always already prescribed by the goal of this analysis.

As far as content goes, phenomenology is the science of the being of beings—ontology. In our elucidation of the task of ontology the necessity arose for a fundamental ontology which would have as its theme that being which is ontologically and ontically distinctive, namely, Dasein. This must be done in such a way that our ontology confronts the cardinal problem, the question of the meaning of being in general.* From the investigation itself we shall see that the methodological meaning of phenomenological description is *interpretation*. The λόγος of the phenomenology of Dasein has the character of ἑρμηνεύειν, through which the proper meaning of being and the basic structures of the very being of Dasein are *made known* to the understanding of being that belongs to Dasein itself. Phenomenology of Dasein is *hermeneutics* in the original signification of that word, which designates the work of interpretation. But since the discovery of the meaning of being and of the basic structures of Dasein in general exhibits the horizon for every further ontological research into beings unlike Dasein, the present hermeneutic is at the same time "hermeneutics" in the sense that it works out the conditions of the possibility of every ontological investigation. Finally, insofar as Dasein has ontological priority over all other beings—as a being in the possibility of existence [Existenz]— hermeneutics, as the interpretation of the being of Dasein, receives the third specific and, philosophically understood, *primary* meaning of an analysis of the existentiality of existence. To the extent that this hermeneutic elaborates the historicity of Dasein ontologically as the ontic condition of the possibility of the discipline of history, it contains the roots of what can be called "hermeneutics" only in a derivative sense: the methodology of the historical humanistic disciplines.

As the fundamental theme of philosophy, being is not a genus of beings; yet it pertains to every being. Its "universality" must be

38

---

* being—not a genus, not being for beings generally; the "in general" = καθόλου = as the whole of: *being of* beings; meaning of difference.

sought in a higher sphere. Being and its structure transcend every being and every possible existent determination of a being. *Being is the transcendens pure and simple.** The transcendence of the being of Dasein is a distinctive one since in it lies the possibility and necessity of the most radical *individuation.* Every disclosure of being as the *transcendens* is *transcendental* knowledge. *Phenomenological truth (disclosedness of being) is veritas transcendentalis.*

Ontology and phenomenology are not two different disciplines which among others belong to philosophy. Both terms characterize philosophy itself, its object and procedure. Philosophy is universal phenomenological ontology, taking its departure from the hermeneutic of Dasein, which, as an analysis of *existence* [*Existenz*],† has fastened the end of the guideline of all philosophical inquiry at the point from which it *arises* and to which it *returns.*

The following investigations would not have been possible without the foundation laid by Edmund Husserl; with his *Logical Investigations* phenomenology achieved a breakthrough. Our elucidations of the preliminary concept of phenomenology show that its essential character does not consist in its *actuality* as a philosophical "movement."‡ Higher than actuality stands *possibility.* We can understand phenomenology solely by seizing upon it as a possibility.[5]

39   With regard to the awkwardness and "inelegance" of expression in the following analyses, we may remark that it is one thing to report narratively about *beings* and another to grasp beings in their *being.* For the latter task not only most of the words are lacking but above all the "grammar." If we may allude to earlier and in their own right altogether incomparable researches on the analysis of being, then we should compare the ontological sections in Plato's *Parmenides* or the fourth chapter of the seventh book of Aristotle's *Metaphysics* with a narrative passage from Thucydides. Then we can see the stunning character of the formulations with which their philosophers challenged

---

* of course not *transcendens*—despite every metaphysical resonance—scholastic and Greek-Platonic κοινόν, rather transcendence as the ecstatic—temporality [Zeitlichkeit]—temporality [Temporalität]; but "horizon"! Beyng [Seyn] has "thought beyond" ["überdacht"] beyngs [Seyendes]. However, transcendence from the truth of beyng [Seyns]: the event [das Ereignis].

† "Existence" fundamental ontologically, i.e., itself related to the truth of being, and only in this way!

‡ i.e., not in the transcendental-philosophic direction of Kantian critical idealism.

5. If the following investigation takes any steps forward in disclosing "the things themselves" the author must above all thank E. Husserl, who by providing his own incisive personal guidance and by very generously turning over his unpublished investigations familiarized the author during his student years in Freiburg with the most diverse areas of phenomenological research.

the Greeks. Since our powers are essentially inferior, and also since the area of being to be disclosed ontologically is far more difficult than that presented to the Greeks, the complexity of our concept-formation and the severity of our expression will increase.

§ 8. *The Outline of the Treatise*

The question of the meaning of being is the most universal and the emptiest. But at the same time the possibility inheres of its most acute individualization in each particular Dasein.* If we are to gain the fundamental concept of "being" and the prescription of the ontologically requisite conceptuality in all its necessary variations, we need a concrete guideline. The "special character" of the investigation does not belie the universality of the concept of being. For we may advance to being by way of a special interpretation of a particular being, Dasein, in which the horizon for an understanding and a possible interpretation of being is to be won. But this being is in itself "historical," so that its most proper ontological illumination necessarily becomes a "historical" interpretation.

The elaboration of the question of being is a two-pronged task; our treatise therefore has two divisions:

Part One: The interpretation of Dasein on the basis of temporality [Zeitlichkeit] and the explication of time as the transcendental horizon of the question of being.

Part Two: Basic features of a phenomenological destruction of the history of ontology along the guideline of the problem of temporality [Temporalität].

The first part is divided into *three divisions*:

1. The preparatory fundamental analysis of Dasein.
2. Dasein and temporality [Zeitlichkeit].
3. Time and being.†

The second part is likewise divided in *three ways*:   40

1. Kant's doctrine of the schematism and of time, as preliminary stage of a problem of temporality [Temporalität].
2. The ontological foundation of Descartes' *"cogito sum"* and the incorporation of medieval ontology in the problem of the *"res cogitans."*
3. Aristotle's treatise on time as a way of discerning the phenomenal basis and the limits of ancient ontology.

---

* authentic: bringing about standing-within the there [Inständigkeit].
† The difference bound to transcendence [transzendenzhafte Differenz]. The overcoming of the horizon as such. The turn back into the source [Herkunft]. The presencing from out of this source.

PART ONE

The Interpretation of Dasein in Terms of
Temporality [Zeitlichkeit]* and the Explication
of Time as the Transcendental Horizon
of the Question of Being†

DIVISION ONE

The Preparatory Fundamental Analysis of Dasein

What is primarily interrogated in the question of the meaning of being   *41*
[Sein] is that being [Seiende] which has the character of Dasein. In
keeping with its uniqueness, the preparatory existential analytic of
Dasein itself needs a prefigurative exposition and delimitation from
investigations which seem to run parallel (Chapter One). Bearing in
mind the point of departure of the investigation, we must analyze a
fundamental structure of Dasein: being-in-the-world [In-der-Welt-sein]
(Chapter Two). This *"a priori"* of the interpretation of Dasein is not
a structure which is pieced together, but rather a structure which is
primordially and constantly whole. It grants various perspectives on
the factors which constitute it. These factors are to be kept constantly
in view, bearing in mind the preceding whole of this structure. Thus,
we have as the object of our analysis: the world in its worldliness
(Chapter Three), being-in-the-world as being a self and being with
others (Chapter Four), being-in as such (Chapter Five). On the basis
of the analysis of this fundamental structure, a preliminary demonstration of the being of Dasein is possible. Its existential meaning is *care*
(Chapter Six).

---

* Only this in the part published.
† Cf. the Marburg lecture, SS 1927 (*Die Grundprobleme der Phänomenologie*).

CHAPTER ONE

The Exposition of the Task of a Preparatory Analysis of Dasein

§ 9. *The Theme of the Analytic of Dasein*

The being whose analysis our task is, is always we* ourselves. The being of this being is always *mine*. In the being of this being it is related to its being.† As the being of this being, it is entrusted to its own being. It is *being*‡ about which this being is concerned. From this characteristic of Dasein two things follow:

1. The "essence" ["Wesen"] of this being lies in its to be.§ The whatness (*essentia*) of this being must be understood in terms of its being (*existentia*) insofar as one can speak of it at all. Here the ontological task is precisely to show that when we choose the word existence for the being of this being, this term does not and cannot have the ontological meaning of the traditional expression of *existentia*. According to the tradition, *existentia* ontologically means *being present* [*Vorhandensein*], a kind of being which is essentially inappropriate to characterize the being which has the character of Dasein. We can avoid confusion by always using the interpretive expression *objective presence* [*Vorhandenheit*] for the term *existentia*, and by attributing existence [Existenz] as a determination of being only to Dasein.

The "essence" ["Wesen"] of Dasein lies in its existence [Existenz]. The characteristics to be found in this being are thus not present "attributes" of an objectively present being which has such and such an "outward appearance," but rather possible ways for it to be, and only this. All being, one way or another, of this being is primarily being [sein]. Thus the term "Dasein," which we use to designate this being,

42

---

* always "I"
† But this is historical being-in-the-world.
‡ Which one? To be [zu sein] the There [Das Da] and thus to persist [bestehen] in beyng [Seyn] as such.
§ that it "has" to be [zu seyn]; definition!

does not express its what—as in the case of table, house, tree—but rather being [Sein].*

2. The being [Sein] which this being [Seienden] is *concerned about* in its being [Sein] is always my own. Thus, Dasein is never to be understood ontologically as a case and instance of a genus of beings objectively present. To something objectively present its being is a matter of "indifference," ["gleichgültig"], more precisely, it "is" in such a way that its being can neither be indifferent nor non-indifferent to it. In accordance with the character of *always-being-my-own-being* [*Jemeinigkeit*], when we speak of Dasein, we must always use the *personal* pronoun along with whatever we say: "I am," "you are."†

Dasein is my own, to be always in this or that way. It has somehow always already decided in which way Dasein is always my own. The being which is concerned in its being about its being is related to its being as its ownmost possibility. Dasein *is* always its possibility. It does not "have" that possibility only as a mere attribute of something objectively present. And because Dasein is always essentially its possibility, it *can* "choose" itself in its being, it can win itself, it can lose itself, or it can never and only "apparently" win itself. It can only have lost itself and it can only have not yet gained itself because it is essentially possible as authentic, that is, it belongs to itself. The two kinds of being [Seinsmodi] of *authenticity* [*Eigentlichkeit*] and *inauthenticity* [*Uneigentlichkeit*]—these expressions are terminologically chosen in the strictest sense of the word—are based on the fact that Dasein is in general determined by always being-mine. But the inauthenticity of Dasein does not signify a "lesser" being or a "lower" degree of being. Rather, inauthenticity can determine Dasein even in its fullest concretion, when it is busy, excited, interested, and capable of pleasure.

43

The two characteristics of Dasein sketched out—on the one hand, the priority of *"existentia"* over *essentia*, and then, always-being-mine—already show that an analytic of this being is confronted with a unique phenomenal region. This being does not have and never has the kind of being of what is merely objectively present within the world. Thus, it is also not to be thematically found in the manner of coming across something objectively present. The correct presentation of it is so little a matter of course that its determination itself constitutes an essential part of the ontological analytic of this being. The possibility of understanding the being of this being stands and falls with the secure accomplishment of the correct presentation of this being. No matter how provisional the analysis may be, it always demands the securing of the correct beginning.

---

* The beyng [Seyn] 'of' the There, 'of': *genitivus objectivus.*
† That is, mineness [Jemeinigkeit] means being appropriated [Übereignetheit].

As a being, Dasein always defines itself in terms of a possibility which it *is*, and that means at the same time that it somehow understands itself in its being. That is the formal meaning of the constitution of the existence of Dasein. But for the *ontological* interpretation of this being, this means that the problematic of its being* is to be developed out of the existentiality of its existence. However, this cannot mean that Dasein is to be construed in terms of a concrete possible idea of existence. At the beginning of the analysis, Dasein is precisely not to be interpreted in the differentiation of a particular existence; rather, it is to be uncovered in the indifferent way in which it is initially and for the most part. This indifference [Indifferenz] of the everydayness of Dasein is *not nothing*; but rather, a positive phenomenal characteristic. All existing is how it is out of this kind of being, and back into it. We call this everyday indifference of Dasein *averageness* [*Durchschnittlichkeit*].

And because average everydayness constitutes the ontic immediacy of this being, it was and will be *passed over* again and again in the explication of Dasein. What is ontically nearest and familiar is ontologically the farthest, unrecognized and constantly overlooked in its ontological significance. Augustine asks: "*Quid autem propinquius meipso mihi?*" ["But what is closer to me than myself?"] and must answer: "*Ego certe laboro hic et laboro in meipso: factus sum mihi terra difficultatis et sudoris nimii*" ["Assuredly I labor here and I labor within myself: I have become to myself a land of trouble and inordinate sweat"].[1] This holds true not only for the ontic and preontological opacity of Dasein, but to a still higher degree for the ontological task of not only not failing to see this being in its phenomenally nearest kind of being, but of making it accessible in its positive characteristics.

But the average everydayness of Dasein must not be understood as a mere "aspect." In it, too, and even in the mode of inauthenticity, the structure of existentiality lies *a priori*. In it, too, Dasein is concerned in a particular way about its being to which it is related in the mode of average everydayness, if only in the mode of fleeing *from* it and of forgetting *it*.

The explication of Dasein in its average everydayness, however, does not just give average structures in the sense of a vague indeterminacy. What *is* ontically in the way of being average can very well be understood ontologically in terms of pregnant structures which are not structurally different from the ontological determinations of an *authentic* being of Dasein.

44

---

* better: of its understanding of being

1. St. Augustine, *Confessions* X.16.

All explications arising from an analytic of Dasein are gained with a view toward its structure of existence. Because these explications are defined in terms of existentiality, we shall call the characteristics of being of Dasein *existentials*. They are to be sharply delimited from the determinations of being of those beings unlike Dasein which we call *categories*. This expression is taken and retained in its primary ontological signification. As the exemplary basis of its interpretation of being, ancient ontology takes the beings which we encounter within the world. νοεῖν or λόγος was regarded as the manner of access to those beings. It is there that beings are encountered. The being of these beings, however, must become comprehensible in a distinctive λέγειν (a letting be seen), so that this being [Sein] is comprehensible from the very beginning as what it is and already is in every being [Seienden]. In the discussion (λόγος) of beings, we have always previously addressed ourselves to being; this addressing is κατηγορεῖσθαι. That means, first of all: to accuse publicly, to say something to someone directly and in front of everyone. Used ontologically, the term means: to say something to a being, so to speak, right in the face, to say what it always already is as a being; that is, to let it be seen for everyone in its being. What is caught sight of in such seeing and what becomes visible are the κατηγορίαι. They include the *a priori* determinations of the beings which can be addressed and spoken about in the λόγος in different ways. Existentials and categories are the two fundamental possibilities of the characteristics of being. The being [Seiende] which corresponds to them requires different ways of primary interrogation. Beings are a *who* (existence) or else a *what* (objective presence in the broadest sense). It is only in terms of the clarified horizon of the question of being that we can treat the connection between the two modes of characteristics of being.

It was intimated in the introduction that a task is furthered in the existential analytic of Dasein, a task whose urgency is hardly less than that of the question of being itself: the exposition of *the a priori* which must be visible if the question "What is human being?" is to be discussed philosophically. The existential analytic of Dasein is *prior* to any psychology, anthropology, and especially biology. By being delimited from these possible investigations of Dasein, the theme of the analytic can become still more sharply defined. Its necessity can thus at the same time be demonstrated more incisively.

§ 10. *How the Analytic of Dasein is to be Distinguished from Anthropology, Psychology, and Biology*

After a theme for investigation has been initially outlined in positive terms, it is always important to show what is to be ruled out, although it can easily become unfruitful to discuss what is not going to happen. We must show that all previous questions and investigations* which aim at Dasein fail to see the real *philosophical* problem, regardless of their factual productivity. Thus, as long as they persist in this attitude, they may not claim to *be able* to accomplish what they are fundamentally striving for at all. In distinguishing the existential analytic from anthropology, psychology, and biology, we shall confine ourselves to what is in principle the fundamental ontological question. Thus, our distinctions will be of necessity inadequate for a "theory of science" simply because the scientific structure of the above-mentioned disciplines (not the "scientific attitude" of those who are working to further them) has today become completely questionable and needs new impulses which must arise from the ontological problematic.

Historiographically, the intention of the existential analytic can be clarified by considering Descartes, to whom one attributes the discovery of the *cogito sum* as the point of departure for all modern philosophical questioning. He investigates the *cogitare* of the *ego*—within certain limits. But the *sum* he leaves completely undiscussed, even though it is just as primordial as the *cogito*. Our analytic raises the ontological question of the being of the *sum*. Only when the *sum* is defined does the manner of the *cogitationes* become comprehensible.

At the same time, it is of course misleading to exemplify the intention of the analytic historiographically in this way. One of our first tasks will be to show that the point of departure from an initially given *ego* and subject totally fails to see the phenomenal content of Dasein. Every idea of a "subject"—unless refined by a previous ontological determination of its basic character—still posits the *subjectum* (ὑποκείμενον) *ontologically* along with it, no matter how energetic one's ontic protestations against the "substantial soul" or the "reification of consciousness." Thingliness itself needs to be demonstrated in terms of its ontological source in order that we can ask what is now to be understood *positively* by the nonreified *being* of the subject, the soul, consciousness, the spirit, the person. All these terms name definite areas of phenomena which can be "developed." But they are never used without a remarkable failure to see the need for inquiring about the being of the beings so designated. Thus we are not being terminologically idiosyncratic when we avoid these terms as well as the expressions "life" ["Leben"] and "human being" ["Mensch"] in designating the beings that we ourselves are.

On the other hand, if we understand it correctly, in any serious and scientifically minded "philosophy of life" ["Lebensphilosophie"] (this expression says about as much as the "botany of plants") there

46

* They did not aim at Dasein at all.

lies an inexplicit tendency toward understanding the being of Dasein.*
What strikes us first of all in such a philosophy (and this is its fundamental lack)† is that "life" itself as a kind of being does not become
a problem ontologically.

    Dilthey's investigations are motivated and sustained by the peren-
nial question of "life." Starting from "life" itself as a whole, he attempts
to understand its "experiences" in their structural and developmental
interconnections. What is philosophically relevant about his "humanistic psychology" is not to be found in the fact that it is no longer oriented toward psychic elements and atoms and no longer tries to piece
together the life of the soul, but rather aims at the "whole of life" and
"gestalt." Rather, it is to be found in the fact that in the midst of all
this he was, *above all*, on the way to the question of "life." It is true that
we can see here very plainly the limits of his problematic and of the
set of concepts with which it had to be expressed. But along with Dilthey and Bergson, all the directions of "personalism" and all tendencies
toward a philosophical anthropology influenced by them share these
limits. The phenomenological interpretation of personality is in principle more radical and transparent; but it does not reach the dimension
of the question of being in Dasein either. Despite all their differences
in questioning, development, and orientation of their worldviews, the
interpretations of personality found in Husserl[2] and Scheler agree in
what is negative. They no longer ask the question about the "*being* of
the person." We choose Scheler's interpretation as an example, not only
because it is accessible in print,[3] but because he explicitly emphasizes
the being of the person as such, and attempts to define it by defin-

47

---

* no!
† Not only that, but the question of truth is totally and essentially inadequate.

2. E. Husserl's investigations on "personality" have not yet been published. The fundamental orientation of the problematic is already evident in the treatise "Philosophie als
strenge Wissenschaft," *Logos* 1 (1910), p. 319. The investigation is extensively furthered in
the second part of *Ideen zu einer Phänomenologie und phänomenologischen Philosophie* (Hus-
serliana IV), the first part of which (cf. this *Jahrbuch*, vol. 1, 1913) presents the problematic
of "pure consciousness" as the basis for investigating the constitution of every possible
reality. The second part gives developmental constitutional analyses and treats in three
sections: 1. The constitution of material nature. 2. The constitution of animal nature. 3.
The constitution of the spiritual world (the personalistic formula in contrast with the
naturalistic one). Husserl begins his presentation with the words: "Dilthey . . . recognized the problems which present the goal, the directions which the work to be done
should take, but he did not press on to the decisive formulations of the problem and the
solutions that are methodologically assured." After this first attempt, Husserl pursued
the problems in a more penetrating way and communicated essential sections of this
in his Freiburg lectures.*
    * But in the goal and result everything is different from what is sought
      and achieved here.
3. Cf. this *Jahrbuch*, vols. 1, 2 (1913) and 2 (1916), especially p. 242ff.

ing the specific being of acts as opposed to everything "psychical." According to Scheler, the person can never be thought of as a thing or a substance. Rather it is "the immediately co-experienced *unity* of experiencing—not just a thing merely thought behind and outside of what is immediately experienced."[4] The person is not a thinglike substantial being. Furthermore, the being of the person cannot consist in being a subject of rational acts that have a certain lawfulness.

The person is not a thing, not a substance, not an object. Here Scheler emphasizes the same thing which Husserl[5] is getting at when he requires for the unity of the person a constitution essentially different from that of things of nature. What Scheler says of the person, he applies to acts as well. "An act is never also an object, for it is the nature of the being of acts only to be experienced in the process itself and given in reflection."[6] Acts are nonpsychical. Essentially the person exists only in carrying out intentional acts, and is thus essentially *not* an object. Every psychical objectification, and thus every comprehension of acts as something psychical, is identical with depersonalization. In any case, the person is given as the agent of intentional acts which are connected by the unity of a meaning. Thus psychical being has nothing to do with being a person. Acts are carried out, the person carries them out. But what is the ontological meaning of "carrying out," how is the kind of being of the person to be defined in an ontologically positive way? But the critical question cannot stop at this. The question is about the being of the whole human being, whom one is accustomed to understand as a bodily-soul-like-spiritual unity. Body, soul, spirit might designate areas of phenomena which are thematically separable for the sake of determinate investigations; within certain limits their ontological indeterminancy might not be so important. But in the question of the being of human being, this cannot be summarily calculated in terms of the kinds of being of body, soul, and spirit which have yet first to be defined. And even for an ontological attempt which is to proceed in this way, some idea of the being of the whole would have to be presupposed. But what obstructs or misleads the basic question of the being of Dasein is the orientation thoroughly colored by the anthropology of Christianity and the ancient world, whose inadequate ontological foundations personalism and the philosophy of life also ignore. Traditional anthropology contains the following:

1. The definition of human being: ζῷον λόγον ἔχον in the interpretation: *animal rationale*, rational life. The kind of being of the

---

4. Ibid., vol. 2, p. 385.
5. Cf. *Logos* 1, op. cit.
6. Ibid., p. 388.

ζῷον, however, is understood here in the sense of being present and occurring. The λόγος is a higher endowment whose kind of being remains just as obscure as that of the being so pieced together.

2. The other guideline for the determination of the being and essence of human being is a *theological* one: καὶ εἶπεν ὁ θεός ποιήσωμεν ἄνθρωπον κατ᾽ εἰκόνα ἡμετέραν καὶ καθ᾽ ὁμοίωσιν; *faciamus hominem ad imaginem nostram et similitudinem nostram.*[7] From this, Christian theological anthropology, taking over the ancient definition, gets an interpretation of the being we call human being. But just as the being [Sein] of God is ontologically interpreted by means of ancient ontology, so is the being of the *ens finitum*, to an even greater extent. The Christian definition was de-theologized in the course of the modern period. But the idea of "transcendence"—that human being is something that goes beyond itself—has its roots in Christian dogma, which can hardly be said to have ever made an ontological problem of the being of human being. This idea of transcendence, according to which the human being is more than a rational being [Verstandeswesen], has elaborated itself in various transformations. We can illustrate its origin with the following quotations: "*His praeclaris dotibus excelluit prima hominis conditio, ut ratio, intelligentia, prudentia, iudicium, non modo ad terrenae vitae gubernationem suppeterent, sed quibus* transscenderet *usque ad Deum et aeternam felicitatem.*"[8] "For the fact that human being *looks toward* God and His word clearly shows that according to his nature he is born closer to God, is more *similar* to God, is somehow *drawn toward* God, that without doubt everything flows solely from the fact that he is created in the *image* of God."[9]

The sources which are relevant for traditional anthropology—the Greek definition and the theological guideline—indicate that, over and above the attempt to determine the essence of "human being" as a being, the question of its being has remained forgotten; rather, this being [Sein] is understood as something "self-evident" in the sense of the *being present* of other created things. These two guidelines intertwine in modern anthropology, where the *res cogitans*, consciousness, and the context of experience, serve as the methodological point of departure. But since these *cogitationes* are also ontologically undetermined, or are again inexplicitly and "self-evidently" taken as something "given" whose "being" is not a matter of question, the anthropological problematic remains undetermined in its decisive ontological foundation.

---

7. Genesis I.26 ["And God said, 'Let us make man in our image, after our likeness.' "]
8. Calvin, *Institutio* I, 15, § 8.
9. Zwingli, *Von klarheit und gewüsse des worts Gottes* (*Deutsche Schriften* I, 58).

This is no less true of *"psychology,"* whose anthropological tendencies are unmistakable today. Nor can the missing ontological foundations be replaced by building anthropology and psychology into a
general *biology*. In the order of possible understanding and interpretation, biology as the "science of life" is rooted in the ontology of Dasein,   50
although not exclusively in it. Life has its own kind of being, but it is
essentially accessible only in Dasein. The ontology of life takes place
by way of a privative interpretation. It determines what must be the
case if there can be anything like just-being-alive. Life is neither sheer
being present, nor is it Dasein. On the other hand, Dasein should never
be defined ontologically by regarding it as life—(ontologically undetermined) and then as something else on top of that.

In suggesting that anthropology, psychology, and biology all fail
to give an unequivocal and ontologically adequate answer to the question of the *kind of being* of this being that we ourselves are, no judgment
is being made about the positive work of these disciplines. But, on the
other hand, we must continually be conscious of the fact that these
ontological foundations can never be disclosed subsequently from
empirical material by the use of hypotheses. Rather, they are always
already "there," even when that empirical material is only *collected*. The
fact that positivistic investigation does not see these foundations and
considers them to be self-evident is no proof of the fact that they do
not lie at the basis of any thesis of positive science and are problematic
in a more radical sense than such science can ever be.[10]

§ 11. *The Existential Analytic and the Interpretation of Primitive
Dasein: The Difficulties in Securing a
"Natural Concept of World"*

The interpretation of Dasein in its everydayness [Alltäglichkeit], however, is not identical with describing a primitive [primitiven] stage of
Dasein, with which we can become acquainted empirically through the
medium of anthropology. *Everydayness is not the same as primitiveness.*
Rather, everydayness is also and precisely a mode of being of Dasein,
even when Dasein moves in a highly developed and differentiated
culture. On the other hand, primitive Dasein also has its possibilities   51
of noneveryday being, and it has *its* own specific everydayness. To
orient the analysis of Dasein toward "the life of primitive peoples" can

---

10. But the discovery of the *a priori* is not an *"a prioristic"* construction. Through Husserl
we have again learned not only to understand the meaning of all genuine philosophical
"empiricism," but we have also learned to use the tools necessary for it. *"A priorism"* is
the method of every scientific philosophy which understands itself. Because *a priorism*
has nothing to do with construction, the investigation of the *a priori* requires the proper
preparation of the phenomenal foundation. The nearest horizon which must be prepared
for the analytic of Dasein lies in its average everydayness.

have a positive methodical significance in that "primitive phenomena" are often less hidden and complicated by extensive self-interpretation on the part of the Dasein in question. Primitive Dasein often speaks out of a more primordial [ürsprunglichen] absorption in "phenomena" (in the pre-phenomenological sense). The conceptuality which perhaps appears to be clumsy and crude to us can be of use positively for a genuine elaboration of the ontological structures of phenomena.

But up until now our information about primitive peoples has been provided by ethnology. And ethnology already moves in certain preliminary concepts and interpretations of human being in general, beginning with the initial "collection" of its materials, its findings and elaborations. We do not know whether commonplace psychology or even scientific psychology and sociology, which the ethnologist brings with him, offer any scientific guarantee for an adequate possibility of access, interpretation, and mediation of the phenomena to be investigated. The situation here is the same as with the disciplines mentioned before. Ethnology itself already presupposes an adequate analytic of Dasein as its guideline. But since the positive sciences neither "can" nor should wait for the ontological work of philosophy, the continuation of research will not be accomplished as "progress"; but, rather, as the *repetition* and the ontologically more transparent purification of what has been ontically discovered.[11]

52   Although the formal differentiation of the ontological problematic as opposed to ontic investigation may seem easy, the development and above all the *approach* [*Ansatz*] of an existential analytic of Dasein is not without difficulties. A need is contained in this task which has made philosophy uneasy* for a long time, but philosophy fails again and again in fulfilling the task: *working out the idea of a "natural concept of world."* The wealth of knowledge of the most far-flung and manifold cultures and forms of existence available today seems favorable

---

* Not at all! The concept of world is not understood at all.

11. Recently E. Cassirer has made mythical Dasein the theme of a philosophical interpretation. Cf. *Philosophie der symbolischen Formen*, part 2, "Das mythische Denken," 1925. Through this investigation more comprehensive guidelines are made available to ethnological investigation. Viewed in terms of the philosophical problematic the question remains whether the foundations of the interpretation are sufficiently transparent, whether especially the architectonic of Kant's *Kritik der reinen Vernunft* and its systematic content are able to offer the possible outline for such a task at all, or whether a new and more primordial beginning is not necessary here. Cassirer himself sees the possibility of such a task as is shown in the footnote on pages 16ff., where Cassirer refers to the phenomenological horizon disclosed by Husserl. In a conversation which the author was able to have with Cassirer on the occasion of a lecture in the Hamburg group of the Kantian Society in December 1923, a lecture on "Aufgaben und Wege der phänomenologischen Forschung," an agreement as to the necessity of an existential analytic which was sketched out in the lecture already became apparent.

to taking up this task in a fruitful way. But that is only an illusion. Fundamentally, this plethora of information seduces us into failing to see the real problem. The syncretistic comparison and classification of everything does not of itself give us genuine essential knowledge. Subjecting the manifold to tabulation does not guarantee a real understanding of what has been ordered. The genuine principle of order has its own content which is never found by ordering, but is rather already presupposed in ordering. Thus the explicit idea of world as such is a prerequisite for the order of world images. And if "world" itself is constitutive of Dasein, the conceptual development of the phenomenon of world requires an insight into the fundamental structures of Dasein.

The positive characteristics and negative considerations of this chapter had the goal of leading the understanding of the basic inclinations and kind of questions in the following interpretation onto the correct path. Ontology can only contribute indirectly to the furtherance of existing positivistic disciplines. It has a goal of its own, provided that the question of being is the spur for all scientific search over and above the acquisition of information about beings.

CHAPTER TWO
Being-in-the-World in General as the Fundamental Constitution of Dasein

§ 12. *The Preliminary Sketch of Being-in-the-World*
[In-der-Welt-sein] in Terms of the Orientation
toward Being-in [In-Sein] as Such

In the preparatory discussion (§ 9) we already profiled characteristics of being which are to provide us with a steady light for our further investigation, but which at the same time receive their structural concretion in this investigation. Dasein is a being which is related understandingly in its being toward that being [Sein]. In saying this we are indicating the formal concept of existence. Dasein exists. Furthermore, Dasein is the being which I myself always am. Mineness [Jemeinigkeit] belongs to existing Dasein as the condition of the possibility of authenticity and inauthenticity. Dasein exists always in one of these modes, or else in the modal indifference [Indifferenz] to them.

53

These determinations of the being [Seinsbestimmungen] of Dasein, however, must now be seen and understood *a priori* on the basis of that constitution of being which we call *being-in-the-world*. The correct point of departure of the analytic of Dasein consists in the interpretation of this constitution.

The compound expression "being-in-the-world" indicates, in the very way we have coined it, that it stands for a *unified* phenomenon. This primary datum must be seen as a whole. But while being-in-the-world cannot be broken up into components that may be pieced together, this does not prevent it from having a multiplicity of constitutive structural factors. The phenomenal fact indicated by this expression actually gives us a threefold perspective. If we pursue it while keeping the whole phenomenon in mind from the outset we have the following:

1. *"In-the-world"*: In relation to this factor, we have the task of questioning the ontological structure of "world" and of defining the idea of *worldliness* as such (cf. Chapter Three of this Division).
2. The *being* [Seiende] which always has being-in-the-world as the way it is. In it we are looking for what we are questioning when we ask about the "who?". In our phenomenological demonstration we should be able to determine who is in the mode of the average everydayness of Dasein (cf. Chapter Four of this Division).
3. *Being in* as such: The ontological constitution of in-ness itself is to be analyzed (cf. Chapter Five of this Division). Any analysis of one of these constitutive factors involves the analysis of the others; that is, each time seeing the whole phenomenon. It is true that being-in-the-world is an *a priori* necessary constitution of Dasein, but it is not at all sufficient to fully determine Dasein's being. Before we thematically analyze the three phenomena indicated individually, we shall attempt to orient ourselves toward a characteristic of the third of these constitutive factors.

What does *being-in* mean? We supplement the expression being-in right away with the phrase "in the world," and are inclined to understand this being-in [In-Sein] as "being in . . ." ["Sein in . . ."]. With this term, the kind of being of a being is named which is "in" something else, as water is "in" the glass, the dress is "in" the closet. By this "in" we mean the relation of being that two beings extended "in" space have to each other with regard to their location in that space. Water and glass, dress and closet, are both "in" space "at" a location in the same way. This relation of being can be expanded; that is, the bench in the lecture hall, the lecture hall in the university, the university in the city, and so on until: the bench in "the cosmos" ["Weltraum"]. These beings whose being "in" one another can be determined in this way all have the same kind of being—that of being present as things occurring "within" the world. Being present "in" something objectively present and the being present together with something having the same kind of being (in the sense of a determinate relation to place) are ontological characteristics which we call *categorial*. They belong to beings whose kind of being is unlike Dasein.

In contrast, being-in designates a constitution of being of Dasein, and is an *existential*. Thus, we cannot understand by this the objective presence of a corporeal thing [Körperding] (the human body [Menschenleib]) "in" a being objectively present. Nor does the term being-in designate a spatial "in one another" of two things objectively present, any more than the word "in" primordially means a spatial relation of this kind.[1] "In" stems from *innan-*, to live, *habitare*, to dwell. *"An"*

_____

1. Cf. Jakob Grimm, *Kleinere Schriften*, vol. 7, p. 247.

means I am used to, familiar with, I take care of something. It has the meaning of *colo* in the sense of *habito* and *diligo*. We characterized this being to whom being-in belongs in this meaning as the being which I myself always am. The expression *"bin"* I connected with *"bei."* "Ich bin" ["I am"] means I dwell, I stay near . . . the world as something familiar in such and such a way. Being* as the infinitive of "I am": that is, understood as an existential, means to dwell near . . . , to be familiar with . . . *Being-in is thus the formal existential expression of the being of Dasein† which has the essential constitution of being-in-the-world.*

"Being together with" ["Sein bei"] the world, in the sense of being absorbed in the world, which must be further interpreted, is an existential which is grounded in being-in. Because we are concerned in these analyses with *seeing* a primordial structure of being of Dasein in accordance with whose phenomenal content the concepts of being must be articulated, and because this structure is fundamentally incomprehensible in terms of the traditional ontological categories, this "being together with" must also be examined more closely. We shall again choose the method of contrasting it with something essentially ontologically different—that is, a categorical relation of being which we express linguistically with the same means. Fundamental ontological distinctions are easily obliterated; and if they are to be envisaged phenomenally in this way, this must be done *explicitly*, even at the risk of discussing something "obvious." The status of the ontological analytic, however, shows that we do not at all have these obvious matters adequately "in our grasp," still less have we interpreted them in the meaning of their being; and we are even farther from possessing the proper structural concepts in a secure form.

As an existential, "being together with" the world never means anything like the being-objectively-present-together of things that occur. There is no such thing as the "being next to each other" of a being called "Dasein" with another being called "world." It is true that, at times, we are accustomed to express linguistically the being together of two objectively present things in such a manner: "The table stands 'next to' ['bei'] the door," "the chair 'touches' the wall." Strictly speaking, we can never talk about "touching," not because in the last analysis we can always find a space between the chair and the wall by examining it more closely, but because in principle the chair can never touch the wall, even if the space between them amounted to nothing. The presupposition for this would be that the wall could be *encountered* "by" the chair. A being [Seiendes] can only touch a being present within the world if that first being fundamentally has the kind of being of being-in—only if with its Dasein something like world is already dis-

---

* Being is also the infinitive of 'is': a being is.
† But not of being in general and not at all of being itself—absolutely.

covered in terms of which beings can reveal themselves through touch and thus become accessible in their being present. Two beings, which are present within the world and are, moreover, *worldless* in themselves, can never "touch" each other, neither can they *"be" "together with"* one another. The supplement "which are moreover worldless" must not be left out, because those beings which are not worldless, for example Dasein itself, are present "in" the world, too. More precisely, they *can* be *understood* within certain limits and with a certain justification as something merely present. To do this, one must completely disregard or just not see the existential constitution of being-in. But with this possible understanding of "Dasein" as something objectively present, and only objectively present, we may not attribute to Dasein its *own* kind of "objective presence." This objective presence does not become accessible by disregarding the specific structures of Dasein, but only in a previous understanding of them. Dasein understands its ownmost being in the sense of a certain "factual objective presence."[2] And yet the "factuality" of the fact of one's own Dasein is ontologically fundamentally different from the factual occurrence of a kind of stone. The factuality of the fact of Dasein, as the way in which every Dasein actually is, we call its *facticity*. The complicated structure of this determination of being is itself comprehensible *as a problem* only in the light of the existential fundamental constitutions of Dasein which we have already worked out. The concept of facticity implies that an "innerworldly" being has being-in-the-world in such a way that it can understand itself as bound up in its "destiny" with the being of those beings which it encounters within its own world.

Initially it is only a matter of seeing the ontological distinction between being-in as an existential and the category of the "insideness" that things objectively present can have with regard to one another. If we define being-in in this way, we are not denying to Dasein every kind of "spatiality." On the contrary. Dasein itself has its own "being-in-space," which in its turn is possible only *on the basis of being-in-the-world in general*. Thus, being-in cannot be clarified ontologically by an ontic characteristic, by saying for example: being-in in a world is a spiritual quality and the "spatiality" of human being is an attribute of its bodiliness which is always at the same time "based on" corporeality. Here again we are faced with a being-objectively-present-together of a spiritual thing thus constituted with a corporeal thing, and the being of the beings thus compounded is more obscure than ever. The understanding of being-in-the-world as an essential structure of Dasein first makes possible the insight into its *existential spatiality*. This insight will keep us from failing to see this structure or from previously cancelling

2. Cf. § 29.

it out, a procedure motivated not ontologically, but "metaphysically" in the naïve opinion that human being is initially a spiritual thing which is then subsequently placed "in" a space.

With its facticity, the being-in-the-world of Dasein has already dispersed itself in definite ways of being-in, perhaps even split itself up. The multiplicity of these kinds of being-in can be indicated by the following examples: to have to do with something, to produce, order and take care of something, to use something, to give something up and let it get lost, to undertake, to accomplish, to find out, to ask about, to observe, to speak about, to determine. . . . These ways of being-in have the kind of being of *taking care of* [*Besorgen*] which we shall characterize in greater detail. The *deficient* modes of omitting, neglecting, renouncing, resting, are also ways of taking care of something, in which the possibilities of taking care are kept to a "bare minimum." The term "taking care" has initially its prescientific meaning and can imply: carrying something out, settling something, "to straighten it out." The expression could also mean to take care of something in the sense of "getting it for oneself." Furthermore, we use the expression also in a characteristic turn of phrase: I will see to it or take care that the enterprise fails. Here "to take care" amounts to apprehensiveness. In contrast to these prescientific ontic meanings, the expression "taking care" is used in this inquiry as an ontological term (an existential) to designate the being of a possible being-in-the-world. We do not choose this term because Dasein is initially economical and "practical" to a large extent, but because Dasein itself is to be made visible as *care* [*Sorge*]. Again, this expression is to be understood as an ontological structure concept (cf. Chapter Six of this Division). The expression has nothing to do with "distress," "melancholy," or "the cares of life" which can be found ontically in every Dasein. These—like their opposites, "carefreeness" and "gaiety"—are ontically possible only because Dasein, *ontologically* understood, is care. Because being-in-the-world belongs essentially to Dasein, its being toward the world is essentially taking care.*

According to what we have said, being-in is not a "property" which Dasein sometimes has and sometimes does not have, *without* which it could *be* just as well as it could with it. It is not the case that human being "is," and then on top of that has a relation of being to the "world" which it sometimes takes upon itself. Dasein is never "initially" a sort of a being which is free from being-in, but which at times is in the mood to take up a "relation" to the world. This taking up of relations to the world is possible only *because*, as being-in-the-world, Dasein is as it is. This constitution of being is not first derived from the fact that besides the being which has the character of Dasein

---

* Human being [Mensch-sein] here equated with Da-sein.

there are other beings which are objectively present and meet up with it. These other beings can only "meet up" "with" Dasein because they are able to show themselves of their own accord within a *world*.

The saying used so often today, "human beings have their environment [Umwelt]," does not say anything ontologically as long as this "having" is undetermined. In its very possibility this "having" has its foundation in the existential constitution of being-in. As a being essentially existing in this way, Dasein can explicitly discover beings which it encounters in the environment, can know about them, can avail itself of them, can *have* the "world." The ontically trivial talk about "having an environment" is ontologically a problem. To solve it requires nothing less than defining the being of Dasein beforehand in an ontologically adequate way. If in biology use has been made of this constitution of being—especially since K. E. von Baer—one must not conclude that its philosophical use implies "biologism." For as a positive science, biology, too, can never find and determine this structure, it must presuppose it and continually make use of it.* This structure itself, however, can be explicated philosophically as the *a priori* condition for the thematic objects of biology only if it is understood beforehand as a structure of Dasein. Only in terms of an orientation toward the ontological structure thus understood can "life" as a constitution of being be defined *a priori* in a privative way. Ontically, as well as ontologically, being-in-the-world has priority as taking care. This structure gets its fundamental interpretation in the analytic of Dasein.

But does not this determination of the constitution of being discussed up to now move exclusively in negative statements? Though this being-in is supposedly so fundamental, we always keep hearing what it is *not*. Indeed. But the prevalence of negative characteristics is no accident. Rather, it makes known what is peculiar to this phenomenon, and is thus positive in a genuine sense—a sense appropriate to the phenomenon itself. The phenomenological demonstration of being-in-the-world has the character of rejecting dissimulations [Verstellungen] and obfuscations [Verdeckungen] *because* this phenomenon is always already "seen" in every Dasein in a certain way. And that is true *because* it makes up a fundamental constitution of Dasein, in that it is always already disclosed, along with its being, for the understanding of being in Dasein. But the phenomenon has mostly been basically misinterpreted, or interpreted in an ontologically inadequate way.† However, this 'seeing in a certain way and yet mostly misinterpreting' is itself based on nothing other than this constitution of being of Dasein itself. In accordance with that constitution, Dasein understands itself

---

* Is one justified in speaking of "world" here at all? Only surroundings [Umgebung]! 'Having' corresponds to this 'giving.' Da-sein never 'has' world.
† Yes! As far as being goes, it *is* not at all.

ontologically—and that means also its being-in-the-world—initially in terms of *those* beings and their being which it itself is *not*, but which it encounters "within" its world.*

Both in Dasein and for it, this constitution of being is always already somehow familiar. If it is now to be recognized, the explicit *cognition* that this task implies takes *itself* (as a knowing of the world) as the exemplary relation of the "soul" to the world. The cognition of world (νοεῖν)—or addressing oneself to the "world" and discussing it (λόγος)—thus functions as the primary mode of being-in-the-world even though being-in-the-world is not understood as such. But because this structure of being remains ontologically inaccessible, yet is ontically experienced as the "relation" between one being (world) and another (soul), and because being is initially understood by taking beings as innerworldy beings for one's ontological support, one tries to conceive the relation between world and soul as grounded in these beings and in the sense of their being; that is, as objective presence. Although it is experienced and known pre-phenomenologically, being-in-the-world is *invisible* if one interprets it in a way that is ontologically inadequate. One is just barely acquainted with this constitution of Dasein only in the form given by an inadequate interpretation—and indeed, as something obvious. In this way it then becomes the "evident" point of departure for the problems of epistemology or a "metaphysics of knowledge." For what is more obvious than the fact that a "subject" is related to an "object" and the other way around? This "subject-object-relation" must be presupposed. But that is a presupposition which, although it is inviolate in its own facticity, is truly fatal, perhaps for that very reason, if its ontological necessity and especially its ontological meaning are left in obscurity.

Thus the phenomenon of being-in has for the most part been represented exclusively by a single exemplar—knowing the world. This has not only been the case in epistemology; for even practical behavior has been understood as behavior which is *not* theoretical and "atheoretical." Because knowing has been given this priority, our understanding of its ownmost kind of being is led astray, and thus being-in-the-world must be delineated more precisely with reference to knowing the world, and must itself be made visible as an existential "modality" of being-in.

§ 13. *The Exemplification of Being-in in a Founded Mode: Knowing the World*

If being-in-the-world is a fundamental constitution of Dasein, and one in which it moves not only in general but especially in the mode of

---

* A subsequent interpretation.

everydayness, it must always already have been experienced ontically. It
60   would be incomprehensible if it remained totally veiled, especially since
Dasein has an understanding of its own being at its disposal, no matter
how indeterminately that understanding functions. However, no sooner
was the "phenomenon of knowing the world" understood than it was
interpreted in an "external," formal way. The evidence for this is the
interpretation of knowledge, still prevalent today, as a "relation between
subject and object" which contains about as much "truth" as it does vacu-
ity. But subject and object are not the same as Dasein and world.*

Even if it were feasible to give an ontological definition of being-
in primarily in terms of being-in-the-world that *knows*, the first task
required would still be the phenomenal characterization of knowing as
a being in and toward the world. If one reflects upon this relation of
being, a being called nature is initially given as that which is known.
Knowing itself is not to be found in this being. If knowing "is" at
all, it belongs solely to those beings which know. But even in those
beings, the things called human beings [Menschending], knowing is
not objectively present. In any case, it cannot be ascertained externally
like corporeal qualities. To the extent that knowing belongs to these
beings and is not an external characteristic, it must be "inside." The
more unequivocally one maintains that knowing is initially and really
"inside," and indeed has by no means the kind of being of physical and psychic beings, the more one believes that one is proceeding
without presuppositions in the question of the essence of knowledge
and of the clarification of the relation between subject and object. For
only then can the problem arise of how this knowing subject comes
out of its inner "sphere" into one that is "other and external," of how
knowing can have an object at all, and of how the object itself is to
be thought so that eventually the subject knows it without having to
venture a leap into another sphere. But in this approach, which has
many variations, the question of the kind of being of this knowing
subject is completely omitted, though its way of being was always
included tacitly in the thematic when one spoke of its knowing. Of
course, one is sometimes assured that the subject's inside and its "inner
sphere" is certainly not to be thought as a kind of "box" or "cabinet."
But what the positive meaning is of the "inside" of immanence in
which knowing is initially enclosed, and how the character of being of
this "being inside" of knowing is founded in the kind of being of the
subject, about this there is silence. However this inner sphere might be
interpreted, if one asks how knowing gets "out" of it and achieves a
61   "transcendence," it becomes evident that the knowing which presents

---

* Certainly not! So little that even putting them together in order to reject this is already
fatal.

such enigmas remains problematic unless one has first clarified how it is and what it is.

With this kind of approach one is blind to what was already implicitly implied in the preliminary thematization of the phenomenon of knowing. Knowing is a mode of being of Dasein as being-in-the-world, and has its ontic foundation in this constitution of being. But if, as we suggest, we thus find phenomenally that *knowing is a kind of being of being-in-the-world*, one might object that with such an interpretation of knowing, the problem of knowledge is annihilated. What is there left to ask about if one *presupposes* that knowing is already together with its world which it is, after all, first supposed to reach in the transcending of the subject? In this last question the "standpoint," which is not demonstrated phenomenally, once again emerges; but, apart from this, what criterion decides *whether* and *in which sense* there is to be a problem of knowledge other than that of the phenomenon of knowing itself and the kind of being of the knower?

If we now ask what shows itself in the phenomenal findings of knowing, we must remember that knowing itself is grounded beforehand in already-being-alongside-the-world, which essentially constitutes the being of Dasein. Initially, this already-being-alongside is not solely a rigid staring at something merely objectively present. Being-in-the-world, as taking care of things, is *taken in by* [*benommen*] the world which it takes care of. In order for knowing to be possible as determining by observation what is objectively present, there must first be a *deficiency* of having to do with the world and taking care of it. In refraining from all production, manipulation, and so on, taking care of things places itself in the only mode of being-in which is left over, in the mode of simply lingering with.... *On the basis* of this kind of being toward the world which lets us encounter beings within the world solely in their mere *outward appearance* (εἶδος), and, *as* a mode of this kind of being, looking explicitly at something thus encountered is possible.* This looking *at* is always a way of assuming a definite direction toward something, a glimpse of what is objectively present. It takes over a "perspective" from the beings thus encountered from the very beginning. This looking itself becomes a mode of independent dwelling together with beings in the world. In this *"dwelling"* [*"Aufenthalt"*]—as refraining from every manipulation and use—the *perception* of what is objectively present takes place. Perception takes place as *addressing* and *discussing* something as something. On the basis of this *interpretation* in the broadest sense, perception becomes *definition*. What   62

---

* Looking at does not occur merely by looking away. Looking at has its own origin and has looking away as its necessary consequence. Observing [Betrachten] has its own primordiality. The look [Blick] to the εἶδος requires something different.

is perceived and defined can be expressed in propositions and as thus *expressed* can be maintained and preserved. This perceptive retention of a proposition about . . . is itself a way of being-in-the-world, and must not be interpreted as a "procedure" by which a subject gathers representations about something for itself which then remain stored up "inside" as thus appropriated, and in reference to which the question can arise at times of how they "correspond" with reality.

In directing itself toward . . . and in grasping something, Dasein does not first go outside of the inner sphere in which it is initially encapsulated, but, rather, in its primary kind of being, it is always already "outside" together with some being encountered in the world already discovered. Nor is any inner sphere abandoned when Dasein dwells together with a being to be known and determines its character. Rather, even in this "being outside" together with its object, Dasein is "inside," correctly understood; that is, it itself is as the being-in-the-world which knows. Again, the perception of what is known does not take place as a return with one's booty to the "cabinet" of consciousness after one has gone out and grasped it. Rather, in perceiving, preserving, and retaining, the Dasein that knows *remains outside as Dasein*. In "mere" knowledge about a context of the being of beings, in "only" representing it, in "solely" "thinking" about it, I am no less outside in the world together with beings than I am when I *originally* grasp them. Even forgetting something, when every relation of being to what was previously known seems to be extinguished, must be understood *as a modification of primordial being-in*, and this holds true for every deception and every error.

The foundational context shown for the mode of being-in-the-world constitutive for the knowledge of the world makes the following clear: in knowing, Dasein gains a new *perspective of being* toward the world always already discovered in Dasein. This new possibility of being can be independently developed. It can become a task, and as scientific knowledge can take over the guidance for being-in-the-world. But knowing neither first *creates* a "commercium" of the subject with the world, nor does this commercium *originate* from an effect of the world on a subject. Knowing is a mode of Dasein which is founded in being-in-the-world. Thus, being-in-the-world, as a fundamental constitution, requires a *prior* interpretation.

The Worldliness of the World

§ 14. *The Idea of the Worldliness of the World [Weltlichkeit der Welt] in General*

First of all, being-in-the-world is to be made visible with regard to the structural factor "world." The accomplishment of this task appears to be easy and so trivial that we still believe we may avoid it. What can it mean, to describe "the world" as a phenomenon? It means letting what shows itself in the "beings" within the world be seen. Thus, the first step is to enumerate the things which are "in" the world: houses, trees, people, mountains, stars. We can *describe* the "outward appearance" of these beings and *tell of* the events occurring with them. But that is obviously a pre-phenomenological "business" which cannot be phenomenologically relevant at all. The description gets stuck in beings. It is ontic. But we are, after all, seeking being. We formally defined "phenomenon" in the phenomenological sense as that which shows itself as being and the structure of being.

Thus, to describe the "world" phenomenologically means to show and to conceptually and categorically determine the being of beings present in the world. Beings within the world are things—natural things and things "having value." Their thingliness becomes a problem; and since the thingliness of the latter is based upon natural thingliness, the being of natural things, nature as such, is the primary theme. The character of being of natural things, of substances, which is the basis of everything, is substantiality. What constitutes its ontological meaning? With this we have given our investigation an unequivocal direction.

But are we asking ontologically about the "world"? The problematic characterized is undoubtedly ontological. But even if it succeeds in the purest explication of the being of nature, in comparison with the fundamental statements made by the mathematical natural sciences about this being, this ontology never gets at the phenomenon of the "world." Nature is itself a being which is encountered within the world and is discoverable on various paths and stages.

Should we accordingly keep to the beings with which Dasein initially and for the most part dwells, to "valuable" things? Do not these things "really" show the world in which we live? Perhaps they do in fact show something like "world" more penetratingly. But these things are, after all, also beings "within" the world.

Neither the ontic description of innerworldly beings nor the ontological interpretation of the being of these beings gets as such at the phenomenon of "world." In both kinds of access to "objective being," "world" is already "presupposed" in various ways.

Can "world" ultimately not be addressed as a determination of the beings mentioned at all? After all, we do say that these beings are innerworldly. Is "world" indeed a character of being of Dasein? And then does every Dasein "initially" have its own world? Does not "world" thus become something "subjective"? Then how is a "common" world still possible "in" which we, after all, *are*? If we pose the question of "world," *which* world is meant? Neither this nor that world, but rather the *worldliness of world in general.* How can we encounter this phenomenon?

"Worldliness" is an ontological concept and designates the structure of a constitutive factor of being-in-the-world. But we have come to know being-in-the-world as an existential determination of Dasein. Accordingly, worldliness is itself an existential. When we inquire ontologically about the "world," we by no means abandon the thematic field of the analytic of Dasein. "World" is ontologically not a determination of *those* beings which Dasein essentially is *not*, but rather a characteristic of Dasein itself. This does not preclude the fact that the path of the investigation of the phenomenon of "world" must be taken by way of innerworldly beings and their being. The task of a phenomenological "description" of the world is so far from obvious that its adequate determination already requires essential ontological clarification.

The multiplicity of meanings of the word "world" is striking now that we have discussed it and made frequent application of it. Unraveling this multiplicity can point toward the phenomenon intended in their various meanings and their connection.

1. World is used as an ontic concept and signifies the totality of beings which can be objectively present within the world.
2. World functions as an ontological term and signifies the being of those beings named in 1. Indeed, "world" can name the region which embraces a multiplicity of beings. For example, when we speak of the "world" of the mathematician, we mean the region of all possible mathematical objects.

3. Again, world can be understood in an ontic sense, but not as beings essentially unlike Dasein that can be encountered within the world; but, rather, as that "*in which*" a factical Dasein "lives" as Dasein. Here world has a pre-ontological, existentiell meaning. There are various possibilities here: world can mean the "public" world of the we or one's "own" and nearest (domestic) surrounding world.

4. Finally, world designates the ontological and existential concept of *worldliness*. Worldliness itself can be modified into the respective structural totality of particular "worlds," but contains in itself the *a priori* of worldliness in general. We shall reserve the expression world as a term for the meaning established in the third meaning of world. If we use it at times in the first meaning, we shall put it in quotation marks.

Thus, terminologically "worldly" means a kind of being of Dasein, never a kind of being of something objectively present "in" the world. We shall call the latter something belonging* to the world or innerworldly.

One look at previous ontology shows us that one *skips over* the phenomenon of worldliness when one fails to see the constitution of Dasein as being-in-the-world. Instead, one tries to interpret the world in terms of the being of the being which is present within the world but has not, however, even been initially discovered, that is, in terms of nature.† Ontologically and categorically understood, nature is a limit case of the being of possible innerworldly beings. Dasein can discover beings as nature in this sense only in a definite mode of its being-in-the-world. This kind of knowledge has the character of a certain "de-worlding" of the world. As the categorial content of structures of being of a definite being encountered in the world, "nature" can never render *worldliness* intelligible.‡ But even the phenomenon "nature," for instance in the sense of the Romantic concept of nature, is ontologically comprehensible only in terms of the concept of world; that is, in terms of an analytic of Dasein.

With regard to the problem of an ontological analysis of the worldliness of the world, traditional ontology is at a dead-end—if it sees the problem at all. On the other hand, an interpretation of the worldliness of Dasein and its possibilities and ways of becoming worldly, must show *why* Dasein skips over the phenomenon of worldliness ontically and ontologically in its way of knowing the world. But at the same   66

---

* It is just Da-sein that *obeys and listens to the world* [*welthörig*].
† "Nature" intended here as Kantian in the sense of modern physics.
‡ Rather, the other way around!

time this fact of skipping over the phenomenon of worldliness indicates that special measures are necessary in order to gain the correct phenomenal point of departure for access to that phenomenon, a point of departure which does not permit any skipping over.

The methodological directive for this has already been given. Being-in-the-world, and thus the world as well, must be the subject of our analytic in the horizon of average everydayness as the *nearest* kind of being of Dasein. We shall pursue everyday being-in-the-world. With it as a phenomenal support, something like world must come into view.

The closest world of everyday Dasein is the *surrounding world* [*Umwelt*]. Our investigation will follow the path from this existential character of average being-in-the-world to the idea of worldliness as such. We shall seek the worldliness of the surrounding world (environmentality [Umweltlichkeit]) by way of an ontological interpretation of those beings initially encountered within the *surroundings*. The expression surrounding world [Umwelt] contains a reference to spatiality in its component "around" ["Um"]. The quality of "around" ["Umherum"] which is constitutive for the surrounding world does not, however, have a primarily "spatial" meaning. Rather, the spatial character which uncontestably belongs to a surrounding world can be clarified only on the basis of the structure of worldliness. Here the spatiality of Dasein mentioned in § 12 becomes phenomenally visible. But ontology has tried precisely to interpret the being of the "world" as *res extensa* on the basis of spatiality. The most extreme tendency toward such an ontology of the "world," an ontology which is oriented in the opposite direction, that is to the *res cogitans* which is neither ontically nor ontologically identical with Dasein, is to be found in Descartes. The analysis of worldliness attempted here becomes clearer if we show how it differs from such an ontological tendency. It has three stages: (A) An analysis of environmentality and worldliness in general. (B) An illustrative contrast between our analysis of worldliness and Descartes' ontology of the "world." (C) The aroundness [Umhafte] of the surrounding world and the "spatiality" of Dasein.

A. Analysis of Environmentality [Umweltlichkeit] and Worldliness [Weltlichkeit] in General

§ 15. *The Being of Beings Encountered in the Surrounding World*

The phenomenological exhibition of the being of beings encountered nearest to us can be accomplished under the guidance of the everyday being-in-the-world, which we also call *dealings in* [*Umgang in*] the world *with* innerworldly beings. Such dealings are already dis-

67

persed in manifold ways of taking care. However, as we showed, the closest kind of dealing is not mere perceptual cognition [Erkennen], but, rather, a handling, using, and taking care which has its own kind of "knowledge" ["Erkenntnis"]. Our phenomenological question is initially concerned with the being of those beings encountered when taking care of something. A methodological remark is necessary to secure the kind of seeing required here.

In the disclosure and explication of being, beings are always our preliminary and accompanying theme. The real theme is being. What shows itself in taking care in the surrounding world constitutes the pre-thematic being in the domain of our analysis. This being is not the object of a theoretical "world"-cognition; it is what is used, produced, and so on. As a being thus encountered, it comes pre-thematically into view for a "knowing" which, as a phenomenological knowing, primarily looks toward being and on the basis of this thematization of being thematizes actual beings as well. Thus, this phenomenological interpretation is not a cognition of existent qualities of beings, but rather a determination of the structure of their being. But as an investigation of being it independently and explicitly brings about the understanding of being which always already belongs to Dasein and is "alive" in every dealing with beings. Phenomenologically pre-thematic beings, what is used and produced, become accessible when we put ourselves in the place of taking care in the world. Strictly speaking, to talk of putting ourselves in the place of taking care is misleading. We do not first need to put ourselves in the place of this way of being in dealing with and taking care. Everyday Dasein always already *is* in this way; for example, in opening the door, I use the doorknob. Gaining phenomenological access to the beings thus encountered consists rather in rejecting the interpretive tendencies crowding and accompanying us which cover over the phenomenon of "taking care" in general, and thus even more so beings *as* they are encountered of their own accord *in* taking care. These insidious mistakes become clear when we ask: Which beings are to be our preliminary theme and established as a pre-phenomenal basis?

One answers: things. But perhaps we have already missed the pre-phenomenal basis we are looking for with this self-evident answer. For an unexpressed anticipatory ontological characterization is contained in addressing beings as "things" (*res*). An analysis which starts with such beings and goes on to inquire about being comes up with thingliness and reality. Ontological explication thus finds, as it proceeds, characteristics of being such as substantiality, materiality, extendedness, side-by-sidedness.... But the beings encountered and taken care of are also pre-ontologically hidden at first in this being [Sein]. When one designates things as the beings that are "initially

68

given" one goes astray ontologically, although one means something else ontically. What one really means remains indefinite. Or else one characterizes these "things" as "valuable" things. What does value mean ontologically? How is this "having" value and being involved with value to be understood categorically? Apart from the obscurity of this structure of having value, is the phenomenal character of being of what is encountered and taken care of in association thus attained?

The Greeks had an appropriate term for "things": πράγματα, that is, that with which one has to do in taking care in dealings (πρᾶξις). But the specifically "pragmatic" character of the πράγματα is just what was left in obscurity and "initially" determined as "mere things."* We shall call the beings encountered in taking care *useful things* [Zeug]. In our dealings we find utensils for writing, utensils for sewing, utensils for working, driving, measuring. We must elucidate the kind of being of useful things. This can be done following the guideline of the previous definition of what makes a useful thing a useful thing: its utility.

Strictly speaking, there "is" no such thing as *a* useful thing. There always belongs to the being of a useful thing a totality of useful things in which this useful thing can be what it is. A useful thing is essentially "something in order to . . .". The different kinds of "in order to" such as serviceability, helpfulness, usability, handiness, constitute a totality of useful things. The structure of "in order to" ["um-zu"] contains a *reference* [Verweisung] of something to something. Only in the following analyses can the phenomenon indicated by this word be made visible in its ontological genesis. At this time, our task is to bring a multiplicity of references phenomenally into view. In accordance with their character of utility, useful things always are *in terms of* their belonging to other useful things: writing utensils, pen, ink, paper, desk blotter, table, lamp, furniture, windows, doors, room. These "things" never show themselves initially by themselves, in order then to fill out a room as a sum of real things. What we encounter as closest to us, although we do not grasp it thematically, is the room, not as what is "between the four walls" in a geometrical, spatial sense, but rather as something useful for living. On the basis of this an "organization" shows itself, and in this organization any "individual" useful thing shows itself. A totality of useful things is always already discovered *before* the individual useful thing.

Dealings geared to useful things, which show themselves genuinely only in such dealings, for example, hammering with the hammer, neither *grasps* these beings thematically as occurring things, nor does

---

* Why? εἶδος–μορφή–ὕλη!, after all, come from τέχνη, thus from an "artistic" interpretation! if μορφή is not interpreted as εἶδος, ἰδέα!

such using even know the structure of useful things as such. Hammering does not just have a knowledge of the useful character of the hammer; rather, it has appropriated this utensil in the most adequate way possible. In such useful dealings, taking care subordinates itself to the in-order-to constitutive for the particular utensil in our dealings; the less we just stare at the thing called hammer, the more we take hold of it and use it, the more original our relation to it becomes and the more undisguisedly it is encountered as what it is, as a useful thing. The act of hammering itself discovers the specific "handiness" ["Handlichkeit"] of the hammer. We shall call the useful thing's kind of being in which it reveals itself by itself *handiness* [*Zuhandenheit*]. It is only because useful things have *this* "being-in-themselves" ["An-sich-sein"], and do not merely occur, that they are handy in the broadest sense and are at our disposal. No matter how keenly we just *look* at the "outward appearance" of things constituted in one way or another, we cannot discover handiness. When we just look at things "theoretically," we lack an understanding of handiness. But a dealing which makes use of things is not blind; it has its own way of seeing which guides our operations and gives them their specific certainly. Our dealings with useful things are subordinate to the manifold of references of the "in-order-to." The kind of seeing of this accommodation to things is called *circumspection* [*Umsicht*].

"Practical" behavior is not "atheoretical" in the sense of a lack of seeing. The difference between it and theoretical behavior lies not only in the fact that in one case we observe and in the other instance we *act*, and that action must apply theoretical cognition if it is not to remain blind. Rather, observation is a kind of taking care just as primordially as action has *its own* kind of seeing. Theoretical behavior is just looking, noncircumspectly. Because it is noncircumspect, looking is not without rules; its canon takes shape in *method*.

That which is handy is not grasped theoretically at all, nor is it itself initially a circumspective theme for circumspection. What is peculiar to what is initially at hand is that it withdraws, so to speak, in its character of handiness in order to be really handy. What everyday dealings are initially busy with is not tools themselves, but the work. What is to be produced in each case is what is primarily taken care of and is thus also what is at hand. The work bears the totality of references in which useful things are encountered. 70

As the *what-for* [*Wozu*] of the hammer, plane, and needle, the work to be produced has in its turn the kind of being of a useful thing. The shoe to be produced is for wearing (footgear), the clock is made for telling time. The work which we primarily encounter when we deal with things and take care of them—what we are at work with—always already lets us encounter the what-for of *its* usability in the usability which essentially belongs to it. The work that has been ordered exists in its turn only on the basis of its use and the referential context of beings discovered in that use.

But the work to be produced is not just useful for . . . ; production itself is always a using *of* something for something. A reference to "materials" is contained in the work at the same time. The work is dependent upon leather, thread, nails, and similar things. Leather in its turn is produced from hides. These hides are taken from animals which were bred and raised by others. We also find animals in the world which were not bred and raised and even when they have been raised these beings produce themselves in a certain sense. Thus beings are accessible in the surrounding world which in themselves do not need to be produced and are always already at hand. Hammer, tongs, nails in themselves refer to—they consist of—steel, iron, metal, stone, wood. "Nature" is also discovered in the use of useful things, "nature" in the light of products of nature.

But nature must not be understood here as what is merely objectively present, nor as the *power of nature*. The forest is a forest of timber, the mountain a quarry of rock, the river is water power, the wind is wind "in the sails." As the "surrounding world" is discovered, "nature" thus discovered is encountered along with it. We can abstract from nature's kind of being as handiness; we can discover and define it in its mere objective presence [Vorhandenheit]. But in this kind of discovery of nature, nature as what "stirs and strives," what overcomes us, entrances us as landscape, remains hidden. The botanist's plants are not the flowers of the hedgerow, the river's "source" ascertained by the geographer is not the "source in the ground."

The work produced refers not only to the what-for [Wozu] of its usability and the whereof [Woraus] of which it consists. The simple conditions of craft contain a reference to the wearer and user at the same time. The work is cut to his figure; he "is" there as the work emerges. This constitutive reference is by no means lacking when wares are produced by the dozen; it is only undefined, pointing to the random and the average. Thus not only beings which are at hand are encountered in the work but also beings with the kind of being of Dasein for whom what is produced becomes handy in its taking care. Here the world is encountered in which wearers and users live, a world which is at the same time our world. The work taken care of in each case is not only at hand in the domestic world of the work-shop, but rather in the *public world*. Along with the public world, the *surrounding world of nature* is discovered and accessible to everyone. In taking care, nature is discovered as having some definite direction on paths, streets, bridges, and buildings. A covered railroad platform takes bad weather into account, public lighting systems take darkness into account, the specific change of the presence and absence of daylight, the "position of the sun." Clocks take into account a specific constellation in the world system. When we look at the clock, we tacitly use the "position of the sun" according to which the official astronomical regulation of time is carried out. The surrounding world of nature is also at hand in the usage of clock equipment which is at first inconspicuously at hand. Our absorption in taking care of things in the work-world nearest to us has the function of discovering; depending upon the way we are absorbed, innerworldly beings that are brought along together with their constitutive references are discoverable in varying degrees of explicitness and with a varying attentive penetration.

The kind of being of these beings is handiness [Zuhandenheit]. But it must not be understood as a mere characteristic of interpretation,* as if such "aspects" were discursively forced upon "beings" which we initially encounter, as if an initially objectively present world-stuff were "subjectively colored" in this way. Such an interpretation overlooks the fact that in that case beings would have to be understood beforehand and discovered as purely objectively present, and would thus have priority and take the lead in the order of discovering and appropriating dealings with the "world." But this already goes against the ontological meaning of cognition which we showed to be a *founded* mode of being-in-the-world. To expose what is merely objectively present, cognition must first penetrate *beyond* things at hand being taken care of. *Handiness is the ontological categorial definition of beings as they are "in themselves."* But "there are" handy things, after all, only on the basis of what is objectively present. Admitting this thesis, does it then follow that handiness is ontologically founded in objective presence?   72

But if, in our continuing ontological interpretation, handiness proves to be the kind of being of beings first discovered within the world, if its primordiality can ever be demonstrated over and against mere objective presence, does what we have explained up to now contribute in the least to an ontological understanding of the phenomenon of world? We have, after all, always "presupposed" world in our interpretation of these innerworldly beings. Joining these beings together does not result as a sum in something like "world." Is there then any path at all leading from the being of these beings to showing the phenomenon of world?[1]

---

* But only as a characteristic of being encountered.

1. The author would like to remark that he has repeatedly communicated the analysis of the surrounding world and the "hermeneutic of the facticity" of Dasein in general in his lecture courses ever since the winter semester of 1919–20.

§ 16. *The Worldly Character of the Surrounding World Announcing Itself in Innerworldly Beings*

World itself is not an innerworldly being, and yet it determines inner-worldly beings to such an extent that they can only be encountered and discovered and show themselves in their being insofar as "there is" world. But how "is there" world? If Dasein is ontically consti-tuted by being-in-the-world and if an understanding of the being of its self belongs just as essentially to it, even if that understanding is quite indeterminate, does it not then have an understanding of world, a pre-ontological understanding which lacks and can dispense with explicit ontological insights? Does not something like world show itself to being-in-the-world taking care of the beings encountered within the world, that is, their innerworldliness? Does not this phenomenon come into view pre-phenomenologically; is it not always in view without requiring a thematically ontological interpretation? In the scope of its heedful absorption in useful things at hand, does not Dasein have a possibility of being in which, together *with* the innerworldly beings taken care of, their worldliness is illuminated in a certain way?

If such possibilities of being of Dasein can be shown in its heedful dealings, a path is opened to pursue the phenomenon thus illuminated and to attempt, so to speak, to "place" it and interrogate the structures evident in it.

73   Modes of taking care belong to the everydayness of being-in-the-world, modes which let the beings taken care of be encountered in such a way that the worldly character of innerworldly beings appears. Beings nearest at hand can be met up with in taking care of things as unusable, as improperly adapted for their specific use. Tools turn out to be damaged, their material unsuitable. In any case, *a useful thing* of some sort is at hand here. But we discover the unusability not by looking and ascertaining properties, but rather by paying attention to the dealings in which we use it. When we discover its unusability, the thing becomes conspicuous. *Conspicuousness* presents the thing at hand in a certain unhandiness. But this implies that what is unus-able just lies there; it shows itself as a thing of use which has this or that appearance and which is always also objectively present with this or that outward appearance in its handiness. Mere objective presence makes itself known in the useful thing only to withdraw again into the handiness of what is taken care of, that is, of what is being put back into repair. This presence of what is unusable still does not lack all handiness whatsoever; the useful thing *thus* present is still not a thing which just occurs somewhere. The damage to the useful thing is still not a mere change in the thing, a change of qualities simply occurring in something present.

But such heedful dealings do not just come up against unusable things *within* what is already at hand. They also find things which are missing, which are not only not "handy," but not "at hand" at all. When we come upon something unhandy, our missing it in this way again discovers what is at hand in a kind of only being present. When we notice its unhandiness, what is at hand enters the mode of *obtrusiveness*. The more urgently we need what is missing and the more truly it is encountered in its unhandiness, all the more obtrusive does what is at hand become, such that it seems to lose the character of handiness. It reveals itself as something merely present, which cannot be budged without the missing element. As a deficient mode of taking care of things, the helpless way in which we stand before it discovers the mere being present of what is at hand.

In dealing with the world taken care of, what is unhandy can be encountered not only in the sense of something unusable or completely missing, but as something unhandy which is *not* missing at all and *not* unusable, but "gets in the way" of taking care of things. That to which taking care cannot turn, for which it has "no time," is something *un*handy in the way of not belonging there, of not being complete. Unhandy things are disturbing and make evident the *obstinacy* of what is initially to be taken care of before anything else. With this obstinacy the presence of what is at hand makes itself known in a new way as the being of what is still present and calls for completion.

The modes of conspicuousness, obtrusiveness, and obstinacy have the function of bringing to the fore the character of objective presence in what is at hand. What is at hand is not thereby *observed* and stared at simply as something present. The character of objective presence making itself known is still bound to the handiness of useful things. These still do not disguise themselves as mere things. Useful things become "things" in the sense of what one would like to throw away. But in this tendency to throw things away, what is at hand is still shown as being at hand in its unyielding objective presence.

But what does this reference to the modified way of encountering what is at hand, a way in which its objective presence is revealed, mean for the clarification of the *phenomenon of world*? In the analysis of this modification, too, we are still involved with the being of innerworldly beings. We have not yet come any closer to the phenomenon of world. We have not yet grasped that phenomenon, but we now have the possibility of catching sight of it.

In its conspicuousness, obtrusiveness, and obstinacy, what is at hand loses its character of handiness in a certain sense. But this handiness is itself understood, although not thematically, in dealing with what is at hand. It does not just disappear, but bids farewell, so to speak, in the conspicuousness of what is unusuable. Handiness shows

74

itself once again, and precisely in doing so the worldly character of what is at hand also shows itself, too.

The structure of being of what is at hand as a useful thing is determined by references [Verweisungen]. The peculiar and self-evident "in-itself" ["An-sich"] of the nearest "things" is encountered when we take care of things, using them but not paying specific attention to them, while bumping into things that are unusable. Something is unusable. This means that the constitutive reference of the in-order-to to a what-for has been disrupted. The references themselves are not observed, rather they are "there" in our heedful adjustment to them. But in a *disruption of reference*—in being unusable for . . .—the reference becomes explicit. It does not yet become explicit as an ontological structure, but ontically for our circumspection which gets annoyed by the damaged tool. This circumspect noticing of the reference to the particular what-for makes the what-for visible and with it the context of the work, the whole "workshop" as that in which taking care has always already been dwelling. The context of useful things is lit up, not as a totality never seen before, but as a totality that has continually been seen beforehand in our circumspection. But with this totality, world makes itself known.

Similarly, when something at hand is missing whose everyday presence was so much a matter of course that we never even paid attention to it, this constitutes a *breach* in the context of references [Verweisungszusammenhänge] discovered in circumspection. Circumspection comes up with emptiness and now sees for the first time *what* the missing thing was at hand *for* [*wofür*] and at hand *with* [*womit*]. Once again the surrounding world makes itself known. What appears in this way is not itself one thing at hand among others and certainly not something *objectively present* which lies at the basis of the useful thing at hand. It is "there" before anyone has observed or ascertained it. It is itself inaccessible to circumspection insofar as circumspection concentrates on beings, but it is always already disclosed for that circumspection. "To disclose" ["Erschließen"] and "disclosedness" are used as technical terms in what follows and mean "to unlock" ["aufschließen"]—"to be open" ["Aufgeschlossenheit"]. Thus "to disclose" never means anything like "obtaining something indirectly by inference."

That the world does not "consist" of what is at hand can be seen from the fact (among others) that when the world appears in the modes of taking care which we have just interpreted, what is at hand becomes deprived of its worldliness so that it appears as something merely present. In order for useful things at hand to be encountered in their character of "being-in-itself" in our everyday taking care of the "surrounding world," the references and referential contexts in

75

which circumspection is "absorbed" must remain nonthematic for that circumspection and all the more so for a noncircumspect, "thematic" abstract comprehension. The condition for the possibility of what is at hand not emerging from its inconspicuousness is that the world *not announce itself*. And this is the constitution of the phenomenal structure of the being-in-itself of these beings.

Privative expressions such as inconspicuousness, unobtrusiveness, and nonobstinacy tell of a positive phenomenal character of the being of what is initially at hand. These negative prefixes express the character of keeping to itself of what is at hand. That is what we have in mind with being-in-itself which, however, we "initially" typically ascribe to things objectively present as that which can be thematically ascertained. When we are primarily and exclusively oriented toward that which is objectively present, the "in itself" cannot be ontologically explained at all. However, we must demand an interpretation if the talk about "in-itself" is to have any ontological importance. Mostly one appeals ontically and emphatically to this in-itself of being, and with phenomenal justification. But this *ontic* appeal does not already fulfill the claim of the *ontological* statement presumably given in such an appeal. The foregoing analysis already makes it clear that the being-in-itself of innerworldly beings is ontologically comprehensible only on the basis of the phenomenon of world.

If, however, world can appear in a certain way, it must be disclosed in general. World is always already predisclosed for circumspect heedfulness together with the accessibility of innerworldly beings at hand. Thus, world is something "in which" ["worin"] Dasein as a being always already *was*, and world is that to which Dasein can always only come back whenever it explicitly moves toward something in some way.

According to our foregoing interpretation, being-in-the-world signifies the unthematic, circumspect absorption in the references constitutive for the handiness of the totality of useful things. Taking care of things always already occurs on the basis of a familiarity with the world. In this familiarity Dasein can lose itself in what it encounters within the world and be numbed by it. With what is Dasein familiar? Why can the worldly character of innerworldly beings appear? How is the referential totality [Verweisungsganzheit] in which circumspection "moves" to be understood more precisely? When this totality is broken, the presence of beings is thrust to the fore.

In order to answer these questions which aim at working out the phenomenon *and problem* of worldliness, a concrete analysis of the structures in whose context our questions are being asked is required.

§ 17. *Reference [Verweisung] and Signs [Zeichen]*

In our preliminary interpretation of the structure of being of things at hand ("useful things"), the phenomenon of reference became visible, but in such a sketchy fashion that, at the same time, we emphasized the necessity of uncovering the phenomenon merely indicated with regard to its ontological origin. Moreover, it became clear that reference and the referential totality were in some sense constitutive of worldliness itself. Until now we have seen the world lit up only in and for particular ways of taking care of what is at hand in the surrounding world, together *with* its handiness. Thus the further we penetrate into the understanding of the being of innerworldly beings, the broader and more secure the phenomenal basis for the uncovering the phenomenon of world becomes.

We shall again take our point of departure with the being of what is at hand with the intention of grasping the phenomenon of *reference* more precisely. For this purpose we shall attempt an ontological analysis of the kind of useful thing in terms of which "references" can be found in a manifold sense. Such a "useful thing" can be found in *signs*. This word names many things. It names not only different *kinds* of signs, but being-a-sign-for something can itself be formalized to a *universal kind of relation* so that the sign structure itself yields an ontological guideline for "characterizing" any being whatsoever.

But signs are themselves initially useful things whose specific character as useful things consists of *indicating [Zeigen]*. Such signs are signposts, boundary-stones, the mariner's storm-buoy, signals, flags, signs of mourning, and the like. Indicating can be defined as a "kind" of referring. Taken in an extremely formal sense, to refer means to *relate [Beziehen]*. But relation does not function as the genus for "species" of reference which are differentiated as sign, symbol, expression, and signification. Relation is a formal definition which can be directly read off by way of "formalization" from every kind of context, whatever its subject matter or way of being.[2]

Every reference is a relation, but not every relation is a reference. Every "indicating" ["Zeigung"] is a reference, but not every reference is an indicating. This means that every "indicating" is a relation, but not every relation is an indicating. Thus the formal, universal character of relation becomes apparent. If we investigate such phenomena as reference, sign, or even signification, nothing is to be gained* by characterizing them as relations. Finally, we must even show that "rela-

---

* This is fundamental for demonstrating the possibility for the claim of logistics.

2. Cf. E. Husserl, *Ideen*, vol. 1, sect. 10ff, and idem, *Logische Untersuchungen*, vol. 1, chapter 11. For the analysis of sign and signification, see ibid., vol. 2, first investigation.

tion" itself has its ontological origin in reference *because* of its formal, universal character.

If this analysis is limited to an interpretation of the sign as distinct from the phenomenon of reference, even within this limitation, the full multiplicity of possible signs cannot be adequately investigated. Among signs there are symptoms, signs pointing backward as well as forward, marks, hallmarks whose way of indicating is different regardless of what it is that serves as a sign. We should differentiate these signs from the following: traces, residues, monuments, documents, certificates, symbols, expressions, appearances, significations. These phenomena can easily be formalized on the basis of their formal relational character. We are especially inclined today to subject all beings to an "interpretation" following the guideline of such a "relation," an interpretation which is always "correct" because it basically says nothing, no more than the facile scheme of form and content.

As an example of a sign, we choose one which we shall see again in a later analysis, though in a different regard. Motor cars are equipped with an adjustable red arrow whose position indicates which direction the car will take, for example, at an intersection. The position of the arrow is regulated by the driver of the car. This sign is a useful thing which is at hand not only for the heedfulness (steering) of the driver. Those who are not in the car—and they especially—make use of this useful thing in that they yield accordingly or remain standing. This sign is handy within the world in the totality of the context of useful things belonging to vehicles and traffic regulations. As a useful thing, this pointer is constituted by reference. It has the character of in-order-to, its specific serviceability; it is there in order to indicate. The indicating of this sign can be taken as a kind of "referring." But here we must note that this "referring" as indicating is not the ontological structure of the sign as a useful thing.

As indicating, "referring" is rather grounded in the structure of being of useful things, in serviceability for. The latter does not automatically make something a sign. The useful thing "hammer" is also characterized by serviceability, but it does not thus become a sign. The "referral" of indicating is the ontic concretion of the what-for of serviceability, and determines a useful thing for that what-for. The referral "serviceability for," on the other hand, is an ontological, categorical determination of the useful thing *as* useful thing. The fact that the what-for of serviceability gets its concretion in indicating is accidental to the constitution of the useful thing as such. The distinction between referral as serviceability and referral as indicating became roughly apparent in the example of the sign. The two coincide so little that their unity first makes possible a particular kind of useful thing. But just as surely as indicating is fundamentally different from referral as the constitution of a useful thing, it is just as incontestable that signs have a peculiar and even distinctive relation to the kind of being of the totality of useful things present in the surrounding world and their worldly character. Useful things which indicate have an *eminent* use in heedful dealings. However, it cannot suffice ontologically simply to ascertain this fact. The ground and meaning of this pre-eminence must be clarified.

What does the indicating of a sign mean? We can only answer this by defining the appropriate way of dealing with things that indicate. In doing this we must also make their handiness genuinely comprehensible. What is the appropriate way of dealing with signs? Taking our orientation toward the above example (the arrow), we must say that the corresponding behavior [Verhalten] (being [Sein]) toward the sign encountered is "yielding" or "remaining still" with reference to the approaching car which has the arrow. As a way of taking a direction, yielding belongs essentially to the being–in-the-world of Dasein. Dasein is always somehow directed and underway. Standing and remaining are only boundary instances of this directed being "underway." A sign is *not* really "comprehended" when we stare at it and ascertain that it is an indicating thing that occurs. Even if we follow the direction which the arrow indicates and look at something which is objectively present in the region thus indicated, even then the sign is not really encountered. The sign applies to the circumspection of heedful dealings in such a way that the circumspection which follows its direction brings the aroundness of the surrounding world in every instance into an explicit "overview" in that compliance. Circumspect overseeing does not *comprehend* what is at hand; instead, it acquires an orientation within the surrounding world. Another possibility of experiencing useful things lies in encountering the arrow as a useful thing belonging to the car. Here the arrow's specific character of being a useful thing need not be discovered. What and how it is to indicate can remain completely undetermined, and yet what is encountered is not a mere thing. As opposed to the nearest finding of a multiply undetermined manifold of useful things, the experience of a thing requires its own *definiteness*.

Signs such as we have described let what is at hand be encountered, more precisely, they let their context become accessible in such a way that heedful dealings get and secure for themselves an orientation. Signs are not things which stand in an indicating relationship to another thing; rather, they are *useful things which explicitly bring a totality of useful things to circumspection so that the worldly character of what is at hand makes itself known at the same time.* In symptoms and preliminary indications "what is coming" "shows itself," but not in the same sense of something merely occurring which is added to what is already objectively present. "What is coming" is something which we expect or "didn't expect" insofar as we were busy with other things. What has happened and occurred becomes accessible to our circumspection through signs after it has already happened. Signs indicate what is actually "going on." Signs always indicate primarily "wherein" we live, what our heedfulness is concerned with, what relevance it has.

The peculiar character of useful things as signs becomes especially clear in "establishing a sign." This happens in and through a circumspect anticipation which needs the possibility at hand of letting the actual surrounding world make itself known for circumspection through something at hand at any time. But the character of not emerging and keeping to itself, which we described, belongs to the being of innerworldy beings at hand closest to us. Thus circumspect dealings in the surrounding world needs a useful thing at hand which, in its character of being a useful thing, takes over the "work" of *letting* things at hand become *conspicuous*. Accordingly, the production of such useful things (signs) must take their conspicuousness into consideration. But even as conspicuous things, they are not taken as objectively present arbitrarily, rather they are "set up" in a definite way with a view toward easy accessibility.

But establishing signs does not necessarily have to come about in such a way that a useful thing at hand which was not yet present at all is produced. Signs also originate when something already at hand is *taken as a sign*. In this mode establishing a sign reveals a still more primordial meaning. Indicating not only creates the circumspectly oriented availability of a totality of useful things and the surrounding world in general; establishing a sign can even discover something for the first time. What is taken as a sign first becomes accessible through its handiness. For example, when the south wind is "accepted" by the famer as a sign of rain, this "acceptance" or the "value attached" to this being is not a kind of bonus attached to something already objectively present, that is, the movement of the wind and a certain geographical direction. As this mere occurrence which is meteorologically accessible, the south wind is *never initially* something objectively present that occasionally takes on the function of an omen. Rather, the   *81* farmer's circumspection first discovers the south wind in its being by taking the lay of the land into account.

But, one will protest, *what* is taken as a sign must, after all, first have become accessible in itself and must be grasped *before* the sign is established. To be sure, it must already be there in some way or another. The question simply remains *how* beings are discovered in this preliminary encounter, whether as something merely occurring and not rather as an uncomprehended kind of useful thing, a thing at hand which one did not know "what to do with" up to now, which accordingly veiled itself to circumspection. *Here again, one must also not interpret the character of useful things at hand which have not been discovered by circumspection as mere thingliness presented for the comprehension of something merely objectively present.*

The being-handy [Zuhandensein] of signs in everyday dealings, and the conspicuousness which belongs to signs and can be produced with varying intentions and in different ways, do not only document the inconspicuousness constitutive for what is at hand nearest to us; the sign itself takes its conspicuousness from the inconspicuousness of the totality of useful things at hand in everydayness as a "matter of course," for example, the well-known "string on one's finger" as a reminder. What it is supposed to indicate is always something to be taken care of within the purview of everydayness. This sign can indicate many things of the most diverse sort. The narrowness of intelligibility and use corresponds to the breadth of what can be indicated in such signs. Not only is it mostly at hand as a sign only for the person who "establishes" it, it can become inaccessible to that same person so that a second sign is necessary for the possible circumspect applicability of the first one. The string which cannot be used as a sign does not thus lose its sign character, but rather acquires the disturbing obtrusiveness of something near at hand.

One could be tempted to illustrate the distinctive role of signs in everyday heedfulness for the understanding of the world itself by citing the extensive use of "signs," such as fetishism and magic, in primitive Dasein. Certainly the establishment of signs that underlies such use of signs does not come about with theoretical intent and by way of theoretical speculation. The use of signs remains completely within an "immediate" being-in-the-world. But when one looks more closely, it becomes clear that the interpretation of fetishism and magic under the guideline of the idea of signs is not sufficient at all to comprehend the kind of "handiness" of beings encountered in the primitive world. With regard to the phenomenon of signs, we might give the following interpretation: for primitive people the sign coincides with what it indicates. The sign itself can represent what it indicates not only in the sense of replacing it, but in such a way that the sign itself always *is* what is indicated. This remarkable coincidence of the sign with what is indicated does not, however, mean that the sign-thing has already undergone a certain "objectification," that it has been experienced as a mere thing and been transposed together with what is signified to the same region of being of objective presence. The "coincidence" is not an identification of hitherto isolated things, but rather the sign has not yet become free from that for which it is a sign. This kind of use of signs is still completely absorbed in the being of what is indicated so that a sign as such cannot be detached at all. The coincidence is not based

82

on a first objectification, but rather upon the complete lack of such an objectification. But this means that signs are not at all discovered as useful things, that ultimately what is "at hand" in the world does not have the kind of being of useful things at all. Perhaps this ontological guideline (handiness and useful things), too, can provide nothing for an interpretation of the primitive world, and certainly for an ontology of thingliness. But if an understanding of being is constitutive for primitive Dasein and the primitive world in general, it is all the more urgent to develop the "formal" idea of worldliness; namely, of a phenomenon which can be modified in such a way that all ontological statements which assert that in a given phenomenal context something is *not yet* or *no longer* such and such may acquire a *positive* phenomenal meaning in terms of what it is *not*.

The foregoing interpretation of signs should simply offer phenomenal support for our characterization of reference. The relation between sign and reference is threefold: (1) As a possible concretion of the what-for of serviceability, indicating is based upon the structure of useful things in general, upon the in-order-to (reference). (2) As the character of useful things at hand, the indicating of signs belongs to a totality of useful things, to a referential context. (3) Signs are not just at hand along with other useful things, rather, in their handiness, the surrounding world becomes explicitly accessible to circumspection. *A sign is something ontically at hand which, as this definite useful thing, functions at the same time as something which indicates the ontological structure of handiness, referential totality, and worldliness.* The distinctive characteristic of these things at hand within the surrounding world circumspectly taken care of is rooted here. Thus reference cannot itself be comprehended as a sign if it is ontologically to be the foundation for signs. Reference is not the ontic specification of something at hand since it, after all, constitutes handiness itself. In what sense is reference the ontological "presupposition" of what is at hand, and as this ontological foundation, to what extent is it at the same time constitutive of worldliness in general?

83

§ 18. *Relevance [Bewandtnis] and Significance [Bedeutsamkeit]: The Worldliness of the World*

Things at hand are encountered within the world. The being of these beings, handiness, is thus ontologically related to the world and to worldliness. World is always already "there" in all things at hand. World is already discovered* beforehand together with everything encountered, although not thematically. However, it can also appear in certain ways of dealing with the surrounding world. World is that

---

* cleared [gelichtet]

in terms of which things at hand are at hand for us. How can world let things at hand be encountered? Our analysis showed that what is encountered within the world is freed in its being for heedful circumspection, for taking matters into account. What does this prior freeing mean and how is it to be understood as the ontological distinction of the world? What problems does the question of the worldliness of the world confront?

The constitution of useful things as things at hand has been described as reference. How can world free beings of this kind with regard to their being, why are these beings encountered first? We mentioned serviceability for, impairment, usability, and so forth, as specific kinds of reference. The what-for of serviceability and wherefore of usability prefigure the possible concretion of reference. The "indicating" of signs, the "hammering" of the hammer, however, are not qualities of beings. They are not qualities at all if this term is supposed to designate the ontological structure of a possible determination of things. In any case, things at hand are suited and unsuited for things, and their "qualities" are, so to speak, still bound up with that suitability or unsuitability, just as objective presence, as a possible kind of being of things at hand, is still bound up with handiness. But as the constitution of useful things, serviceability (reference) is also not the suitability of beings, but the condition of the possibility of being for their being able to be determined by suitability. But then what does reference mean? The fact that the being of things at hand has the structure of reference means that they have in themselves the character of *being referred*. Beings are discovered with regard to the fact that they are referred, as those beings which they are, to something. They are relevant *together with* something else. The character of being of things at hand is *relevance*. To be relevant means to let something be together with something else. The relation of "together . . . with . . ." is to be indicated by the term reference.

Relevance is the being [Sein] of innerworldly beings, for which they are always already initially freed. Beings are in each case relevant. That it has a relevance is an *ontological* determination of the being of these beings, not an ontic statement about beings. What the relevance is about is the what-for of serviceability, the wherefore of applicability [Verwendbarkeit]. The what-for of serviceability can in turn be relevant. For example, the thing at hand which we call a hammer has to do with hammering, the hammering has to do with fastening something, fastening something has to do with protection against bad weather. This protection "is" for the sake of providing shelter for Dasein, that is, for the sake of a possibility of its being. *Which* relevance things at hand have is prefigured in terms of the total relevance. The total relevance which, for example, constitutes the things at hand in a workshop in their handiness is "earlier" than any single useful thing,

84

as is the farmstead with all its utensils and neighboring lands. The total relevance itself, however, ultimately leads back to a what-for which *no longer* has relevance, which itself is not a being of the kind of being of things at hand within a world, but is a being whose being is defined as being-in-the-world, to whose constitution of being worldliness itself belongs. This primary what-for is not just another for-that as a possible factor in relevance. The primary "what-for" is a for-the-sake-of-which. But the for-the-sake-of-which always concerns the being of *Dasein* which is essentially concerned *about* this being itself in its being. For the moment we shall not pursue any further the connection indicated which leads from the structure of relevance to the being of Dasein itself as the real and unique for-the-sake-of-which. "Letting something be relevant" first of all requires a clarification which goes far enough to bring the phenomenon of worldliness to *the* kind of definiteness needed in order to be able to ask questions about it in general.

Ontically, to let something be relevant means to let things at hand *be** in such and such a way in factical taking care of things, to let them be *as* they are and *in order that* they be such. We grasp the ontic meaning of this "letting be" in a fundamentally ontological way. Thus we interpret the meaning of the previous freeing of innerworldly beings initially at hand. Previously letting "be" does not mean first to bring something to its being and produce it, but rather to discover something that is already a "being" in its handiness and thus let it be encountered as the being of this being.† This "*a priori*" letting something be relevant is the condition of the possibility that things at hand be encountered so that Dasein in its ontic dealings with the beings thus encountered can let them be relevant in an ontic sense. On the other hand, letting something be relevant, understood in an ontological sense, concerns the freeing of *every* thing at hand as a thing at hand, whether it is relevant in the ontic sense or whether it is such a being which is precisely *not* relevant ontically. Such a being is one that, initially and for the most part, is taken care of, but which we do not let "be" as the discovered being it is, since we work on it, improve it, destroy it.

To have always already let something be freed for relevance is an *a priori perfect*‡ characterizing the kind of being of Dasein itself.

85

---

* Letting-be (*Seyn-lassen*). Cf. "On the Essence of Truth," where letting-be is related in principle and very broadly to *every* kind of being [Seiende]!
† Thus to let it presence in its truth.
‡ In the same paragraph we speak of "previous freeing"—namely (generally speaking) of being for the possible manifestness of beings: "Previously" in this ontological sense means in Latin *a priori*, in Greek πρότερον τῇ φύσει (Aristotle, *Physics*, A 1). More clearly in *Metaphysics* E 1025b29—τὸ τί ἦν εἶναι 'what already was—being [sein],' 'what always already presences in advance,' what has-been, the perfect. The Greek verb εἶναι has no perfect tense; it is named here in ἦν εἶναι. It is not something ontically past, but rather what is always earlier, what we are referred *back* to in the question of beings as such. Instead of *a priori* perfect we could also call it ontological or transcendental perfect (cf. Kant's doctrine of the schematism).

Understood ontologically, letting something be relevant is the previous freeing of beings for their innerworldly handiness. The with-what of relevance is freed in terms of the together-with-what of relevance. It is encountered by heedfulness as this thing at hand. When a *being* shows itself in general to heedfulness, that is, when a being is discovered in its being, it is always already a thing at hand in the surrounding world and precisely not "initially" merely present "world-stuff."

As the being of things at hand, relevance itself is always discovered only on the basis of a totality of relevance [Bewandtnisganzheit] previously discovered, that is, in the things at hand encountered; what we called the worldly character of things at hand thus lies predis-covered. This totality of relevance previously discovered contains an ontological relation to the world. Letting beings be relevant and thus freeing them for a totality of relevance must have already somehow disclosed that for which it is freeing. That for which things at hand in the surrounding world are freed (in such a way that the things at hand first become accessible *as* innerworldly beings) cannot itself be understood as a being with the kind of being [Sein] thus discovered. It is essentially not discoverable if we reserve *discoveredness* [Entdecktheit] as the term for a possibility of being of all beings *un*like Dasein.

But now what does it mean to say that for which innerworldly beings are initially freed must previously be disclosed? An understand-ing of being belongs to the being of Dasein. Understanding has its
being in an act of understanding. If the kind of being of being-in-the-world essentially belongs to Dasein, then the understanding of being-in-the-world belongs to the essential content of its understanding of being. The previous disclosure of that for which the freeing of things encountered in the world ensues is none other than the understanding of world to which Dasein as a being is always already related.

Previous letting something be relevant to . . . with . . . is ground-ed in an understanding of something like letting things be relevant [Bewendenlassen] as well as the in-which and with-which of relevance. These things and what underlies them, such as the what-for to which relevance is related, the for-the-sake-of-which from which every what-for is ultimately derived, all of these must be previously disclosed in a certain intelligibility. And what is that in which Dasein understands itself pre-ontologically as being-in-the-world? In understanding a con-text of relations, Dasein has been referred to an in-order-to in terms of an explicitly or inexplicitly grasped potentiality for being [Seinkön-nen] for the sake of which it is, which can be authentic or inauthentic. This prefigures a what-for as the possible letting something be relevant which structurally allows for relevance *to* something else. Dasein is always in each case already referred in terms of a for-the-sake-of-which to the with-what of relevance. This means that, insofar as it is, it always already lets beings be encountered as things at hand. That *within* which Dasein understands itself beforehand in the mode of self-reference is that *for which* it lets beings be encountered beforehand. *As that for which one lets beings be encountered in the kind of being of relevance, the wherein of self-referential understanding is the phenomenon of world.* And the structure of that to which Dasein is referred is what constitutes the *worldliness* of the world.

Dasein is primordially familiar with that within which it understands itself in this way. This familiarity with the world does not necessarily require a theoretical transparency of the relations constituting the world as world. However, the possibility of an explicit ontological and existential interpretation of these relations is grounded in the familiarity with the world constitutive for Dasein, which, in its turn, constitutes Dasein's understanding of being. This possibility can be explicitly appropriated when Dasein has set as its task a primordial interpretation of its being and the possibilities of that being or, for that matter, of the meaning of being in general.

But as yet our analyses have only first laid bare the horizon within which something akin to world and worldliness is to be sought. For our further reflection, we must first make clear how the context of the self-referral of Dasein is to be understood ontologically.   *87*

Understanding, which will be analyzed with proper penetration in what follows (cf. § 31), holds the indicated relations in a preliminary disclosure. In its familiar being-in-relevance, understanding holds itself *before* that disclosure as that within which its reference moves. Understanding can itself be referred to in and by these relations. We shall call the relational character of these referential relations *signifying* [*be-deuten*]. In its familiarity with these relations, Dasein "signifies" ["bedeutet"] to itself. It primordially gives itself to understand its being and potentiality-of-being with regard to its being-in-the-world. The for-the-sake-of-which signifies an in-order-to, the in-order-to signifies a what-for, the what-for signifies a what-in of letting something be relevant, and the latter a what-with of relevance. These relations are interlocked among themselves as a primordial totality. They are what they are as this signifying in which Dasein gives itself to understand its being-in-the-world beforehand. We shall call this relational totality of signification *significance* [*Bedeutsamkeit*]. It is what constitutes the structure of the world, of that in which Dasein* as such always already is. *In its familiarity with significance Dasein is the ontic condition of the possibility of the discovery of beings with the kind of being of relevance (handiness) which are encountered in a world and that can thus make themselves known in their in-itself.* As such, Dasein always means that a context

---

* The Da-sein in which human being [Mensch] presences.

of things at hand is already essentially discovered with its being. In that it *is*, Dasein has always already referred itself* to an encounter with a "world." This *dependency upon being referred* belongs essentially to its being [Sein].

But the significance itself with which Dasein is always already familiar contains the ontological condition of the possibility that Dasein, as understanding and interpreting, can disclose something akin to "significations" ["Bedeutungen"] which in turn found the possible being of words and language.†

As the existential constitution of Dasein, of its being-in-the-world, this disclosed significance is the ontic condition of the possibility for discovering a totality of relevance.

If we thus define the being of what is at hand (relevance) and even worldliness itself as a referential context, are we not volatizing the "substantial being" of innerworldly beings into a system of relations, and, since relations are always "something thought," are we not dissolving the being of innerworldly beings into "pure thought?"

88

Within the present field of investigation the repeatedly designated differences of the structures and dimensions of the ontological problematic are to be fundamentally distinguished:

1. The being of the innerworldly beings initially encountered (handiness);
2. The being *of* beings (objective presence) that is found and determined by discovering them in their own right in going through beings initially encountered;
3. The being of the ontic condition of the possibility of discovering innerworldly beings in general, the worldliness‡ of the world.

This third kind of being is an *existential* determination of being-in-the-world, that is, of Dasein. The other two concepts of being are *categories* and concern beings unlike Dasein. The referential context that constitutes worldliness as significance can be formally understood in the sense of a system of relations. But we must realize that such formalizations level down the phenomena to the extent that the true phenomenal content gets lost, especially in the case of such "simple" relations as are contained in significance. These "relations" and "relata" of the in-order-to, for-the-sake-of, and with-what of relevance resist any kind of mathematical functionalization in accordance with their phenomenal content. Nor are they something thought, something first posited in

* But not as the egoistic deed of a subject, rather: Dasein and being [Sein].
† Untrue. Language is not imposed, but *is* the primordial essence of truth as there [Da].
‡ Better: the holding sway [Walten] of the world.

"thinking," but rather relations in which heedful circumspection as such already dwells. As constitutive of worldliness, this "system of relations" does not volatize the being of innerworldly beings at all. On the contrary, these beings are discoverable in their "substantial" "in itself" only on the basis of the worldliness of the world. And only when innerworldly beings can be encountered at all does the possibility exist of making what is merely objectively present accessible in the field of these beings. On the basis of their merely objective presence these beings can be determined mathematically in "functional concepts" with regard to their "properties." Functional concepts of this kind are ontologically possible only in relation to beings whose being has the character of pure substantiality. Functional concepts are always possible only as formalized substantial concepts.

In order to delineate the specific ontological problematic of worldliness still more clearly, the interpretation of worldliness is to be clarified in terms of an extreme counter-example before preceding with our analysis.

B. The Contrast Between Our Analysis of Worldliness and Descartes' Interpretation of the World

Our investigation can secure the concept of worldliness and the structures contained in this phenomenon only step by step. Since the interpretation of the world initially starts with an innerworldly being and then never gains sight of the phenomenon of world again, we shall attempt to clarify this point of departure ontologically in what is perhaps its most extreme development. We shall not only give a short presentation of the fundamental features of Descartes' ontology of the "world," but also ask about its presuppositions and try to characterize those presuppositions in the light of what has been clarified up to now. This discussion should tell us on what fundamentally undiscussed ontological "foundations" the interpretations of the world after Descartes, and especially those preceding him, are based.

Descartes sees the fundamental ontological determination of the world as *extensio*. Since extension is a component of spatiality, for Descartes in fact identical with it, and since spatiality is in some sense constitutive of the world, our discussion of the Cartesian ontology of the "world" at the same time offers a negative support for the positive explication of the spatiality of the surrounding world and of Dasein itself. With regard to Descartes' ontology we shall discuss three things:

1. The determination of the "world" as *res extensa* (§ 19).
2. The foundations of this ontological determination (§ 20).

89

3. The hermeneutical discussion of the Cartesian ontology of the "world" (§ 21).

The following reflections can be grounded in more detail only by the phenomenological destruction of the *"cogito sum"* (cf. Part Two, Division Two).*

§ 19. *The Determination of the "World" as* res extensa

Descartes distinguishes the *"ego cogito"* as *"res cogitans"* from the *"res corporea."* From then on this distinction ontologically defines the distinction of "nature and spirit." Although this opposition between nature and spirit is formulated ontically in many variations of content, the unclarity of its ontological fundamentals and even of the poles of this opposition itself have their proximate roots in Descartes' distinction. In what kind of understanding of being did he determine the being of these beings? The term for the being of beings in themselves is *substantia*. This expression sometimes means the *being* of beings as substance, *substantiality*, sometimes beings themselves, *a particular substance*. This ambiguity of *substantia*, already inherent in the ancient concept of οὐσία,† is not accidental.

The ontological determination of the *"res corporea"* requires the explication of substance, that is, of the substantiality of these beings as particular substances. What constitutes the true being-in-itself of the *res corporea*? How is a substance as such, that is, its substantiality, to be understood? *Et quidem ex quolibet attributo substantia cognoscitur; sed una tamen est cuiusque substantiae praecipua proprietas, quae ipsius naturam essentiamque constituit, et ad quam aliae omnes referuntur.*[3] Substances are accessible through their "attributes," and every substance has an eminent property in terms of which the essence of the substantiality of a definite substance can be determined. What is this property with regard to the *res corporea*? *Nempe* extensio *in longum, latum et profundum, substantiae coporeae naturam constituit.*[4] Extension in terms of length, breadth, and depth constitutes the real being of the corporeal substance that we call "world." What gives the *extensio* this distinction? *Nam omne aliud quod corpori tribui potest, extensionem praesupponit.*[5] Extension

90

---

* Never published [Tr.].
† And especially the ὄν; τὸ ὄν: (1) being (beingness) [das Seiend (Seiendsein)], (2) beings [das Seiende].

3. *Principia* I. pr. 53, p. 25 (*Oeuvres*, ed. Adam-Tannery, Vol. VIII).
4. Ibid.
5. Ibid.

is *the* constitution of being of the beings under discussion, a constitution which must already "be" before other determinations of being in order for the latter to be able to "be" what they are. Extension must primarily be "attributed" to the corporeal thing. Accordingly, the proof for the extension and the substantiality of the "world" characterized by that extension is accomplished by showing how all other properties of this substance, above all *divisio, figura, motus* can only be conceived as modes of *extensio,* and that *extensio,* conversely, remains intelligible *sine figura vel motu.*

Thus a corporeal being can maintain its total extension and yet change the distribution of the extension in many ways and in various dimensions and still present itself as one and the same thing in manifold shapes. *Atque unum et idem corpus, retinendo suam eandem quantitatem, pluribus diversis modis potest extendi: nunc scilicet magis secundum longitudinem, minusque secundum latidudinem vel profunditatem, ac paulo post e contra magis secundum latitudinem, et minus secundum longitudinem.*[6]

Gestalt is a mode of *extensio,* and motion as well: for *motus* is comprehended only *si de nullo nisi locali cogitemus, ac de vi a qua excitatur non inquiramus.*[7] If motion is an existent property of *res corporea,* then it must be understood in terms of the being of this being itself, in terms of *extensio,* that is, as mere change of location in order to be experienced in its being. Something like "force" adds nothing to the determination of the *being* of this being. Properties such as *durities* (hardness), *pondus* (weight), *color* (color) can be removed from matter, yet matter remains what it is. These properties do not constitute its true being, and insofar as they *are,* they turn out to be modes of *extensio.* Descartes attempts to show this in detail with regard to "hardness": *Nam, quantum ad duritiem, nihil aliud de illa sensus nobis indicat, quam partes durorum corporum resistere motui manuum nostrarum, cum in illas incurrunt. Si enim, quotiescunque manus nostrae versus aliquam partem moventur, corpora omnia ibi existentia recederent eadem celeritate qua illae accedunt, nullam unquam duritiem sentiremus. Nec ullo modo potest intelligi, corpora quae sic recederent, idcirco naturam corporis esse amissura; nec proinde ipsa in duritie consistit.*[8] Hardness is experienced by touch. What does the sense of touch "tell" us about hardness? The parts of

91

---

6. Ibid., pr. 64, p. 31 ["And that one body, retaining the same size, may be extended in many different ways, sometimes being greater in length and less in breadth or depth, and sometimes on the contrary greater in breadth and less in length"].
7. Ibid., pr. 65, p. 32 ["If we inquire only about locomotion, without taking into account the force that produces it"].
8. Ibid., *Principia* II, pr. 4, p. 42.

the hard thing "resist" the motion of the hands, for instance in wanting to push something away. But if the hard bodies, those that do not
give way, changed their location with the same speed as the hand
"approaching" the bodies, nothing would ever be touched. Hardness
would not be experienced and thus would never *be*. But it is in no way
comprehensible that bodies that give way with such velocity should
thus forfeit any of their corporeal being. If they were to retain this even
under a change of velocity which makes it impossible for anything
like "hardness" to be, then hardness does not belong to the being of
these beings either. *Eademque ratione ostendi potest, et pondus, et colorem,
et alias omnes eiusmodi qualitates, quae in materia corporea sentiuntur, ex
ea tolli posse, ipsa integra remanente: unde sequitur, a nulla ex illis eius (sc.
extensionis) naturam dependere.*[9] Thus, what constitutes the being of the
res corporea is *extensio*, the *omnimodo divisibile, figurabile et mobile*, what
can change in every kind of divisibility, shape, and motion, the *capax
mutationum*, what persists throughout all these changes, *remanet*. In a
corporeal being what is capable of such a *remaining constant* is what is
genuinely that being, indeed, it is this in such a way that it characterizes the substantiality of this substance.

§ 20. *The Fundaments of the Ontological Definition of the "World"*

The idea of being from which the ontological characteristics of the *res
extensa* are derived is substantiality. *Per substantiam nihil aliud intelligere
possumus, quam rem quae ita existit, ut nulla alia re indigeat ad existendum.*
By substance we can understand nothing other than a being which *is* in
such a way that it needs no other being in order *to be*.[10] The being of a
"substance" is characterized by not needing anything. Whatever in its
being absolutely needs no other being, satisfies the idea of substance in
the true sense. This being is the *ens perfectissimum. Substantia quae nulla
plane re indigeat, unica tantum potest intelligi, nempe Deus.*[11] Here, "God" is
a purely ontological term when he is understood as *ens perfectissimum.*
At the same time, the "self-evident" connotation of the concept of God
makes possible an ontological interpretation of the constitutive factor of
substantiality, that of not needing anything. *Alias vero omnes (res), non
nisi ope concursus Dei existere posse percipimus.*[12] All beings other than
God need to be produced in the broadest sense and to be sustained.

---

9. Ibid. ["The same reason shows us that weight, color, and all the other qualities of the
kind that are perceived in corporeal matter, may be taken from it, it remaining meanwhile entire: it thus follows that the nature of body depends on none of these"].
10. Ibid., *Principia* I, pr. 51.
11. Ibid. ["Nothing but God answers to this description as being that which is absolutely
self-sustaining"].
12. Ibid.

The production of what is objectively present and the lack of need for production constitute the horizon within which "being" is understood. Every being other than God is *ens creatum*. The being which belongs to one of these entities is "infinitely" different from that which belongs to the other; yet we still consider what is created and the creator alike *as beings*. We thus use being in such a broad sense that its meaning encompasses an "infinite" distinction. Thus we can also call created beings substances with a certain justification. It is true that these beings need to be produced and sustained, relative to God, but within the region of created beings, of the "world" in the sense of the *ens creatum*, there are beings which are "in need of no other being" relative to creaturely production and sustenance—for instance, human beings. There are two such substances: *res cogitans* and *res extensa*.

The being of *that* substance whose eminent *proprietas* is *extensio*   93 is thus definable in principle ontologically when the *meaning* of being *"common"* to the three substances, the one infinite and the two finite ones, is clarified. But *nomen substantiae non convenit Deo et illis* univoce, *ut dici solet in Scholis, hoc est . . . quae Deo et creaturis sit communis.*[13] Here Descartes touches upon a problem which occupied medieval ontology in many ways, namely, the question in what way the meaning of being signifies the being under consideration. In statements such as "God is" and "the world is" we predicate being. But this word "is" cannot signify the being in question in the same sense (συνωνύμως, *univoce*)* when, after all, there is an *infinite* distinction of being between the two beings. If the significance of "is" were univocal, the creature would be understood as the uncreated or else the uncreated would be degraded to being a creature. But "being" does not simply function as the same name; rather, in both cases "being" is understood. Scholasticism understands the positive sense of the significance of "being" as an "analogous" meaning in contradistinction to the univocal or merely homonymous one. Following Aristotle, in whom the problem is prefigured as it is in the point of departure of Greek ontology in general, various kinds of analogy were established according to which the "schools" differ in their interpretation of the functional significance of being. With regard to the ontological development of the problem, Descartes is far behind the scholastics;[14] he actually evades the question. *Nulla eius* [*substantiae*] *nominis significatio potest distincte intelligi, quae*

---

* in a consistent sense

13. Ibid. ["That is why the word substance does not pertain *univoce* to God and to other things, as the Scholastics say, that is, no common signification for this appellation which will apply equally to God and to them can be distinctly understood"].
14. Cf. *Opuscula omnia Thomae de Vio Caietani Cardinalis*, Lugduni (1580), III.5, "de nominum analogia," pp. 211–19.

Deo et creaturis sit communis.[15] This evasion means that Descartes leaves the meaning of being contained in the idea of substantiality and the character of the "universality" of this meaning unexplained. Medieval ontology left the question of what being itself means just as unquestioned as did ancient ontology. Thus it is not surprising if a question such as that of the kinds of significations of being gets nowhere as long as it is to be discussed on the foundation of an unclarified meaning of being which the signification "expresses." The meaning was unclarified because it was held to be "self-evident."*

94   Descartes not only completely evades the ontological question of substantiality, he emphasizes explicitly that substance as such, that is, its substantiality, is in and for itself inaccessible from the very beginning. *Verumtamen non potest substantia primum animadverti ex hoc solo, quod sit res existens, quia hoc solum per se nos non afficit.*[16] "Being" itself does not "affect" us, therefore it cannot be perceived. "Being is not a real predicate"† according to Kant, who is only repeating Descartes' statement. Thus the possibility of a pure problematic of being is renounced in principle, and a way out is sought for arriving at the definitions of substances designated above. Because "being" is in fact not accessible *as a being*, it is expressed by existing [seiende] definite qualities of the beings in question, by attributes. Not, however, by arbitrary qualities, but by those that most purely satisfy the meaning of being and substantiality tacitly presupposed. *Extensio* is the primarily necessary "attribute" in the *substantia finita* as *res corporea. Quin et facilius intelligimus substantiam extensam, vel substantiam cogitantem, quam substantiam solam, omisso eo quod cogitet vel sit extensa;*[17] for substantiality is *ratione tantum*, it is not detachable *realiter,*‡ nor is it to be found like substantial beings themselves.

Thus the ontological foundations for the definition of "world" as *res extensa* have become clear: they are found in the idea of substantiality which is not only unexplained in the meaning of its being, but is also declared to be inexplicable, and which is presented by way of a detour around the most distinctive substantial attribute of the substance in question. In the definition of substance in terms of a substantial being one also sees the reason why the term substance is ambiguous. What

* and was content with intelligibility.
† "Real" belonging to whatness [Sachheit], to the what, which alone can concern us in this or that way.
‡ The content of the what [wasgehaltlich].

15. Descartes, *Principia* I, pr. 51, p. 24. ["No signification of this name (substance) which would be common to God and his creation can be distinctly understood"].
16. Ibid., pr. 52, p. 25. ["Yet substance cannot be first discovered merely from the fact that it is a thing that exists, for that fact alone is not observed by us"].
17. Ibid., pr. 63, p. 31. ["It is moreover easier to know a substance that thinks, or an extended substance, than substance alone, without regarding whether it thinks or is extended"].

is intended is substantiality and it is understood in terms of an existent quality of substance. Because something ontic is made to underlie the ontological, the expression *substantia* functions sometimes in an ontological, sometimes in an ontic meaning, but mostly in a meaning which shifts about in a hazy mixture of the two. But behind this slight difference of meaning lies hidden the failure to master the fundamental problem of being.* Its development requires "tracking down" the equivocations in the *right way*. Whoever tries this sort of thing is not "occupied" with "mere verbal meanings," but must venture forth to the most primordial problematic of the "things themselves" to get such "nuances" straightened out.

95

§ 21. *The Hermeneutical Discussion of the Cartesian Ontology of the "World"*

The critical question now arises: Does this ontology of the "world" see the phenomenon of world at all, and if not, does it at least define innerworldly beings to the extent that their worldly character can be made visible? *To both questions we must answer "No!"* The being which Descartes is trying to grasp ontologically and in principle with the *extensio* is rather of such a nature that it can be discovered initially only through an innerworldly being initially at hand. But if this, along with the ontological characteristic of *this* particular innerworldly being (nature)—the idea of substantiality, as well as the meaning of *existit* and *ad existendum*, contained in its definition—leads to obscurity, the possibility nonetheless still exists that through an ontology grounded in the radical separation of God, ego, "world," the ontological problem of the world is in some sense raised and further advanced. But even if this possibility does not exist, we must show explicitly that Descartes not only goes amiss ontologically in his definition of the world, but that his interpretation and its foundations led him to *pass over* the phenomenon of world, as well as the being of innerworldly beings initially at hand.

In our exposition of the problem of worldliness (§ 14) we referred to the importance of gaining proper access to this phenomenon. Thus in our critical discussion of the Cartesian point of departure we must ask which kind of being of Dasein we should fix upon as the appropriate kind of access to *that* being with whose being as *extensio* Descartes equates the being of the "world." The sole, genuine access to this being is knowing, *intellectio*, in the sense of the kind of knowledge we get in mathematics and physics. Mathematical knowledge is regarded as the one way of apprehending beings which can always be certain of the secure possession of the being of the beings which it apprehends.

---

* Ontological difference.

Whatever has the kind of being adequate to the being accessible in mathematical knowledge *is* in the true sense. This being is *what always is what*
96   *it is*. Thus what can be shown to have the character of *constantly remain-*
ing, as *remanens capax mutationum*, constitutes the true being of beings
which can be experienced in the world. What enduringly remains truly
is. This is the sort of thing that mathematics knows. What *mathematics*
makes accessible in beings constitutes their being. Thus the being of
the "world" is, so to speak, dictated to it in terms of a definite idea of
being, which is embedded in the concept of substantiality, and in terms
of an idea of knowledge which cognizes beings *in this way*. Descartes
does not allow the kind of being of innerworldly beings to present
itself, but rather prescribes to the world, so to speak, its "true" being
on the basis of an idea of being (being = constant presence) the source
of which has not been revealed and the justification of which has not
been demonstrated. Thus it is not primarily his dependence upon a science, mathematics, which just happens to be especially esteemed, that
determines his ontology of the world; rather, his ontology is determined
by a basic ontological orientation toward being as constant objective
presence, which mathematical knowledge is exceptionally well suited
to grasp.* In this way Descartes explicitly switches over philosophically
from the development of traditional ontology to modern mathematical
physics and its transcendental foundations.

Descartes does not need to raise the problem of the appropriate access to innerworldly beings. Under the unbroken dominance of
traditional ontology, the way to get a grasp of what truly is has been
decided in advance. That way lies in νοεῖν, "intuition" ["Anschauung"]
in the broadest sense, of which διανοεῖν, "thinking," is just a derivative
form. It is in terms of this basic ontological orientation that Descartes
gives his "critique" of the possible intuitive-perceptive access to beings,
of *sensatio* (αἴσθησις) as opposed to *intellectio*.

Descartes knows very well that beings do not initially show themselves in their true being. What is given "initially" is this waxen thing
which is colored, flavored, hard, cold, and resonant in a definite way.
But this is not important ontologically, nor, in general, is anything which
is given through the senses. *Satis erit, si advertamus sensuum perceptiones*
non referri, nisi ad istam corporis humani cum mente coniunctionem, et nobis
quidem ordinarie exhibere, quid ad illam externa corpora prodesse possint aut
nocere.[18] The senses do not enable us to know any being in its being;

---

* but orientation to the mathematical as such, μάθημα and ὄν.

18. Ibid., II, pr. 3, p. 41. ["It will be sufficient for us to observe that the perceptions
of the senses are related simply to the intimate union which exists between body and
mind, and that while by their means we are made aware of what in external bodies
can profit or hurt this union, they do not present them to us as they are in themselves
unless occasionally and accidentally"].

they merely make known the usefulness and harmfulness of "external" innerworldly things for human beings encumbered with bodies. *Nos non docent, qualia (corpora) in seipsis existant;*[19] they tell us nothing at all about beings in their being. *Quod agentes, percipiemus naturam materiae, sive corporis in universum spectati, non consistere in eo quod sit res dura, vel ponderosa, vel colorata, vel alio aliquo modo sensus afficiens: sed tantum in eo, quod sit res extensa in longum, latum et profundum.*[20]   *97*

If we subject Descartes' interpretation of the experience of hardness and resistance to a critical analysis, it will be plain how unable he is to let what shows itself in sensation present itself in its own kind of being, let alone determine its character. (Cf. § 19.)

Hardness is understood as resistance. But neither hardness nor resistance is understood in a phenomenal sense, as something experienced in itself and determinable in such experience. For Descartes, resistance amounts to no more than not yielding place, that is, not undergoing any change of location. A thing's resistance means that it stays in a definite place, relative to another thing changing its place, or else that it changes its own location with a velocity that permits the thing to "catch up" with it. But when the experience of hardness is interpreted in this way, the kind of being that belongs to sensory perception is obliterated, and with it the possibility of grasping the being of those beings encountered in such perception. Descartes translates the kind of being of the perception of something into the only kind of being that he knows; the perception of something becomes a definite objective presence of two objectively present *res extensa* next to each other. The relation of their movements is itself a mode of *extensio* that primarily characterizes the objective presence of the corporeal thing. It is true that the possible "fulfillment" of the act of touching requires a distinctive "nearness" of what is touchable. But that does not mean that touching and the hardness made known in touching consist, ontologically understood, in different velocities of two corporeal things. Hardness and resistance do not show themselves at all unless there is a being which has the kind of being of Dasein, or at least of a living being.

Thus Descartes' discussion of the possible kinds of *access to* innerworldly beings is dominated by an idea of being which is patterned after a particular region of these beings themselves.

---

19. Ibid., II, pr. 3, pp. 41–42. ["That the perceptions of the senses do not teach us what is really in things, but merely that whereby they are useful or hurtful to man's composite nature"].

20. Ibid., pr. 4, p. 42. ["In this way we shall ascertain that the nature of matter or of body in its universal aspect does not consist in its being hard, or heavy, or colored, or one that affects our senses in some other way, but solely in the fact that it is a substance extended in length, breadth, and depth"].

98   The idea of being as constant presence not only motivates an
extreme definition of the being of innerworldly beings and their
identification with the world as such. At the same time, it blocks the
possibility of bringing to view attitudes of Dasein in a way which is
ontologically appropriate. But thus the road is completely blocked to
seeing the founded character of all sensuous and intellective apprehension, and to understanding them as a possibility of being-in-the-world.
But Descartes understands the being of "Dasein," to whose basic constitution being-in-the-world belongs, in the same way as the being of
res extensa, as substance.

        But with these criticisms have we not foisted upon Descartes a
task altogether beyond his horizon, and then "demonstrated" that he
failed to solve it? How could Descartes identify a definite innerworldly
being and its being with the world if he does not know the phenomenon of world at all and thus something akin to innerworldliness?

        In the realm of controversy over principles, one must not only
attach oneself to theses which can be grasped doxographically, rather
one must take the objective tendency of the problematic as an orientation, even if it does not go beyond a rather common version of
that problematic. The fact that Descartes not only *wanted to raise* the
question of "ego and world" with his doctrine of *res cogitans* and *res
extensa*, but claimed to give a radical solution, becomes clear in his
Meditations (cf. especially I and IV). The preceding discussion should
have demonstrated that the basic ontological orientation toward the
tradition, devoid of any positive criticism, made it impossible for him
to clear the way for a primordial ontological problematic of Dasein,
and necessarily distorted his view of the phenomenon of the world and
forced the ontology of the "world" into the ontology of a particular
innerworldly being.

        One might object, however, that even if the problem of the world
and also the being of beings encountered in the surrounding world
indeed remain obscured, Descartes nonetheless laid the foundation for
the ontological characteristic of *that* innerworldly being which in its
being is the basis for every other being, material nature.* The other
strata of innerworldly reality are based upon it, the fundamental stratum. The definite properties which show themselves as qualities, but
which are "basically" quantitative modifications of the modes of *extensio* itself, initially have their basis in the extended thing as such. Specific
99 qualities such as beautiful, not beautiful, fitting, unfitting, usable, unusable then find a footing in these qualities which are themselves further
reducible. Those specific qualities must be understood in a primary

---

* Critique of Husserl's development of "ontologies!" just as the whole critique of Descartes is inserted here with this intention!

orientation to thingliness as nonquantifiable value predicates through which the thing, initially merely material, gets stamped as something good. But with this stratification we come, after all, to the being that we characterized ontologically as the useful thing at hand. Thus the Cartesian analysis of the "world" first makes possible a secure erection of the structure of what is initially at hand. It only needs to round out the natural thing so that it becomes in every sense a thing of use, a task easily accomplished.

But apart from the specific problem of the world, is the being of what we initially encounter in the world ontologically attainable in this way? When we speak of material thingliness, do we not tacitly posit a kind of being—the constant objective presence of a thing—which is so far from being rounded out ontologically by subsequently outfitting beings with value predicates that these value characters themselves rather remain mere ontic qualities of a being which has the kind of being of a thing? The addition of value predicates is not in the least able to tell us anything new about the being of goods, *but rather only again presupposes for them the kind of being of mere objective presence.* Values are *objectively present* determinations of a thing. In the end, values have their ontological origin solely in the previous point of departure of the reality of the thing as the fundamental stratum. But pre-phenomenological experience already shows something about the being supposed to be a thing which is not fully intelligible through thingliness. Thus thinglike being needs a supplement. What, then, does the being of values or their "validity," which Lotze understood as a mode of "affirmation," mean ontologically? What does this "inherence" of values in things mean ontologically? As long as these matters remain obscure, the reconstruction of a thing of use in terms of a thing of nature is an ontologically questionable undertaking, not to speak of the fundamental distortion of the problematic. And does not this reconstruction of the initially "stripped" thing of use always need the *previous, positive view of the phenomenon whose totality is to be reestablished in the reconstruction*? But if its ownmost constitution of being of the phenomenon is not adequately explicated, are we not building the reconstruction without a plan? Insofar as this reconstruction and "rounding out" of the traditional ontology of the "world" results in our reaching the same *being* from which the above analysis of the handiness of useful things and totality of relevance took its point of departure, it seems as if the *being* of that being were indeed clarified or had at least become a *problem*. Just as Descartes cannot grasp the being of substance with *extensio* as *proprietas*, the flight to "valuable" qualities cannot even catch sight of being as handiness, let alone make it ontologically thematic.

Descartes narrowed down the question of the world to that of the thingliness of nature as that innerworldly being which is initially

100

accessible. He strengthened the opinion that the supposedly strictest ontic *knowledge* of a being is also the possible access to the primary being of the being discovered in such knowledge. But we must at the same time realize that the "roundings-out" of an ontology of things are fundamentally on the same dogmatic basis as that of Descartes.

We have already intimated (§ 14) that passing over the world and those beings initially encountered is not a matter of chance, not an oversight which we could simply make up for, but rather is grounded in the essential kind of being of Dasein itself. When our analytic of Dasein has made the most important basic structures of Dasein transparent in the scope of this problematic, when we have assigned to being in general the horizon of its possible intelligibility,* thus first making handiness and objective presence ontologically and primordially intelligible, too, only then can the critique of the Cartesian ontology of the world, which basically is still customary today, claim its philosophical justification.

To do this, we must show several things (cf. Part I, Division 3):

1. Why was the phenomenon of world passed over at the beginning of the ontological tradition decisive for us, explicitly in Parmenides; where does the constant recurrence of this passing over come from?
2. Why do innerworldly beings take the place of the phenomenon thus passed over as the ontological theme?
3. Why are these beings initially found in "nature"?
4. Why does the rounding out of such an ontology of the world, experienced as necessary, take place with the help of the phenomenon of value?

In the answers to these questions a positive understanding of the *problematic* of the world will be reached for the first time, the source of our failure to recognize it will be demonstrated, and the justification for rejecting the traditional ontology of the world will have been demonstrated.

101   The world, Dasein, and innerworldly beings are the ontologically constitutive states nearest to us; but we have no guarantee that we achieve the basis for encountering them phenomenally by the seemingly obvious procedure of starting with the things of this world, still less by taking our orientation from what is supposedly the most rigorous knowledge of beings. Our remarks about Descartes should have brought us this insight.

But if we recall that spatiality also manifestly constitutes innerworldly beings, it is, after all, possible to "salvage" the Cartesian

---

* Sic! Of course, "intelligibility" is based on understanding as projection and this projection as ecstatic temporality.

analysis of the "world" in the long run. With his radical exposition of *extensio* as the *praesuppositum* for every quality of the *res corporea*, Descartes prepared the way for the understanding of an *a priori* whose content Kant then made precise with greater penetration. Within certain limits, the analysis of *extensio* remains independent of his neglecting to provide an explicit interpretation of the being of extended beings. Taking *extensio* as the basic determination of the "world" has its phenomenal justification, although in recourse to it neither the spatiality of the world nor the spatiality initially discovered of beings encountered in the surrounding world, nor even the spatiality of Dasein itself, can be conceived ontologically.

C. The Aroundness [Umhafte] of the Surrounding World [Umwelt] and the Spatiality of Dasein

In connection with our first preliminary sketch of being-in (cf. § 12), Dasein had to be contrasted with a way of being in space which we call insideness. This means that a being which is itself extended is enclosed [umschlossen] by the extended boundaries of something extended. The being which is inside and what surrounds it are both present in space. Our rejection of such an insideness of Dasein in a spatial container should not, however, basically exclude all spatiality of Dasein, but only keep the way clear for seeing the kind of spatiality which is essential for Dasein. This must now be set forth. But since inner*worldly* beings are also in space, their spatiality has an ontological connection with the world.* Thus we must determine in what sense space is constitutive for the world which in turn was characterized as a structural factor of being-in-the-world. We must especially show how the aroundness of the surrounding world, the specific spatiality of the beings encountered in the surrounding world, is grounded in the worldliness of the world, and not the other way around, that is, we cannot say that the world in its turn is present in space. Our study of the spatiality of Dasein and the spatial definiteness of the world takes its point of departure from an analysis of the innerworldly things at hand in space. We shall consider three stages: (1) The spatiality of innerworldly things at hand (§ 22). (2) The spatiality of being-in-the-world (§ 23). (3) The spatiality of Dasein and space (§ 24).

102

§ 22. *The Spatiality of Innerworldly Things at Hand*

If space constitutes the world in a sense which we have yet to determine, it cannot be surprising that in our foregoing ontological characterization of the being of what is within the world we already had to

---

* Thus world is also spatial.

regard that being as something in space too. This spatiality of things at hand has not yet been grasped phenomenally in an explicit way and its interconnection with the structures of being of what is at hand has not yet been demonstrated. That is now the task.

To what extent have we already bumped up against this spatiality in our characterization of what is at hand? We spoke of what is *initially* [*zunächst*] at hand. This means not only beings which we encounter *first* [*zuerst*] before others, but means at the same time beings that are "near by" ["in der Nähe"]. The things at hand of everyday dealings have the character of *nearness* [*Nähe*]. To be exact, this nearness of useful things is already hinted at in the term which expresses their being, in "handiness." Beings "at hand" have their various nearnesses which are not ascertained by measuring distances. Their nearness is determined by the handling and use that circumspectly "calculate." The circumspection of taking care of things at the same time establishes what is thus near with respect to the direction in which useful things are always accessible. The structured nearness of useful things means that they do not simply have a place in space, objectively present somewhere, but as useful things are essentially installed, put in their place, set up, and put in order. Useful things have their *place*, or else they "lie around," which is fundamentally different from merely occurring in a random spatial position. The actual place is defined as the place of this useful thing for . . . in terms of a totality of the interconnected places of the context of useful things at hand in the surrounding world. Place and the multiplicity of places must not be interpreted as the where of a random being present of things. Place is always the definite "over there" and the "there" of a useful thing *belonging there*. In each and every case belonging there corresponds to the useful character of what is at hand, that is, to its relevant belonging to a totality of useful things. But a whereto in general, in which the positional totality is referred to a context of useful things, underlies the positional belonging somewhere of a totality of useful things as the condition of their possibility. This whereto of the possible belonging somewhere of useful things, which is circumspectly held in view in advance and in heedful dealings, we call the *region*.

"In the region of" means not only "in the direction of," but also in the orbit of something that lies in that direction. The kind of place which is constituted by direction and remoteness—nearness is only a mode of the latter—is already oriented toward a region and within that region. Something akin to a region must already be discovered if there is to be any possibility of referring and finding the places of a totality of useful things available to circumspection. This regional orientation of the multiplicity of places of what is at hand constitutes the aroundness [Umhafte], the being around [Um-uns-herum] us of beings encountered initially in the surrounding world [umweltlich]. There is never a three-dimensional multiplicity of possible positions

103

initially given which is then filled out with objectively present things. This dimensionality of space is still veiled in the spatiality of what is at hand. The "above" is what is "on the ceiling," the "below" is what is "on the floor," the "behind" is what is "at the door." All these wheres are discovered and circumspectly interpreted on the paths and ways of everyday dealings; they are not ascertained and catalogued by the observational measurement of space.

Regions are not first formed by things present together, but are always already at hand in individual places. The places themselves are assigned to what is at hand in the circumspection of taking care of things, or else we come across them as such. Thus things constantly at hand, with which circumspect being-in-the-world reckons from the outset, have their place. The where of their handiness is taken account of in taking care and is oriented toward other handy things. Thus the sun whose light and warmth we make use of every day has its circumspectly discovered, eminent places in terms of the changing usability of what it gives us: sunrise, noon, sunset, midnight. The places of these things, which are constantly at hand in various ways and yet uniformly, become accentuated "indicators" of the regions contained in them. These regions of the sky which do not yet need to have any geographical meaning at all, give beforehand the whereto for every particular development of regions which can be occupied by places. The house has its sunny side and its shady side. This provides the orientation for dividing up the "rooms" and "arranging" them according to their useful character. Churches and graves, for example, are laid out according to the rising and setting of the sun—the regions of life and death which determine Dasein itself with regard to its ownmost possibilities of being [Seinsmöglichkeiten] in the world. In taking care, Dasein, which is in its very being concerned about that being, discovers beforehand the regions which are each in a decisive relevance. The discovery of regions beforehand is determined by the totality of relevance for which what is at hand is set free as something encountered.

The handiness which belongs to each region beforehand has as the being of what is at hand the *character of inconspicuous familiarity* in a more primordial sense. The familiarity itself becomes visible in a conspicuous manner only when what is at hand is discovered circumspectly in the deficient mode of taking care. When we do not find something in *its* place, the region of that place often becomes explicitly accessible as such for the first time. Space, which is discovered in circumspect being-in-the-world as the spatiality of a totality of useful things, belongs to beings themselves as their place. Bare space is still veiled. Space is split up into places.* But this spatiality has its own unity by virtue of the worldlike totality of relevance of what is

104

---

* No, rather a peculiar unity of places that are not split up.

spatially at hand. The "surrounding world" does not arrange itself in a previously given space, but rather its specific worldliness articulates in its significance the relevant context of a totality of places that have been circumspectly assigned. In each case the world discovers the spatiality of space belonging to it. The fact that what is at hand can be encountered in its space of the surrounding world is ontically possible only because Dasein itself is "spatial" with regard to its being-in-the-world.

§ 23. *The Spatiality of Being-in-the-World*

When we attribute spatiality to Dasein, this "being [Sein] in space" must evidently be understood in terms of the kind of being of this being. The spatiality of Dasein, which is essentially not objective presence, can mean neither something like being found in a position in "world space," nor being at hand in a place. Both of these are kinds of being belonging to beings encountered in the world. But Dasein is "in" the world in the sense of a familiar and heedful dealing with the beings encountered within the world. Thus when spatiality is attributed to it in some way, this is possible only on the basis of this being-in. But the spatiality of being-in shows the character of *de-distancing* and *directionality*.

105

By de-distancing as a kind of being of Dasein with regard to its being-in-the-world, we do not understand anything like remoteness (nearness) or even being at a distance. We use the expression de-distancing in an active and transitive sense. It means a constitution of being of Dasein, of which de-distancing something, putting it away, is only a definite, factical mode. De-distancing means *making distance disappear*,* making the being at a distance of something disappear, bringing it near. Dasein is essentially de-distancing. As the being that it is, it lets beings be encountered in nearness.† De-distancing discovers remoteness. Remoteness, like distance, is a categorial determination of beings unlike Dasein. De-distancing, on the other hand, must be kept in mind as an existential. Only because beings in general are discovered by Dasein in their remoteness, do "distances" and intervals among innerworldly beings become accessible in relation to other things. Two points are as little remote from each other as two things in general because neither of these beings can de-distance in accordance with its kind of being. They merely have a measurable distance between them which is encountered in de-distancing.‡

Initially and for the most part, de-distancing is a circumspect approaching, a bringing near as supplying, preparing, having at hand.

---

* Where does the distance come from that is de-distanced?
† Nearness and *presence* [*Anwesenheit*], not the extent of the distance, is essential.
‡ De-distancing is more precise [*schärfer*] than nearing.

But particular kinds of the purely cognitive discovery of beings also have the character of bringing near. *An essential tendency toward nearness lies in Dasein.** All kinds of increasing speed which we are more or less compelled to go along with today push for overcoming distance. With the "radio," for example, Dasein is bringing about today a de-distancing of the "world," which is unforeseeable in its meaning for Dasein, by way of expanding and destroying the everyday surrounding world.

De-distancing does not necessarily imply an explicit estimation of the farness of things at hand in relation to Dasein. Above all, remoteness is never understood as measurable distance. If farness is estimated, this is done relative to the de-distancing in which everyday Dasein is involved. In the calculative sense these estimations may be imprecise and variable, but they have their *own* thoroughly intelligible *definiteness* in the everydayness of Dasein. We say that to go over there is a good walk, a stone's throw, as long it takes to smoke a pipe. These measures express the fact that they not only do not intend to "measure," but that the estimated remoteness belongs to a   *106* being which one approaches in a circumspect, heedful way. But even when we use more exact measures and say "it takes half an hour to get to the house," this measure must be understood as an estimation. "Half an hour" is not thirty minutes, but a duration which does not have any "length" in the sense of a quantitative stretch. This duration is always interpreted in terms of familiar, everyday "activities." Even where "officially" calculated measurements are familiar, remoteness is initially estimated circumspectly. Because what is de-distanced is at hand in such estimates, it retains its specifically innerworldly character. This even implies that the paths we take in our dealings with remote beings are of different lengths every day. What is at hand in the surrounding world is, after all, not objectively present for an eternal spectator exempt from Dasein, but is encountered in the circumspect, heedful everydayness of Dasein. On these paths Dasein does not traverse, like an objectively present corporeal thing, a stretch of space, it does not "eat up kilometers"; nearing and de-distancing are always a heedful being toward what is approached and de-distanced. An "objectively" long path can be shorter than an "objectively" much shorter path which is perhaps an "onerous one" and strikes one as infinitely long. *When it "strikes" one thus, however, the actual world is first truly at hand.* The objective distances of objectively present things do not coincide with the remoteness and nearness of what is at hand within the world. The former may be exactly known, but this knowledge is blind. It does not have the function of the circumspectly discovering approach to the surrounding world. One uses such knowledge only in

---

* To what extent and why? Being qua constant presence [beständige Anwesenheit] has priority, making present.

and for a heedful being which does not measure stretches and which is related to the world which "concerns" us.

When there is a prior orientation toward "nature" and the "objectively" measured distances of things, one is inclined to consider such interpretations and estimates of remoteness "subjective." However, that is a "subjectivity" which perhaps discovers what is most real about the "reality" of the world, which has nothing to do with "subjective" arbitrariness and the subjectivistic "conceptions" of beings which are "in themselves" otherwise. *The circumspect de-distancing of everyday Dasein discovers the being-in-itself of the "true world" of beings with which Dasein as existing is always already together.*

The primary and even exclusive orientation toward remoteness as measured distances obscures the primordial spatiality of being-in. What is supposedly "nearest" is by no means that which has the smallest distance "from us." What is "near" ["Nächste"] lies in that which is in the circle of an average reach, grasp, and look. Since Dasein is essentially spatial in the manner of de-distancing, its dealings always take place in a "surrounding world" which is remote from it in a certain leeway. Thus we initially always overlook and fail to hear what is measurably "nearest" to us. Seeing and hearing are senses of distance not because of their scope, but because Dasein, de-distancing, predominantly lives in them. For someone who, for example, wears spectacles which are distantially so close to him that they are "sitting on his nose," this useful thing is further, in being used, further away in the surrounding world than the picture on the wall across the room. This useful thing has so little nearness [Nähe] that it is often not even to be found at all initially. Useful things for seeing, and those for hearing, for example, the telephone receiver, have the inconspicuousness of what is initially at hand which we characterized. That is also true, for example, of the street, the useful thing for walking. When we walk, we feel it with every step and it seems to be what is nearest and most real about what is generally at hand, it slides itself, so to speak, along certain parts of our body—the soles of one's feet. And yet it is more remote than the acquaintance one meets while walking in the "distance," twenty steps away "on the street." Circumspect heedfulness decides about the nearness and farness of what is initially at hand in the surrounding world. Whatever this heedfulness dwells in from the beginning is what is nearest, and regulates our de-distancing.

When Dasein in taking care brings something near, this does not mean that it fixes upon something at a position in space which has the least measurable distance from a point of its body. To be near means to be in the range of what is initially at hand for circumspection. Bringing near is not oriented toward the I-thing encumbered with a body, but rather toward heedful being-in-the-world, that is, what that being-

107

in-the-world initially encounters. Neither is the spatiality of Dasein determined by citing the position where a corporeal thing is objectively present. It is true that we also say of Dasein that it occupies a place. But this "occupying" is to be fundamentally differentiated from being at hand at a place in terms of a region. Occupying a place must be understood as de-distancing what is at hand in the surrounding world in a region previously discovered circumspectly. Dasein understands its here in terms of the over there of the surrounding world. The here does not mean the where of something objectively present, but the where of de-distancing being with . . . together with this de-distancing. In accordance with its spatiality, Dasein is initially never here, but over there. From this over there it comes back to its here, and it does this only by interpreting its heedful being toward something in terms of what is at hand over there. This becomes quite clear from a phenomenal peculiarity of being-in which has the structure of de-distancing.

     108

As being-in-the-world, Dasein essentially dwells in de-distancing. This de-distancing, the farness from itself of what is at hand, is something that Dasein can *never cross over*. It is true that Dasein can take the remoteness of something at hand from Dasein to be distance if that remoteness is determined in relation to a thing which is thought of as being objectively present at a place which Dasein has already occupied. Dasein can subsequently traverse the "between" of this distance, but only in such a way that the distance itself becomes de-distanced. So little has Dasein crossed over its de-distancing that it rather has taken it along and continues to do so *because it is essentially de-distancing, that is, it is spatial*. Dasein cannot wander around in the current range of its de-distancings, it can only change them. Dasein is spatial by way of circumspectly discovering space so that it is related to beings thus spatially encountered by constantly de-distancing.

As being-in which de-distances, Dasein has at the same time the character of *directionality*. Every bringing near has always taken a direction in a region beforehand from which what is de-distanced approaches so that it can be discovered with regard to its place. Circumspect heedfulness is a directional de-distancing. In this heedfulness, that is, in the being-in-the-world of Dasein itself, the need for "signs" is already present. As useful things, signs take over the giving of directions in a way which is explicit and easily handled. They explicitly keep the circumspectly used regions open, the actual whereto of belonging, going, bringing, fetching. If Dasein *is*, it always already has, as directing and de-distancing, its discovered region. As modes of being of being-in-the-world, directing and de-distancing are guided beforehand *by the circumspection* of heedfulness.

The firm directions of right and left originate out of this directionality. Dasein continually takes these directions along together with its de-distancing. The spatialization of Dasein in its "corporeality," which contains a problematic of its own not to be discussed here, is also marked out in accordance with these directions. Thus things at hand and in use for the body, such as gloves, for example, that must go along with the hands' movement, must be oriented in terms of right and left. Tools, however, which are held in the hand and moved with it, do not go along with the specifically "handlike" movement of the hand. Thus there are no right- and left-handed hammers, even though they are held with the hand as gloves are.

But we must observe the fact that the directionality that belongs to de-distancing is grounded in being-in-the-world. Left and right are not something "subjective" for which the subject has a feeling, but they are directions of orientation in a world which is always already at hand. I could never find my way around in a world "through a mere feeling of a difference between my two sides."[21] The subject with the "mere feeling" of this difference is a construct posited without regard to the true constitution of the subject, namely that whenever Dasein has this "mere feeling" it is always already in a world *and must be* in order to be able to orient itself at all. This becomes clear in the example with which Kant tries to clarify the phenomenon of orientation.

Let us assume that I enter a familiar but dark room which has been rearranged during my absence in such a way that everything which was on the right-hand side is now on the left-hand side. If I am to get oriented, the "mere feeling of the difference" between my two sides does not help at all as long as I do not apprehend some particular object "whose position," as Kant casually remarks, "I have in mind." But what else does this mean except that I necessarily orient myself in and from already being in a "familiar"* world. The context of useful things in a world must already be given to Dasein. The fact that I am always already in a world is no less constitutive for the possibility of orientation than the feeling for right and left. That this constitution of being of Dasein is obvious does not justify suppressing it in its onto-logically constitutive role. Kant does not suppress it either, any more than any other interpretation of Dasein. Continual use of this constitution does not, however, exempt us from giving an adequate ontological explication, but rather requires it. The psychological interpretation that the ego has something "in mind" fundamentally refers to the existential constitution of being-in-the-world. Because Kant did not see this structure, he failed to understand the full context of the constitution of a possible orientation. Directedness toward the right or the left is

---

* From a familiar belongingness that I hold before myself and vary *accordingly*.

21. Immanuel Kant: "Was Heisst: Sich im Denken orientieren?" (1786) *Werke* (Akademie Ausgabe), vol. VIII, pp. 131–47.

grounded in the essential directionality of Dasein in general, which in turn is essentially determined by being-in-the-world. However, Kant is not interested in a thematic interpretation of orientation, either. He only wishes to show that all orientation needs a "subjective principle." But "subjective" means here *a priori*. The *a priori* of directionality in terms of right and left, however, is grounded in the "subjective" *a priori* of being-in-the-world, which has nothing to do with a determinate character restricted beforehand to a worldless subject.

As constitutive characteristics of being-in, de-distancing and directionality determine the spatiality of Dasein, for its being heedfully and circumspectly in discovered innerworldly space. Our previous explication of the spatiality of innerworldly things at hand and the spatiality of being-in-the-world first lays out the presuppositions for working out the phenomenon of the spatiality of the world and for formulating the ontological problem of space.

§ 24. *The Spatiality of Dasein and Space*

As being-in-the-world, Dasein has always already discovered a "world." We characterized this discovering, which is founded in the worldliness of the world, as the freeing of beings for a totality of relevance. Freeing something and letting it be relevant occur by way of circumspect self-reference which is grounded in a previous understanding of signification. We have now shown that circumspect being-in-the-world is spatial. And only because Dasein is spatial by way of de-distancing and directionality can things at hand in the surrounding world be encountered in their spatiality. The freeing of a totality of relevance is equiprimordially a letting something be relevant in a region which de-distances and gives direction. It is a freeing of the spatial belongingness of things at hand. The essential disclosure of space lies in the significance with which Dasein as heedful being-in is familiar.

Space that is disclosed with the wordliness of the world does not yet have the characteristic of a pure manifold of three dimensions. In this proximate disclosedness, space is still hidden as the pure wherein [reine Worin] in which points are ordered by measurement and the positions of things are determined. With the phenomenon of the region we have already indicated that for which space is discovered beforehand in Dasein. We understand the region as that to which the context of useful things at hand possibly belongs, a context which can be encountered as something directional, that is, containing places and as de-distanced. The belongingness is determined by the significance constitutive for the world and articulates the here and there within the possible whereto. The whereto in general is prefigured by the referential totality established in a for-the-sake-of-which

111

of heedfulness. Freeing and letting something be relevant is referred within this totality. *With* that which is encountered as at hand there is always relevance in a region. A regional spatial relevance belongs to the totality of relevance which constitutes the being of things at hand in the surrounding world. On the basis of this relevance, things at hand can be found and determined according to form and direction. In accordance with the possible transparency of heedful circumspection, innerworldly things at hand are de-distanced and oriented with the factical being of Dasein.

Letting innerworldly beings be encountered, which is constitutive for being-in-the-world, is "giving space." This "giving space," which we call *making room*, frees things at hand for their spatiality. As a way of discovering and presenting a possible totality of places relevantly determined, making room makes actual factical orientation possible. As circumspect taking care of things in the world, Dasein can change things around, remove them or "make room" for them only because making room—understood as an existential—belongs to its being-in-the-world. But neither the previously discovered region nor the actual spatiality in general are explicitly in view. In itself, it is present in the inconspicuousness of things at hand being taken care of by a circumspection absorbed in them for that circumspection. Space is initially discovered in this spatiality with being-in-the-world. On the basis of the spatiality thus discovered, space itself becomes accessible to cognition.

Space is neither in the subject nor is the world in space. Rather, space is "in" the world since the being-in-the-world constitutive for Dasein has disclosed space. Space is not in the subject, nor does that subject observe the world "as if" it were in space. Rather, the "subject," correctly understood ontologically, Dasein, is spatial in a primordial sense. And because Dasein is spatial in the way described, space shows itself as *a priori*. This term does not mean something like belonging beforehand to an initially worldless subject which spins a space out of itself. Here, apriority means the previousness of encountering space (as region) in the actual encountering of things at hand in the surrounding world.

The spatiality of what is initially circumspectly encountered can itself become thematized and so the task of calculation and measurement for circumspection, as for example, in building a house and surveying land. With this predominantly circumspect thematization of the spatiality of the surrounding world, space in itself already comes to view in a way. The space which thus shows itself can be studied by purely looking at it, if that which was formerly the sole possibility of access to space, namely, circumspect calculation, is given up. The "formal intuition" of space discovers pure possibilities of spatial relations. Here there is a series of stages laying bare pure, homogeneous

112

space, going from the pure morphology of spatial shapes to *Analysis Situs* and finally to the purely metrical science of space. In this present study we shall not consider how these are all interconnected.[22] In our problematic we wish solely to establish ontologically the phenomenal basis for the thematic discovery and working out of pure space.

Where space is discovered non-circumspectly by just looking at it, the regions of the surrounding world get neutralized to pure dimensions. The places and the totality of places of useful things at hand, which are circumspectly oriented, are reduced to a multiplicity of positions for random things. The spatiality of innerworldly things at hand thus loses its character of relevance. The world loses its specific character of aroundness, the surrounding world becomes the natural world. "The world" as a totality of useful things at hand is spatialized to become a context of extended things which are merely present. The homogeneous space of nature shows itself only when the beings we encounter are discovered in such a way that the worldly character of what is at hand gets specifically *deprived of its worldliness*.

In accordance with its being-in-the-world, Dasein has always already been pre-given its discovered space, even if unthematically. On the other hand, space in itself is initially still obscured with regard to the mere possibilities of the pure spatial being of something contained in it. The fact that space essentially *shows* itself *in a world* does not tell us anything about its kind of being. It need not have the kind of being of something itself at hand in space or objectively present. Nor does the being of space have the kind of being of Dasein. From the fact that the being of space itself cannot be conceived as the kind of being of *res extensa*, it follows neither that it must be ontologically determined as "phenomenon" of this *res*—it would not be distinguished from that *res* in its being—nor that the being of space can be equated with that of the *res cogitans* and be conceived as something merely "subjective," quite apart from the questionability of the *being* of this subject.

113

The perplexity still present today with regard to the interpretation of the being of space is grounded not so much in an inadequate knowledge of the factual constitution of space itself as in the lack of a fundamental transparency of the possibilities of being in general and of their ontologically conceived interpretation. What is decisive for the understanding of the ontological problem of space lies in freeing the question of the being of space from the narrowness of the accidentally available and, moreover, undifferentiated concepts of being, and, with respect to the phenomenon itself, in moving the problematic of the

---

22. Cf. O. Becker, "Beiträge zur phänomenologischen Bergründung der Geometrie und ihrer physikalischen Anwendungen," Jahrbuch, vol. 6 (1923), pp. 385ff.

being of space and the various phenomenal spatialities in the direction of clarifying the possibilities of being in general.

The primary ontological character of the being of innerworldly beings is not found in the phenomenon of space, either as unique or as one among others. Still less does space constitute the phenomenon of world. Space can only be understood by going back to the world. Space does not become accessible only by depriving the surrounding world of its worldliness. Spatiality can be discovered only on the basis of world; indeed, space *co*-constitutes the world in accordance with the essential spatiality of Dasein itself with regard to its fundamental constitution of being-in-the-world.

CHAPTER FOUR

Being-in-the-World as Being-with and Being a Self: The "They"

The analysis of the worldliness of the world continually brought the whole phenomenon of being-in-the-world into view without thereby delimiting all of its constitutive factors with the same phenomenal clarity as the phenomenon of world itself. The ontological interpretation of the world which discussed innerworldly things at hand came first not only because Dasein in its everydayness is in a world in general and remains a constant theme with regard to that world, but because it relates itself to the world in a predominant mode of being. Initially and for the most part, Dasein is taken over [benommen] by its world. This mode of being absorbed in the world, and in general the being-in which underlies it, essentially determine the phenomenon which we shall now pursue with the question: *Who* is it that Dasein is in everydayness? All of the structures of being of Dasein, thus also the phenomenon that answers to this question of who, are ways of its being [Sein]. Their ontological characteristic is an existential one. Thus, we need to pose the question correctly and outline the procedure for bringing into view a broader phenomenal domain of the everydayness of Dasein. By taking our investigation in the direction of the phenomenon which allows us to answer the question of the who, we are led to structures of Dasein which are equiprimordial with *being-in-the-world*: *being-with* [*Mitsein*] and *Dasein-with* [*Mitdasein*]. In this kind of being the mode of everyday being a self is grounded; the explication of this mode makes visible what we might call the "subject" of everydayness, the *they* [*das Man*]. This chapter on the "who" of average Dasein thus has the following structure:

    (1) The approach to the existential question of the who of Dasein (§ 25). (2) The Dasein-with of others and everyday being-with (§ 26). (3) Everyday being a self and the they (§ 27).

114

111

§ 25. *The Approach to the Existential Question of the Who of Dasein*

The answer to the question of who this being actually is (Dasein) seems to have already been given with the formal indication of the basic characteristics of Dasein (cf. § 9). Dasein is a being which I myself am, its being is in each case mine. This determination *indicates* an *ontological* constitution, but no more than that. At the same time, it contains an *ontic* indication, albeit an undifferentiated one, that an I is always this being, and not others. The who is answered in terms of the I itself, the "subject," the "self." The who is what maintains itself as an identity throughout changes in behavior and experiences, and in this way relates itself to this multiplicity. Ontologically, we understand it as what is always already and constantly present in a closed region and for that region, as that which lies at its basis in an eminent sense, as the *subjectum*. As something self-same in manifold otherness, this subject has the character of the *self*. Even if one rejects a substantial soul, the thingliness of consciousness, and the objectivity of the person, ontologically one still posits something whose being [Sein] retains the meaning of objective presence, whether explicitly or not. Substantiality is the ontological clue for the determination of beings in terms of which the question of the who is answered. Dasein is tacitly conceived in advance as objective presence. In any case, the indeterminacy of its being always implies this meaning of being. However, objective presence is the mode of being of beings unlike Dasein.

115

The ontic obviousness of the statement that it is I who is in each case Dasein must not mislead us into supposing that the way for an ontological interpretation of what is thus "given" has been unmistakably prescribed. It is even questionable whether the ontic content of the above statement reaches the phenomenal content of everyday Dasein. It could be the case that the who of everyday Dasein is precisely *not* I myself.

If, in coming to ontic and ontological statements, our phenomenal demonstration in terms of the mode of being of beings is to retain priority over the most obvious and usual answers as well as the problems arising from these, then the phenomenological interpretation of Dasein must be protected from a perversion [Verkehrung] of its problematic with regard to the question to be raised now.

But does it not go against the rules of a sound method when the approach to a problematic does not stick to the givens that are evident within the thematic realm? And what is less dubious than the givenness of the I? And, for the purpose of working this givenness out in a primordial way, does it not direct us to abstract from everything else that is "given," not only from an existing "world," but also from the being of other "I"s? Perhaps what this kind of giving gives—

this simple, formal, reflective perception of the I—is indeed evident. This insight even opens access to an independent phenomenological problematic which has its fundamental significance in the framework known as "formal phenomenology of consciousness."

In the present context of an existential analytic of factical Dasein, the question arises whether the I's mode of givenness which we mentioned discloses Dasein in its everydayness, if it discloses it at all. Is it then *a priori* self-evident that the access to Dasein must be a simple perceiving reflection of the I of acts? What if this kind of "self-giving" of Dasein were to lead our existential analytic astray and do so in a way grounded in the being of Dasein itself? Perhaps when Dasein addresses itself in the way which is nearest to itself, it always says it is I, and finally says this most loudly when it is "not" this being. What if the fact that Dasein is so constituted that it is in each case mine were the reason for the fact that Dasein *is*, initially and for the most part, *not itself*? What if, with the approach mentioned above, the existential analytic fell into the trap, so to speak, of starting with the givenness of the I for Dasein itself and its obvious self-interpretation? What if it should turn out that the ontological horizon for the determination of what is accessible in simple giving should remain fundamentally undetermined? We can probably always correctly say ontically of this being that "I" am it. However, the ontological analytic which makes use of such statements must have fundamental reservations about them. The "I" must be understood only in the sense of a noncommittal *formal indication* of something which perhaps reveals itself in the actual phenomenal context of being as that being's "opposite." Then "not I" by no means signifies something like a being which is essentially lacking "I-hood," but means a definite mode of being of the "I" itself; for example, having lost itself.*

But even the positive interpretation of Dasein that has been given up to now already forbids a point of departure from the formal givenness of the I if the intention is to find a phenomenally adequate answer to the question of the who. The clarification of being-in-the-world showed that a mere subject without a world "is" not initially and is also never given. And, thus, an isolated I without the others is in the end just as far from being given initially.[1] But if "the others" *are* always already *there with us* in being-in-the-world, ascertaining this phenomenally, too, must not mislead us into thinking that the *ontological* structure of what is thus "given" is self-evident and not in need of

116

---

* or else genuine selfhood as opposed to miserable egotism.

1. Cf. the phenomenological elucidations of M. Scheler, *Zur Phänomenologie und Theorie der Sympathiegefühle*, 1913, addendum, pp. 118ff. Also, the second edition under the title *Wesen und Formen der Sympathie*, 1923, pp. 244ff.

an investigation. The task is to make this Dasein-with [Mitdasein] of the nearest everydayness phenomenally visible and to interpret it in an ontologically appropriate way.

Just as the ontic, self-evident character of the being-in-itself of innerworldly beings misleads us by convincing us of the ontologically self-evident character of the meaning of this being [Sein] and makes us overlook the phenomenon of world, so too does the ontically self-evident character that Dasein is always my own harbor the possibility that the ontological problematic inherent in it might be led astray. *Initially* the who of Dasein is not only a problem *ontologically*, it also remains concealed *ontically*.

117   But, then, is the existential analytical answer to the question of the who without any clues at all? By no means. To be sure, of the formal indications of the constitution of being of Dasein given above (§§ 9 and 12), it is not so much the ones which we discussed thus far which function as such a clue, but rather, the one according to which the "essence" ["Essenz"] of Dasein is grounded in its existence. *If the "I" is an essential determination of Dasein, then it must be interpreted existentially.* The question of the who can then be answered only by a phenomenal demonstration of a definite kind of being of Dasein. If Dasein is always only its self *in existing*, the constancy of the self as well as its possible "inconstancy" require an existential-ontological kind of questioning as the only adequate access to the problematic.

But if the self is conceived "only" as a way of the being of this being, then that seems tantamount to volatizing the authentic "core" of Dasein. But such fears are nourished by the distorted presumption that the being in question really has, at bottom, the kind of being of something objectively present, even if one avoids attributing to it the solidifying element of a corporeal thing. However, the *"substance"* of human being is not spirit as the synthesis of body and soul; it is rather *existence*.

§ 26. *The Dasein-with of Others and Everyday Being-with*

The answer to the question of the who of everyday Dasein is to be won through the analysis of *the* kind of being in which Dasein, initially and for the most part, maintains itself. Our investigation takes its orientation from being-in-the-world. This fundamental constitution of Dasein codetermines every mode of its being. If we justifiably stated that all other structural factors of being-in-the-world already came into view by means of the previous explication of the world, then the answer to the question of the who must also be prepared by that explication.

The "description" of the surrounding world closest to us, for example, the work-world of the handworker, showed that together with the useful things found in work, others are "also encountered" for whom the "work" is to be done. In the kind of being of these things at hand, that is, in their relevance, there lies an essential reference to possible wearers for whom they should be "made to measure." Similarly, the producer or "supplier" is encountered in the material used as one who "serves" well or badly. The field, for example, along which we walk "outside" shows itself as belonging to such and such a person who keeps it in good order, the book which we use is bought at such and such a place, given by such and such a person, and so on. The boat anchored at the shore refers in its being-in-itself to an acquaintance who undertakes his voyages with it, but even as a "boat which is unknown to us," it still points to others. The others who are "encountered" in the context of useful things in the surrounding world at hand are not somehow added on in thought to an initially merely objectively present thing, but these "things" are encountered from the world in which they are at hand for the others. This world is always already from the outset my own. In our previous analysis, the scope of what is encountered in the world was initially narrowed down to useful things at hand, or nature objectively present, thus to beings of a character unlike Dasein. This restriction was not only necessary for the purpose of simplifying the explication, but, above all, because the kind of being of the existence of others encountered within the surrounding world is distinct from handiness and objective presence. The world of Dasein thus frees beings which are not only completely different from tools and things, but which themselves in accordance with their kind of being *as Dasein* are themselves "in" the world as being-in-the-world in which they are at the same time encountered. These beings are neither objectively present nor at hand, but they *are like* the very Dasein which frees them—*they are there, too, and there with it*. So, if one wanted to identify the world in general with innerworldly beings, one would have to say the "world" is also Dasein.

But the characteristic of encountering *others* is, after all, oriented toward one's *own* Dasein. Does not it, too, start with the distinction and isolation of the "I," so that a transition from this isolated subject to others must then be sought? In order to avoid this misunderstanding, we must observe in what sense we are talking about "others." "Others" does not mean everybody else but me—those from whom the I distinguishes itself. Others are, rather, those from whom one mostly does *not* distinguish oneself, those among whom one also is. This being-there-too with them does not have the ontological character of being objectively present "with" them within a world. The "with" is of the character of Dasein, the "also" means the sameness of being [Sein] as circumspect, heedful being-in-the-world. "With" and "also" are to be understood *existentially*, not categorially. On the basis of this *with-bound* [*mithaften*] being-in-the-world, the world is always already

118

the one that I share with others. The world of Dasein is a *with-world* [*Mitwelt*]. Being-in is *being-with* [*Mitsein*] others. The innerworldly being-in-itself of others is *Dasein-with* [*Mitdasein*].

119   Others are not encountered by grasping and discriminating beforehand one's own subject, initially objectively present, from other subjects also present. They are not encountered by first looking at one-self and then ascertaining the opposite pole of a distinction. They are encountered from out of the *world* in which Dasein, heedful and cir-cumspect, essentially dwells. As opposed to the theoretically concocted "explanations" of the presence of others, which easily urge themselves upon us, we must hold fast to the phenomenal fact which we have pointed out, namely, that they are encountered in the *surrounding world*. This nearest and elemental way of Dasein encountering the world goes so far that even one's *own* Dasein *initially* becomes "discoverable" by *looking away* from its "experiences" and the "center of its actions," or by not yet "seeing" them at all. Dasein initially finds "itself" in *what* it does, needs, expects, has charge of, in the things at hand which it initially *takes care of* in the surrounding world.

And even when Dasein explicitly addresses itself as "I here," the locative personal designation must be understood in terms of the existential spatiality of Dasein. When we interpreted this (§ 23), we already intimated that this I-here does not refer to an eminent point of an I-thing; rather, it understands itself, as being-in, in terms of the over there of the world at hand where Dasein dwells in *taking care*.

W. v. Humboldt[2] has alluded to certain languages which express the "I" by "here," the "thou" by "there," and the "he" by "over there," thus rendering the personal pronouns by locative adverbs, to put it grammatically. It is disputed whether the primordial meaning of loca-tive expressions is adverbial or pronominal. This dispute loses its basis if one notes that locative adverbs have a relation to the I qua Dasein. The "here," "over there," and "there" are not primarily pure locative designations of innerworldly beings objectively present at positions in space, but, rather, characteristics of the primordial spatiality of Dasein. The supposedly locative adverbs are determinations of Dasein; they have primarily an existential, not a categorial, meaning. But they are not pronouns, either. Their significance is prior to the distinction of locative adverbs and personal pronouns. The true spatial meaning of these expressions of Dasein, however, documents the fact that the theo-
120   retically undistorted interpretation of Dasein sees the latter immedi-ately in its spatial "being-together-with" [Sein bei] the world taken care of, spatial in the sense of de-distancing and directionality. In the "here"

---

2. "Über die Verwandtschaft der Ortsadverbien mit dem Pronomen in einigen Sprachen" (1829). *Gesammelte Schriften*, ed. Preuß. Akad. der Wissenschaften, vol. VI, part 1, pp. 304–30.

Dasein, absorbed in its world, does not address itself, but speaks away from itself, in circumspection, to the "over there" of something at hand and means, however, *itself* in its existential spatiality.

Dasein understands itself, initially and for the most part, in terms of its world, and the Dasein-with of others is frequently encountered from innerworldly things at hand. But when others become, so to speak, thematic in their Dasein, they are not encountered as objectively present thing-persons, rather we meet them "at work," that is, primarily in their being-in-the-world. Even when we see the other "just standing around," he is never understood as a human-thing objectively present. "Standing around" is an existential mode of being, the lingering with everything and nothing which lacks heedfulness and circumspection. The other is encountered in his Dasein-with in the world.

But, after all, the expression "Dasein" clearly shows that this being [Seiende] is "initially" unrelated to others, that it can, of course, also be "with" others subsequently. But we must not overlook the fact that we are also using the term *Dasein-with* as a designation *of* the being to which the existing others are freed within the world. The Dasein-with of others is disclosed only within the world, and so too for beings who are Daseins with us, because Dasein in itself is essentially being-with. The phenomenological statement that Dasein is essentially being-with has an existential-ontological meaning. It does not intend to ascertain ontically that I am factically not objectively present alone, rather that others of my kind also are [vorkommen]. If the statement that the being-in-the-world of Dasein is essentially constituted by being-with meant something like this, being-with would not be an existential attribute that belongs to Dasein of itself on the basis of its kind of being, but something which occurs at times on the basis of the existence of others. Being-with existentially determines Dasein even when an other is not factically present and perceived. The being-alone of Dasein, too, is being-with in the world. The other can be *lacking* only *in* and *for* a being-with. Being-alone is a deficient mode of being-with, its possibility is a proof for the latter. On the other hand, factical being alone is not changed by the fact that a second instance of a human being is "next to" me, or by ten such human beings. Even when these and still more are present, Dasein can be alone. Thus, being-with and the facticity of being-with-one-another are not based on the fact that several "subjects" are physically there together. Being alone "among" many, however, does not mean with respect to the being of others that they are simply objectively present. Even in being "among them," they are *there with* [*mit-da*]. Their Dasein-with is encountered in the mode of indifference and being alien. Lacking and "being away" are modes of Dasein-with and are possible only because Dasein as being-with lets the Dasein of others be encountered in its world. Being-with is an attribute of one's own Dasein. Dasein-with characterizes the Dasein of others in that it is freed for a being-with [Mitsein] by the world of that being-with. Only because it has the essential structure of being-with, is one's own Dasein encounterable by others as Dasein-with.

If Being-with* remains existentially constitutive for being-in-the-world, it must be interpreted, as must also circumspect dealings with the innerworldly things at hand which we characterized by way of anticipation as taking care [Besorgen], in terms of the phenomenon of *care* [Sorge] which we used to designate the being of Dasein in general. (Cf. Chapter Six of this Division.) Taking care of things is a character of being which being-with cannot have as its own, although this kind of being is a *being toward* [Sein zu] beings encountered in the world, as is taking care of things. The being [Seiende] to which Dasein is related as being-with does not, however, have the kind of being of useful things at hand; it is itself Dasein. This being is not taken care of, but is a matter of *concern* [Fürsorge].

Even "taking care" of food and clothing, and the nursing of the sick body is concern. But we understand this expression in a way which corresponds to our use of taking care as a term for an existential. For example, "welfare work" [Fürsorge] as a factical social institution is based on the constitution of being of Dasein as being-with. Its factical urgency is motivated by the fact that Dasein initially and, for the most part, lives in the deficient modes of concern. Being for-, against-, and without-one-another, passing-one-another-by, not-mattering-to-one-another, are possible ways of concern. And precisely the last named modes of deficiency and indifference characterize the everyday and average being-with-one-another. These modes of being show the characteristics of inconspicuousness and obviousness which belong to everyday innerworldly Dasein-with of others, as well as to the handiness of useful things taken care of daily. These indifferent modes of being-with-one-another tend to mislead the ontological interpretation into initially interpreting this being as the pure objective presence of several subjects. It seems as if only negligible variations of the same kind of being lie before us, and yet ontologically there is an essential distinction between the "indifferent" being together of

122   arbitrary things and the not-mattering-to-one-another of beings who are with one another.

With regard to its positive modes, concern has two extreme possibilities. It can, so to speak, take the other's "care" away from him and put itself in his place in taking care, it can *leap in* for him. Concern takes over what is to be taken care of for the other. The other is thus displaced, he steps back so that afterwards, when the matter has

---

* Earlier editions have Dasein-with here. Tr.

been attended to, he can take it over as something finished and available or disburden himself of it completely. In this concern, the other can become someone who is dependent and dominated even if this domination is a tacit one and remains hidden from him. This kind of concern which does the job and takes away "care" is, to a large extent, determinative for being-with-one another and pertains, for the most part, to our taking care of things at hand.

In contrast to this, there is the possibility of a concern which does not so much leap in for the other as *leap ahead* of him in his existentiell potentiality-of-being, not in order to take "care" away from him, but rather to authentically give it back as such. This concern which essentially pertains to authentic care—that is, it pertains to the existence of the other, and not to a *what* which it takes care of—helps the other to become transparent to himself *in* his care and *free for* it.

Concern proves to be constitutive of the being of Dasein which, in accordance with its different possibilities, is bound up with its being toward the world taken care of and also with its authentic being toward itself. Being-with-one-another is based initially and often exclusively on what is taken care of together in such being. A being-with-one-another which arises from one's doing the same thing as someone else not only keeps for the most part within outer limits but enters the mode of distance and reserve. The being-with-one-another of those who are employed for the same thing often thrives only on mistrust. On the other hand, when they devote themselves to the same thing in common, their doing so is determined by the Dasein that each has grasped as his own. This *authentic* alliance first makes possible the proper kind of objectivity [Sachlichkeit] which frees the other for himself in his freedom.

Between the two extremes of positive concern—the one which does someone's job for him and dominates him, and the one which is in advance of him and frees him—everyday being-with-one-another maintains itself and shows many mixed forms whose description and classification lie outside of the limits of this investigation.

Just as *circumspection* [*Umsicht*] belongs to taking care of things   *123*
as a way of discovering things at hand, concern is guided by *considerateness* [*Rücksicht*] and *tolerance* [*Nachsicht*]. With concern, both can go through the deficient and indifferent modes up to the point of *inconsiderateness* and the tolerance which is guided by indifference [Gleichgültigkeit].

The world frees not only things at hand as beings encountered within the world but also Dasein and others in their Dasein-with. But in accordance with its own meaning of being, this being which is freed in the surrounding world is being-in in the same world in which, as encounterable for others, it is there with them. Worldliness was interpreted (§ 18) as the referential totality of significance. In being familiar with this significance and previously understanding it, Dasein lets things at hand be encountered as things discovered in their relevance. The referential context of significance is anchored in the being [Sein] of Dasein toward its ownmost being—it cannot be relevance, it is rather being [Sein] *for the sake of which* [*worumwillen*] Dasein itself is as it is.

But, according to the analysis which we have now completed, being-with with others [Mitsein mit Anderen] belongs to the being of Dasein, with which it is concerned in its very being. As being-with, Dasein "is" essentially for the sake of others. This must be understood as an existential statement as to its essence. But even when actual, factical Dasein does *not* turn to others and thinks that it does not need them, or does without them, it *is* in the mode of being-with. In being-with as the existential for-the-sake-of-others, these others are already disclosed in their Dasein. This previously constituted disclosedness of others together with being-with thus helps to constitute significance, that is, worldliness. As this worldliness, disclosedness is anchored in the existential for-the-sake-of-which. Hence the worldliness of the world thus constituted in which Dasein always already essentially is, lets things at hand be encountered in the surrounding world in such a way that the Dasein-with of others is encountered at the same time with them as circumspectly taken care of. The structure of the worldliness of the world is such that others are not initially present as unattached subjects along with other things, but show themselves in their heedful being in the surrounding world in terms of the things at hand in that world.

The disclosedness of the Dasein-with of others which belongs to being-with means that the understanding of others already lies in the understanding of being of Dasein because its being is being-with. This understanding, like all understanding, is not a knowledge derived from cognition, but a primordially existential kind of being which first makes knowledge and cognition possible. Knowing oneself is grounded in being-with which primordially understands. It operates initially in accordance with the nearest kind of being of being-together-in-the-world [mitseienden In-der-Welt-seins] in the understanding knowledge of what Dasein circumspectly finds and takes care of with others. Concernful taking care of things is understood in terms of what is taken care of and with an understanding of them. Thus the other is initially disclosed in the taking care of concern.

But because concern, initially and for the most part, dwells in the deficient or at least indifferent modes—in the indifference of passing-one-another-by—the closest and essential knowing oneself is in need of a getting-to-know-oneself. And when even knowing oneself loses itself in aloofness, concealing oneself and misrepresenting oneself, being-with-one-another requires special ways in order to get close to others or to "see through them."

But just as opening oneself up or closing oneself off are grounded in the actual mode of being of being-with-one-another, indeed it *is* nothing besides this mode itself, so too does the explicit disclosure of the other in concern only grow out of one's primary being-with with him. Such a disclosure of the other which is indeed *thematic*, but not in the mode of theoretical psychology, easily becomes the phenomenon that first comes into view for the theoretical problematic of understanding the "psychical life of others." What "initially" presents phenomenally a way of being-with-one-another that understands—is at the same time, however, taken to mean that which "originally" and primordially makes possible and constitutes being toward others. This phenomenon, which is none too happily designated as *"empathy,"* is then supposed, as it were, to provide the first ontological bridge from one's own subject, initially given by itself, to the other subject, which is initially quite inaccessible.

To be sure, being-toward-others is ontologically different from being toward objectively present things. The "other" being itself has the kind of being of Dasein. Thus, in being with and toward others, there is a relation of being from Dasein to Dasein. But, one would like to say, this relation is after all already constitutive for one's own Dasein, which has an understanding of its own being and is thus related to Dasein. The relation of being to others then becomes a projection of one's own being toward oneself "into an other." The other is a duplicate [Dublette] of the self.

But it is easy to see that this seemingly obvious deliberation has little ground to stand on. The presupposition which this argument makes use of—that the being of Dasein toward itself is a being toward another—is incorrect. As long as the presupposition has not been demonstrated clearly in its legitimacy, it remains puzzling how the relation of Dasein to itself should thus be disclosed to the other as other.

Being toward others is not only an autonomous, irreducible relation of being, as being-with it already exists with the being of Dasein. Of course, it is indisputable that a lively mutual acquaintanceship on the basis of being-with often depends on how far one's own Dasein has actually understood itself, but this only means that it depends upon how far it has made one's essential being with others transparent and not disguised it. This is possible only if Dasein as being-in-the-world is always already with others. "Empathy" does not first constitute being-with, but is first possible on its basis, and is motivated by the prevailing deficient modes of being-with in their inevitability.

But the fact that "empathy" is not an original existential phenomenon, any more than is knowing in general, does not mean that there is no

125

problem here. Its special hermeneutic will have to show how the various possibilities of being of Dasein themselves mislead and obstruct being-with-one-another and its self-knowledge, so that a genuine "understand-ing" is suppressed and Dasein takes refuge in surrogates; this positive existential condition presupposes a correct understanding of the stranger for its possibility. Our analysis has shown that being-with is an existential constituent of being-in-the-world. Dasein-with has proved to be a manner of being which beings encountered within the world have as their own. Insofar as Dasein *is* at all, it has the kind of being of being-with-one-another. Being-with-one-another cannot be understood as an accumulative result of the occurrence of multiple "subjects." Encountering a number of "subjects" itself is possible only by treating the others encountered in their Dasein-with merely as "numbers." Such numbers are discovered only by a distinctive being-with and being-toward-one-another. "Inconsiderate" being-with "reckons" with others without seriously "counting on them" or even wishing "to have anything to do" with them.

One's own Dasein, like the Dasein-with of others, is encountered initially and for the most part in terms of the surrounding world taken care of that is shared [Mitwelt]. In being absorbed in the world of taking care of things, that is, at the same time in being-with toward others, Dasein is not itself. *Who* is it, then, who has taken over being as everyday being-with-one-another?

§ 27. *Everyday Being a Self and the They*

The *ontologically* relevant result of the foregoing analysis of being-with is the insight that the "subject character" of one's own Dasein and of the others is to be defined existentially, that is, in terms of certain ways to be. In what is taken care of in the surrounding world, others are encountered as what they are; they *are* what they do.

In taking care of the things which one has taken hold of, for, and against others, there is constant care as to the way one differs from them, whether this difference is to be equalized, whether one's own Dasein has lagged behind others and wants to catch up in relation to them, whether Dasein in its priority over others is intent on suppress-ing them. Being-with-one-another is, unknown to itself, disquieted by the care about this distance. Existentially expressed, being-with-one-another has the character of *distantiality*. The more inconspicuous this kind of being is to everyday Dasein itself, all the more stubbornly and primordially does it work itself out.

But this distantiality which belongs to being-with is such that, as everyday being-with-one-another, Dasein stands in *subservience* to others. It itself *is* not; the others have taken its being away from it. The everyday possibilities of being of Dasein are at the disposal of the whims of others. These others are not *definite* others. On the contrary, any other can represent them. What is decisive is only the inconspicuous domination by others that Dasein as being-with has already taken over unawares. One belongs to the others oneself, and entrenches their power. "The others," whom one designates as such in order to cover over one's own essential belonging to them, are those who *are there* initially and for the most part in everyday being-with-one-another. The who is not this one and not that one, not oneself, not some, and not the sum of them all. The "who" is the neuter, *the they*.

We have shown earlier how the public "surrounding world" is always already at hand and taken care of in the surrounding world nearest to us. In utilizing public transportation, in the use of information services such as the newspaper, every other is like the next. This being-with-one-another dissolves one's own Dasein completely into the kind of being of "the others" in such a way that the others, as distinguishable and explicit, disappear more and more. In this inconspicuousness and unascertainability, the they unfolds its true dictatorship. We enjoy ourselves and have fun the way *they* enjoy themselves. We read, see, and judge literature and art the way *they* see and judge. But we also withdraw from the "great mass" the way *they* withdraw, we find "shocking" what *they* find shocking. The they, which is nothing definite and which all are, though not as a sum, prescribes the kind of being of everydayness.

The they has its own ways to be. The tendency of being-with which we called distantiality is based on the fact that being-with-one-another as such creates *averageness*. It is an existential characteristic of the they. In its being, the they is essentially concerned with averageness. Thus, the they maintains itself factically in the averageness of what belongs to it, what it does and does not consider valid, and what it grants or denies success. This averageness, which prescribes what can and may be ventured, watches over every exception which thrusts itself to the fore. Every priority is noiselessly squashed. Overnight, everything that is original is flattened down as something long since known. Everything won through struggle becomes something manageable. Every mystery loses its power. The care of averageness reveals, in turn, an essential tendency of Dasein, which we call the *leveling down* of all possibilities of being.

Distantiality, averageness, and leveling down, as ways of being of the they, constitute what we know as "publicness." Publicness initially controls every way in which the world and Dasein are interpreted, and it is always right, not because of an eminent and primary relation of being to "things," not because it has an explicitly appropriate transparency of Dasein at its disposal, but because it does not get to "the heart of the matter," because it is insensitive to every difference of level and genuineness. Publicness obscures everything, and then claims that what has been thus covered over is what is familiar and accessible to everybody.

The they is everywhere, but in such a way that it has always already stolen away when Dasein presses for a decision. However, because the they presents every judgment and decision as its own, it takes the responsibility of Dasein away from it. The they can, as it were, manage to have "them" constantly invoking it. It can most easily be responsible for everything because no one has to vouch for anything. The they always "did it," and yet it can be said that "no one" did it. In the everydayness of Dasein, most things happen in such a way that we must say "no one did it."

In this way, the they *disburdens* Dasein in its everydayness. Not only that; but disburdening it of its being, the they accommodates Dasein in its tendency to take things easily and make them easy. And since the they constantly accommodates Dasein by disburdening its being [Seinsentlastung], it retains and entrenches its stubborn dominance.

Everyone is the other, and no one is himself. The *they*, which supplies the answer to the *who* of everyday Dasein, is the *nobody* to whom every Dasein has always surrendered itself, in its being-among-one-another.

In these characteristics of being which we have laid out—everyday being-among-one-another, distantiality, averageness, leveling down, publicness, disburdening of one's being, and accommodation—lies the initial "constancy" of Dasein. This constancy pertains not to the enduring objective presence of something, but to the kind of being of Dasein as being-with. Existing in the modes we have mentioned, the self of one's own Dasein and the self of the other have neither found nor lost themselves. One is in the manner of dependency and inauthenticity. This way of being does not signify a lessening of the facticity of Dasein, just as the they as the nobody is not nothing. On the contrary, in this kind of being Dasein is an *ens realissimum*, if by "reality" we understand being [Sein] that has the character of Dasein.

Of course, the they is as little objectively present as Dasein itself. The more openly the they behaves, the more slippery and hidden it is, but then too the less is it nothing at all. To the unprejudiced ontic-ontological "eye," it reveals itself as the "most real subject" of everydayness. And if it is not accessible like an objectively present stone, that is not the least decisive with regard to its kind of being. One may neither decree prematurely that this they is "really" nothing, nor profess the opinion that the phenomenon has been interpreted ontologically if one "explains" it as the result of the objective presence of several subjects which one has put together in hindsight. On the contrary, the elaboration of the concepts of being must be guided by these indubitable phenomena.

Nor is the they something like a "universal subject" which hovers over a plurality of subjects. One could understand it this way only if the being of "subjects" is understood as something unlike Dasein, and if these are regarded as factually objectively present cases of an existing genus. With this approach, the only possibility ontologically is to understand everything which is not a case of this sort in the sense of genus and species. The they is not the genus of an individual Dasein, nor can it be found in this being as an abiding quality. That traditional logic also fails in the face of these phenomena, cannot surprise us if we consider that it has its foundation in an ontology of objective presence—an ontology which is still crude. Thus, it fundamentally cannot be made more flexible no matter how many improvements and expansions might be made. These reforms of logic, oriented toward the "humanistic sciences," only increase the ontological confusion.

The they is an existential and belongs as a primordial phenomenon to the positive constitution of Dasein. It has, in turn, various possibilities of concretion as a characteristic of Dasein. The extent to which its dominance becomes penetrating and explicit may change historically.

The self of everyday Dasein is the *they-self*, which we distinguish from the *authentic self*, that is, the self which has explicitly grasped itself. As the they-self, Dasein is *dispersed* in the they and must first find itself. This dispersion characterizes the "subject" of the kind of being which we know as heedful absorption in the world encountered as closest. If *Dasein* is familiar with itself as the they-self, this also means that the they prescribes the nearest interpretation of the world and of being-in-the-world. The they itself, for the sake of which Dasein is every day, articulates the referential context of significance. The world of Dasein frees the beings encountered for a totality of relevance which is familiar to the they in the limits which are established with the averageness of the they. *Initially*, factical Dasein is in the with-world, discovered in an average way. *Initially*, "I" "am" not in the sense of my own self, but I am the others in the mode of the they. In terms of the they, and as the they, I am initially "given" to "myself." Initially, Dasein is the they and for the most part it remains so. If Dasein explicitly discovers the world and brings it near, if it discloses its authentic being to itself, this discovering of "world" and disclosing of Dasein always comes about by clearing away coverings and obscurities, by breaking up the disguises with which Dasein cuts itself off from itself.

With this interpretation of being-with and being one's self in the they, the question of the who in the everydayness of being-with-one-another is answered. These considerations have at the same time given us a concrete understanding of the basic constitution of Dasein. Being-in-the-world became visible in its everydayness and averageness.

130   Everyday Dasein derives the pre-ontological interpretation of its being from the closest kind of being of the they. The ontological interpretation initially follows this tendency of interpretation, it understands Dasein in terms of the world and finds it there as an innerworldly being. But that is not all: the "closest" ontology of Dasein takes the meaning of being on the basis of which these existing "subjects" are understood also in terms of the "world." But since the phenomenon of world itself is passed over in this absorption in the world, it is replaced by objective presence in the world, by things. The being of beings, which *is there too*, comes to be understood as objective presence. Thus, by showing the positive phenomenon of closest everyday being-in-the-world, we have made possible an insight into the basic reason why the ontological interpretation of this constitution of being is lacking. It itself, in its everyday kind of being, is what initially misses itself and covers itself over.

If the being of everyday being-with-one-another, which seems ontologically to approach mere objective presence, is really fundamentally different from that kind of presence, then the being of the authentic self can be understood still less as objective presence. *Authentic being a self* is not based on an exceptional state of the subject, detached from the they, *but is an existentiell modification of the they as an essential existential.*

But, then, the sameness of the authentically existing self is separated ontologically by a gap from the identity of the I maintaining itself in the multiplicity of its experiences.

CHAPTER FIVE
Being-in as Such

§ 28. *The Task of a Thematic Analysis of Being-in*

In the preparatory stage of the existential analytic of Dasein we have for our leading theme this being's basic constitution, being-in-the-world. Our first aim is to bring into relief phenomenally the unitary primordial structure of the being of Dasein by which its possibilities and ways "to be" are ontologically determined. Until now, the phenomenal characterization of being-in-the-world has been directed toward the structural moment of the world and has attempted to provide an answer to the question of the who of this being in its everydayness. But in first sketching out the tasks of a preparatory fundamental analysis of Dasein we already provided an orientation to *being-in as such*[1] and demonstrated it by the concrete mode of knowing the world.[2]

    The anticipation of this sustaining structural moment arose out of the intention of enclosing the analysis of individual moments, from the outset, within a steadfast anticipation of the structural whole, and with the intention of guarding against any disruption and fragmentation of the unitary phenomenon. Now, keeping in mind what has been achieved in the concrete analysis of world and who, we must turn our interpretation back to the phenomenon of being-in. By considering this more penetratingly, however, we shall not only get a new and more certain phenomenological view of the structural totality of being-in-the-world, but shall also pave the way to grasping the primordial being of Dasein itself, care.

    But what more is there to point out in being-in-the-world beyond the essential relations of being together with the world (taking care), being-with (concern), and being a self (who)? There is still the

131

---

1. Cf. § 12.
2. Cf. § 13.

possibility of broadening the analysis by comparing the variations of
taking care and its circumspection, of concern and its considerateness,
and of distinguishing Dasein from all beings unlike Dasein by a more
precise explication of the being of all possible innerworldly beings.
Without question, there are unfinished tasks in this direction. What we
have set forth so far needs to be supplemented in many ways with
respect to a full elaboration of the existential *a priori* of philosophical
anthropology. But this is not the aim of our investigation. *Its aim is that
of fundamental ontology.* If we thus inquire into being-in thematically,
we cannot be willing to nullify the primordiality of the phenomenon
by deriving it from others, that is, by an inappropriate analysis in the
sense of dissolving it. But the fact that we cannot derive something
primordial does not exclude a multiplicity of characteristics of being
constitutive for it. If these characteristics show themselves, they are
existentially equiprimordial. The phenomenon of the *equiprimordiality* of constitutive factors has often been disregarded in ontology on
account of a methodologically unrestrained tendency to derive everything and anything from a simple "primordial ground."

132   In which direction must we look for the phenomenal characteristics of being-in as such? We get the answer to this question by recall-
ing what we were charged with keeping in view phenomenologically
when we pointed out this phenomenon: being-in in contradistinction
to the objectively present insideness of something objectively present
"in" an other; being-in not as an attribute of an objectively present
subject effected or even just triggered by the objective presence of
the "world"; rather, being-in essentially as the kind of being of this
being itself. But then what else presents itself with this phenomenon
other than the objectively present *commercium between* an objectively
present subject and an objectively present object? This interpretation
would come closer to the phenomenal content if it stated that Dasein
is the being of this "between." Nonetheless, the orientation toward the
"between" would still be misleading. It colludes unawares with the
ontologically indefinite approach that there are beings between which
this between as such "is." The between is already understood as the
result of the *convenientia* of two objectively present things. But this
kind of approach always already *splits* the phenomenon beforehand,
and there is no prospect of ever again putting it back together from
the fragments. Not only do we lack the "cement," even the "schema"
according to which this joining together is to be accomplished has been
split apart, or never as yet unveiled. What is ontologically decisive
is to avoid splitting the phenomenon beforehand, that is, to secure
its positive phenomenal content. The fact that extensive and complicated preparations are necessary for this, only shows that something

ontically self-evident in the traditional treatment of the "problem of knowledge" was ontologically distorted [verstellt] in many ways to the point of becoming invisible.

The being which is essentially constituted by being-in-the-world *is* itself always its "there." According to the familiar meaning of the word, "there" points to "here" and "over there." The "here" of an "I-here" is always understood in terms of an "over there" at hand in the sense of being toward it which de-distances, is directional, and takes care. The existential spatiality of Dasein which determines its "place" for it in this way is itself based upon being-in-the-world. The over there is the determinateness of something encountered within the *world*. "Here" and "over there" are possible only in a "there," that is, when there is a being which as the being of the "there" has disclosed spatiality. This being [Seiende] bears in its ownmost being [Sein] the character of not being closed off [Unverschlossenheit]. The expression "there" means this essential disclosedness [Erschlossenheit]. Through disclosedness this being (Dasein) is "there" for itself together with the there-being [Da-sein] of the world.

When we talk in an ontically figurative way about the *lumen naturale* in human being, we mean nothing other than the existential-ontological structure of this being, the fact that it *is* in such a way as to be its there [sein Da zu sein]. To say that it is "illuminated" means that it is cleared* in itself *as* being-in-the-world, not by another being, but in such a way that it *is* itself the clearing [Lichtung].† Only for a being thus cleared existentially do objectively present things become accessible in the light or concealed in darkness. By its very nature, Dasein brings its there along with it. If it lacks its there, it is not only factically not, but is in no sense, the being [Seiende] which is essentially Dasein. *Dasein is‡ its disclosedness.*

We must set forth the constitution of this being. But since the nature of this being is existence, the existential statement that "Dasein *is* its disclosedness" means: the being [Sein] about which this being [Seienden] is concerned in its being is its "there," which it is to be. In addition to characterizing the primary constitution of the being of disclosure, we must, in accordance with the character of our analysis, interpret the kind of being in which this being is its there in an *everyday way.*

133

---

* Ἀλήθεια—openness—clearing, light, shining.
† But not produced.
‡ Dasein exists, and it alone. Thus existence is standing out, into and enduring, the openness of the there: Ek-sistence.

This chapter, which undertakes the explication of being-in as such, that is, of the being of the there, has two parts: (A) The existential constitution of the there. (B) The everyday being of the there and the entanglement* of Dasein.

We see the two equiprimordially constitutive ways to be the there in *attunement* [*Befindlichkeit*] and *understanding* [*Verstehen*]. Their analysis obtains the necessary phenomenal confirmation through an interpretation of a concrete mode which is important for the subsequent problematic. Attunement and understanding are equiprimordially determined by *discourse*.

Under Part A (the existential constitution of the there) we shall treat Da-sein as attunement (§ 29), fear as a mode of attunement (§ 30), Da-sein as understanding (§ 31), understanding and interpretation (§ 32), statement as a derivative mode of interpretation (§ 33), Da-sein, discourse, and language (§ 34).

The analysis of the characteristics of the being of *Da-sein* is an existential one. This means that the characteristics are not properties of something objectively present, but essentially existential ways to be. Thus, their kind of being in everydayness must be brought out.

Under Part B (the everyday being of the there and the entanglement of Dasein), we shall analyze idle talk (§ 35), curiosity (§ 36), and ambiguity (§ 37) as existential modes of the everyday being of the there: we shall analyze them as corresponding to the constitutive phenomenon of discourse, the vision which lies in understanding, and the interpretation (meaning) belonging to that understanding. In these phenomena a fundamental kind of being of the there becomes visible which we interpret as *entanglement* [*Verfallen*]. This "falling" ["Fallen"] exhibits a movement which is existentially its own (§ 38).

134

A. The Existential Constitution of the There

§ 29. *Da-sein as Attunement*

What we indicate *ontologically* with the term *attunement* is *ontically* what is most familiar and an everyday kind of thing: mood, being in a mood. Prior to all psychology of moods, a field which, moreover, still lies fallow, we must see this phenomenon as a fundamental existential and outline its structure.

---

* "Verfallen" has several senses in ordinary speech. It refers to something that is "decayed," "wasting away," or "ruined" [eine verfallene Burg], as well as to "addiction" [dem Alkohol verfallen] or to "fall for someone." There is a sense of something succumbing to something else that unites many of the various meanings of the word. There is also the sense of a kind of movement that does not go anywhere [er ist dem Wahnsinn verfallen]. Heidegger's use of the word emphasizes the sense of getting caught up in something. "Entanglement" and "falling prey" are both used as translations here. [TR]

Both the undisturbed equanimity and the inhibited discontent of everyday heedfulness, the way we slide over from one to another or slip into bad moods, are by no means nothing ontologically although these phenomena remain unnoticed as what is supposedly the most indifferent and fleeting in Dasein. The fact that moods can be spoiled and change only means that Dasein is always already in a mood. The often persistent, smooth, and pallid lack of mood, which must not be confused with a bad mood, is far from being nothing. Rather, in this Dasein becomes tired of itself. The being of the there has, in such a bad mood, become manifest as a burden [Last].*† One does not *know* why. And Dasein cannot know why because the possibilities of disclosure belonging to cognition fall far short of the primordial disclosure of moods in which Dasein is brought before its being as the there. Furthermore, an elevated mood can alleviate the manifest burden of being. But the possibility of this mood, too, discloses the burdensome character of Dasein even when it alleviates that burden. Mood makes manifest "how one is and is coming along." In this "how one is" being in a mood brings being to its "there."

In being in a mood, Dasein is always already disclosed in accordance with its mood as *that* being to which Dasein was delivered over in its being as the being which it, existing, has to be. To be disclosed does not, as such, mean to be known. And even in the most indifferent and harmless everydayness the being of Dasein can burst forth as the naked "that it is and has to be." The pure "that it is" shows itself, the whence and whither remain obscure. The fact that Dasein normally does not "give in" to such everyday moods, that is, does not pursue what they disclose and does not allow itself to confront what has been disclosed, is no evidence *against* the phenomenal fact of the moodlike disclosure of the being of the there in its that, but is rather evidence for it. For the most part Dasein evades the being that is disclosed in moods in an *ontic* and existentiell way. *Ontologically* and existentially this means that, even in that to which such a mood pays no attention, Dasein is unveiled in its being delivered over to the there. In the evasion itself the there *is* something disclosed.

We shall call this character of being of Dasein which is veiled in its whence and whither, but in itself all the more openly disclosed, this "that it is," the *thrownness* [*Geworfenheit*] of this being into its there; it is thrown in such a way that it is the there as being-in-the-world. The expression thrownness is meant to suggest the *facticity of its being*

135

---

* 'Burden': what bears [das Zu-tragende]; human being is delivered to Da-sein, appropriated by it. To bear [Tragen]: to take over something from out of belonging to being itself.
† In earlier editions this sentence read: "Being has become manifest as a burden." [TR].

delivered over [*Überantwortung*]. The "that it is and has to be" disclosed in the attunement of Dasein is not the "that" which expresses ontologically and categorially the factuality belonging to objective presence; the latter is accessible only when we ascertain it by looking at it. Rather, the that disclosed in attunement must be understood as an existential attribute of *that* being which is in the mode of being-in-the-world. *Facticity is not the factuality of the* factum brutum *of something objectively present, but is a characteristic of the being of Dasein taken on in existence, although initially thrust aside.* The that of facticity is never to be found by looking.

Beings of the character of Dasein are their there in such a way that they find themselves in their thrownness, whether explicitly or not. In attunement, Dasein is always already brought before itself, it has always already found itself, not as perceiving oneself to be there, but as one finds one's self in attunement. As a being which is delivered over to its being, it is also delivered over to the fact that it must always already have found itself, found itself in a finding which comes not from a direct seeking, but from a fleeing. Mood does not disclose in the mode of looking at thrownness, but as turning toward and away from it. For the most part, mood does not turn itself toward the burdensome character of Dasein manifest in it, it does this least of all in an elevated mood in which this burden is lifted. This turning away is always what it is in the mode of attunement.

Phenomenally, *what* mood discloses and *how* it discloses would be completely misunderstood if what has been disclosed were conflated with that which attuned Dasein "at the same time" is acquainted with, knows, and believes. Even when Dasein is "sure" of its "whither" in faith or thinks it knows about its whence in rational enlightenment, all of this makes no difference in the face of the phenomenal fact that moods bring Dasein before the that of its there, which stares directly at it with the inexorability of an enigma. Existentially and ontologically there is not the slightest justification for minimizing the "evidence" of attunement by measuring it against the apodictic certainty of the theoretical cognition of something merely objectively present. But *the* falsification of the phenomena, which banishes them to the sanctuary of the irrational, is no better. Irrationalism, as the counterpart of rationalism, talks about the things to which rationalism is blind, but only with a squint.

That a Dasein factically can, should, and must master its mood with knowledge and will may signify a priority of willing and cognition in certain possibilities of existing. But that must not mislead us into ontologically denying mood as a primordial kind of being of Dasein in which it is disclosed to itself *before* all cognition and willing and *beyond* their scope of disclosure. Moreover, we never master a mood by being free of a mood, but always through a counter mood. The *first* essential ontological characteristic of attunement is: *attunement*

discloses Dasein in its thrownness and, initially and for the most part, in the
mode of an evasive turning away.

From this we can already see that attunement is far removed
from anything like finding a psychical condition. Far from having the
character of an apprehension which first turns itself around and then
turns back, all immanent reflection can find "experiences" only because
the there is already disclosed in attunement. "Mere mood" discloses
the there more primordially, but it also *closes* it *off* more stubbornly,
than any *not*-perceiving.

Bad moods [Verstimmung] show this. In bad moods, Dasein
becomes blind to itself, the surrounding world of heedfulness is veiled,
the circumspection of taking care is led astray. Attunement is so far
from being reflected upon that, in the unreflected devotion to and giving in to the "world" of its heedfulness, it assails Dasein. Mood assails.
It comes neither from "without" nor from "within," but rises from
being-in-the-world itself as a mode of that being. But thus by negatively contrasting attunement with the reflective apprehension of the
"inner," we arrive at a positive insight into its character of disclosure.
Mood has always already disclosed being-in-the-world as a whole and first   137
makes possible directing oneself toward something. Being attuned is not
initially related to something psychical, it is itself not an inner condition which then in some mysterious way reaches out and leaves its
mark on things and persons. In this the *second* essential characteristic
of attunement shows itself. It is a fundamental existential mode of the
equiprimordial disclosedness of world, Dasein-with and existence because
this disclosure itself is essentially being-in-the-world.

Besides these two essential determinations of attunement just
explicated, the disclosure of thrownness and the actual disclosure of
the whole of being-in-the-world, we must notice a *third*, which above
all contributes to a more penetrating understanding of the worldliness
of the world. We said earlier[3] that the world already disclosed lets
innerworldly things be encountered. This prior disclosedness of the
world which belongs to being-in is also constituted by attunement.
Letting something be encountered is primarily *circumspective*, not just
a sensation or staring out at something. Letting things be encountered
in a circumspect heedful way has—we can see this now more precisely
in terms of attunement—the character of being affected or moved. But
being affected by the unserviceable, resistant, and threatening character
of things at hand is ontologically possible only because being-in as
such is existentially determined beforehand in such a way that what
it encounters in the world can *matter* to it in this way. This mattering
to it is grounded in attunement, and as attunement it has disclosed

---

3. Cf. § 18.

the world, for example, as something by which it can be threatened. Only something which is the attunement of fearing, or fearlessness, can discover things at hand in the surrounding world as being threatening. The moodedness of attunement constitutes existentially the openness to world of Dasein.

And only because the "senses" belong ontologically to a being whose kind of being is an attuned being-in-the-world can they be "touched" and "have a sense" for something so that what touches them shows itself in an affect. Something like an affect would never come about under the strongest pressure and resistance, resistance would be essentially undiscovered, if attuned being-in-the-world were not already related to having things in the world matter to it in a way prefigured by moods. *In attunement lies existentially a disclosive submission to world out of which things that matter to us can be encountered.* Indeed, we must *ontologically* in principle leave the primary discovery of the world to "mere mood." Pure beholding [reines Anschauen], even if it penetrated into the innermost core of the being of something objectively present, would never be able to discover anything like what is threatening.

The fact that everyday circumspection goes wrong on account of attunement, which is primarily disclosive and is vastly subject to deception, is, gauged by the idea of an absolute "world"-cognition, a μὴ ὄν. But such ontologically unjustified value judgments completely fail to recognize the existential positivity of the capacity for being deceived. When we see the "world" in an unsteady and wavering way in accordance with our moods, what is at hand shows itself in its specific worldliness, which is never the same on any given day. Theoretical looking at the world has always already flattened it down to the uniformity of what is merely present, although, of course, a new abundance of what can be discovered in pure determination lies within that uniformity. But the purest θεωρία does not abandon all moods, either. Even when we look theoretically at what is merely present, it does not show itself in its pure outward appearance unless this θεωρία lets it come toward us in a *tranquil* lingering . . . in ῥᾳστώνη and διαγωγή.[4] We must not confuse demonstrating the existential-ontological constitution of cognitive determination in the attunement of being-in-the-world with the attempt to surrender science ontically to "feeling."

The various modes of attunement and their interconnected foundations cannot be interpreted within the problematic of this investigation. As phenomena they have long been familiar ontically under the terms of affects and feelings and have always been considered in philosophy. It is not a matter of chance that the first traditional and

138

---

4. Aristotle, *Metaphysics* A, 982b22ff.

systematically developed interpretation of affects is not treated in the scope of "psychology." Aristotle investigated the πάϑη in the second book of his *Rhetoric*. Contrary to the traditional orientation of the concept of rhetoric according to which it is some kind of "discipline," Aristotle's *Rhetoric* must be understood as the first systematic hermeneutic of the everydayness of being-with-one-another. Publicness as the kind of being of the they (cf. § 27) not only has its attunedness, it needs mood and "makes" it for itself. The speaker speaks to it and from out of it. He needs an understanding of the possibilities of mood in order to arouse and direct it in the right way.

The continuation of the interpretation of the affects in the Stoics, as well as their tradition in patristic and scholastic theology down to modern times, are well known. What has not been noted is the fact that the fundamental ontological interpretation of affects has hardly been able to take one step worthy of mention since Aristotle. On the contrary, the affects and feelings fall thematically under the psychic phenomena, functioning as a third class of these, mostly along with representational thinking and willing. They sink to the level of accompanying phenomena.

It is the merit of phenomenological investigation that it has again created a freer view of these phenomena. Not only that: Scheler, adopting the suggestions of Augustine and Pascal,[5] steered the problematic toward the connections between the foundations of "representing" and "interested" acts. Of course, here too, the existential-ontological foundations of the phenomenon of act remain in the dark.

Attunement discloses Dasein not only in its thrownness and dependence on the world already disclosed with its being, it is itself the existential kind of being in which it is continually surrendered to the "world" and lets itself be concerned by it in such a way that it, in a certain sense, evades its very self. The existential constitution of this evasion becomes clear in the phenomenon of entanglement.

Attunement is an existential, fundamental way in which Dasein is its there. It not only characterizes Dasein ontologically, but is at the same time of fundamental methodological significance for the

---

5. Cf. Pascal, *Pensées*: "Et de là vient qu'au lieu qu'en parlant des choses humaines on dit qu'il faut les connaître avant que de les aimer, ce qui a passé en proverbe, les saints au contraire disent en parlant des choses divines qu'il faut les aimer pour les connaître, et qu'on n'entre dans la vérité que par la charité, dont ils ont fait une de leurs plus utiles sentences." ["And thence it comes about that in the case where we are speaking of human things, it is said to be necessary to know them before we can love them, and this has become a proverb; but the saints, on the contrary, when they speak of divine things, say we must love them before we know them, and that we enter into truth only through charity; they have made of this one of their most useful maxims."] Cf. Augustine, *Opera* (Migne, ed., *Patrologiae Latinae*, vol. 8), *Contra Faustum*, book 32, chap. 18; *non intratur in veritatem, nisi per charitatem* ["one does not enter into truth except through charity"].

existential analytic because of its disclosure. Like every ontological interpretation in general, the analytic can only listen in, so to speak, on beings already previously disclosed with regard to their being. And it will keep to the eminent disclosive possibilities of Dasein of the widest scope in order to gain from them information about this being. The phenomenological interpretation must give to Dasein itself the possibility of primordial disclosure and let it, so to speak, interpret itself. It goes along with this disclosure only in order to raise the phenomenal content of disclosure existentially to a conceptual level.

140

With regard to the later interpretation of such an existential-ontologically significant basic attunement of Dasein, anxiety (cf. § 40), the phenomenon of attunement will be demonstrated more concretely in the definite mode of *fear*.

§ 30. *Fear as a Mode of Attunement*[6]

The phenomenon of fear can be considered in three aspects. We shall analyze what we are afraid of, fearing, and that about which are afraid. These possible aspects of fear are not accidental; they belong together. With them, the structure of attunement in general comes to the fore. We shall complete our analysis by alluding to the possible modifications of fear, each of which concerns its various structural factors.

That *before which* [*Wovor*] we are afraid, the "fearsome," is always something encountered within the world, either with the kind of being of something at hand or something objectively present or Dasein-with. We do not intend to report ontically about beings which often and for the most part can be "fearsome," but to determine phenomenally what is fearsome in its fearsome character. What is it that belongs to the fearsome as such which is encountered in fearing? What is feared has the character of being threatening. Here several points must be considered:

1. What is encountered has the relevant nature of harmfulness. It shows itself in a context of relevance.
2. Thus harmfulness aims at a definite range of what can be affected by it. So determined, it comes from a definite region.
3. The region itself and what comes from it is known as something which is "unnerving" ["geheuer"].
4. As something threatening, what is harmful is not yet near enough to be dealt with, but it is coming near. As it approaches, harmfulness radiates and thus has the character of threatening.
5. This approaching occurs within nearness. Something may be harmful in the highest degree and may even be constantly coming nearer,

---

6. Cf. Aristotle, *Rhetoric* B 5.1382a20–1383b11.

but if it is still far off it remains veiled in its fearsome nature. As something approaches in nearness, however, what is harmful is threatening, it can get us, and yet perhaps not. In approaching, this   *141* "it can and yet in the end it may not" gets worse. It is fearsome, we say.

6. This means that what is harmful, approaching near, bears the revealed possibility of not happening and passing us by. This does not lessen or extinguish fearing, but enhances it.

Fearing itself frees what we have characterized as threatening in a way which lets us be concerned with it. It is not that we initially ascertain a future evil (*malum futurum*) and then are afraid of it. But neither does fearing first confirm something approaching us, rather it discovers it beforehand in its fearsomeness. And then fear, in being afraid, can "clarify" what is fearsome by explicitly looking at it. Circumspection sees what is fearsome because it is in the attunement of fear. As a dormant possibility of attuned being-in-the-world, fearing, "fearfulness" has already disclosed the world with regard to the fact that something like a fearful thing can draw near to us from this fearfulness. The ability to draw near is itself freed by the essential, existential spatiality of being-in-the-world.

The *about which* [*Worum*] fear is afraid is the fearful being itself, Dasein. Only a being which is concerned in its being about that being can be afraid. Fearing discloses this being in its jeopardization, in its being left to itself. Although it does so in varying degrees of explicitness, fear always reveals Dasein in the being of its there. When we are afraid for house and home, this is not a counter-example for the above determination of what it is we are fearful about. For as being-in-the-world, Dasein is always a heedful being-with. Initially and for the most part, Dasein *is* in terms of *what* it takes care of. The jeopardization of that is a threat to being with. Fear predominantly discloses Dasein in a privative way. It bewilders us and makes us "lose our heads." At the same time that fear closes off our jeopardized being-in it lets us see it, so that when fear has subsided Dasein has to first find its way about again.

Fear about, as being afraid of, always equiprimordially discloses—whether privatively or positively—innerworldly beings in their possibility of being threatening and being-in with regard to its being threatened. Fear is a mode of attunement.

But fearing about can also involve others, and we then speak of fearing for them. This fear for . . . does not take away the other's fear from him. That is out of the question because the other *for* whom we are afraid does not even have to be afraid on his part. We are afraid *for* the other most of all precisely when *he* is *not* afraid and blunders recklessly into what is threatening. Fearing for . . . is a mode of   *142* co-attunement with others, but it is not necessarily being afraid with them or even being afraid together. One can be afraid for ... without being afraid oneself. But viewed precisely, fearing for ... is, after all, being afraid *oneself*. What is "feared" here is the being-with the other who could be snatched away from us. What is fearsome is not aimed directly at the one who is fearing with. Fearing for ... knows in a way that it is unaffected and yet is affected in the involvement of Dasein-with for whom it is afraid. It is not a matter here of degrees of "feeling tones," but of existential modes. Fearing for ... does not lose its specific genuineness when it is "really" not afraid.

The factors constitutive for the full phenomenon of fear can vary. Thus various possibilities of fear result. Approaching nearby belongs to the structure of encountering what is threatening. When something threatening itself suddenly bursts into heedful being-in-the-world in its character of "not right now, but at any moment," fear becomes *alarm*. Accordingly, we must distinguish in what is threatening the nearest approach of what threatens and the way in which this approach itself is encountered, namely, its suddenness. What we are alarmed about is initially something known and familiar. But when what threatens has the character of something completely unfamiliar, fear becomes *horror*. And when something threatening is encountered in the aspect of the horrible, and at the same time is encountered as something alarming, suddenness, fear becomes *terror*. We are familiar with further variations of fear, such as timidity, shyness, nervousness, misgiving. All modifications of fear as possibilities of attunement point to the fact that Dasein as being-in-the-world is "fearful." This "fearfulness" must not be understood in the ontic sense of a factical, "isolated" tendency, but rather as the existential possibility of the essential attunement of Dasein in general, which is, of course, not the only one.

§ 31. *Da-sein as Understanding*

Attunement is *one* of the existential structures in which the being [Sein] of the "there" dwells. Equiprimordially with it, *understanding* constitutes this being [Sein]. Attunement always has its understanding, even if only by suppressing it. Understanding is always attuned. If we interpret understanding as a fundamental existential,* we see that this phenomenon is conceived as a fundamental mode of the *being* of Dasein. On the other hand, "understanding" in the sense of *one* possible kind of cognition among others, distinguished as "explaining," must be interpreted, like explaining, as an existential derivative of the primary understanding which constitutes the being of the there in general.

143

---

* Fundamental ontology, that is, from the relation of the truth of being.

Our previous inquiry already encountered this primordial understanding, but without explicitly taking it up within the theme under
consideration. To say that Dasein, existing, is its there means: World is
"there"; its *Da-sein* is being-in. Being-in is "there" as that for the sake
of which Dasein is. Existing being-in-the-world as such is disclosed
in the for-the-sake-of-which, and we called this disclosedness understanding.[7] In understanding the for-the-sake-of-which, the significance
grounded therein is also disclosed. The disclosure of understanding, as
that of the for-the-sake-of-which and of significance, equiprimordially
pertains to the entirety of being-in-the-world. Significance is that for
which world as such is disclosed. The statement that the for-the-sake-
of-which and significance are disclosed in Dasein means that Dasein is
a being which, as being-in-the-world, is concerned about itself.

Speaking ontically, we sometimes use the expression "to understand something" to mean "being able to handle it," "being up to it,"
"being able to do something." In understanding as an existential, the
thing we are able to do is not a what, but being [Sein] as existing. The
mode of being of Dasein as a potentiality of being lies existentially in
understanding. Dasein is not something objectively present which then
has in addition the ability to do something, rather it is primarily being-
possible. Dasein is always what it can be and how it is its possibility.
The essential possibility of Dasein concerns the ways of taking care of
the "world" which we characterized, of concern for others and, always
already present in all of this, the potentiality of being itself, for its own
sake. The being-possible, which Dasein always is existentially, is also
distinguished from empty, logical possibility and from the contingency
of something objectively present, where this or that can "happen" to
it. As a modal category of objective presence, possibility means what
is *not yet* real and *not always* necessary. It characterizes what is *only*
possible. Ontologically, it is less than reality and necessity. In contrast,
possibility as an existential is the most primordial and the ultimate   *144*
positive ontological determination of Dasein; as is the case with existentiality, possibility can initially be prepared for solely as a problem.
Understanding, as a potentiality of being that is disclosive, offers the
phenomenal ground to see it at all.

As an existential, possibility does not refer to a free-floating potentiality of being in the sense of the "liberty of indifference" (*libertas indif-
ferentiae*). As essentially attuned, Dasein has always already got itself into
definite possibilities. As a potentiality for being which it *is*, it has let some
go by; it constantly adopts the possibilities of its being, grasps them, and
sometimes fails to grasp them. But this means that Dasein is a being-possible which is entrusted to itself, it is *thrown possibility* throughout. Dasein

---

7. Cf. § 18.

is the possibility of being free *for* its ownmost potentiality of being. Being-possible is transparent for it in various possible ways and degrees.

Understanding is the being of such a potentiality of being which is never still outstanding as something not yet objectively present, but, as something essentially never objectively present, "*is*" together with the being of Dasein in the sense of existence. Dasein is in the way that in each case it understands (or, alternatively, has not understood) to be in this or that way. As this understanding, it "knows" *what* is going on, that is, what its potentiality of being is. This "knowing" does not first grow out of an immanent self-perception, but belongs to the being of the there which is essentially understanding. And only *because* Dasein is its there, understandingly, *can* it go astray and fail to recognize itself. And since understanding is attuned and attunement is existentially surrendered to thrownness, Dasein has always already gone astray and failed to recognize itself. In its potentiality of being, it is thus delivered over to the possibility of first finding itself again in its possibilities.

Understanding is the existential being [Sein] of the ownmost potentiality of being of Dasein itself in such a way that this being [Sein] discloses in itself what its very being is about. The structure of this existential must be grasped more precisely.

As disclosing, understanding always concerns the whole fundamental constitution of being-in-the-world. As a potentiality of being, being-in is always a potentiality of being-in-the-world. Not only is the world, qua world, disclosed in its possible significance, but innerworldly beings themselves are freed, freed for *their own* possibilities. What is at hand is discovered as such in its service*ability*, us*ability*, detriment*ality*. The totality of relevance reveals itself as the categorial whole of a *possibility* of the nexus of things at hand. But the "unity," too, of manifold presence, nature, is discoverable only on the basis of the disclosedness of one of its *possibilities*. Is it a matter of chance that the question of the *being* of nature aims at the "conditions of its *possibility*?" On what is this questioning based? It cannot omit the question: *Why* are beings unlike Dasein understood in their being if they are disclosed in terms of the conditions of their possibility? Kant presupposed something like this, perhaps correctly so. But this presupposition itself cannot be left without demonstrating how it is justified.

Why does understanding always penetrate into possibilities from among all the essential dimensions of what can be disclosed to it? Because understanding in itself has the existential structure which we call *project* [*Entwurf*]. It projects the being of Dasein just as primordially upon its for-the-sake-of-which as upon significance as the worldliness of its particular world. The project character of understanding constitutes being-in-the-world with regard to the disclosedness of its there as the there of a potentiality of being. Project is the existential constitution of being in the realm of factical potentiality of being. And, as thrown, Dasein is thrown into the mode of being of projecting. Projecting has nothing to do with being related to a plan thought out, according to which Dasein arranges its being, but, as Dasein, it has always already projected itself and is, as long as it is, projecting. As long as it is, Dasein always has understood itself and will understand itself in terms of possibilities. Furthermore, the project character of understanding means that understanding does not thematically grasp that upon which it projects, namely, possibilities themselves. Such a grasp precisely takes its character of possibility away from what is projected, it degrades it to the level of a given, intended content, whereas, in projecting, project throws possibility before itself as possibility, and as such lets it *be*. As projecting, understanding is the mode of being of Dasein in which it *is* its possibilities as possibilities.

Because of the kind of being which is constituted by the existential of projecting, Dasein is constantly "more" than it actually is, assuming that one wanted to, and if one could, give an inventory of it as something objectively present in its content of being. But it is never more than it factically is because its potentiality of being belongs essentially to its facticity. But, as being-possible, Dasein is also never less. It is existentially that which it is *not yet* in its potentiality of being. And only because the being of the there gets its constitution through understanding and its character of project, only because it *is* what it becomes or does not become, can it say understandingly to itself: "become what you are!"*

Project always concerns the complete disclosedness of being-in-the-world. As a potentiality of being, understanding itself has possibilities which are prefigured by the scope of what can be essentially disclosed to it. Understanding *can* turn primarily to the disclosedness of the world, that is, Dasein can understand itself initially and for the most part in terms of the world. Or else understanding throws itself primarily into the for-the-sake-of-which, which means Dasein exists as itself.† Understanding is either authentic, originating from its own self as such, or else inauthentic. The "in" ["*in*authentic] does not mean that Dasein cuts itself off from itself and understands "only" the world. World belongs to its being a self as being-in-the-world. Again, authentic as well as inauthentic understanding *can* be either genuine or not genuine. As a potentiality of being understanding is altogether permeated with possibility. Turning to one of these fundamental possibilities of understanding, however, does not dispense with the other. *Rather,*

146

---

* But who are "you"? The one who lets *go*—and *becomes*.
† But not qua subject and individual or qua person.

because understanding always has to do with the complete disclosedness of
Dasein as being-in-the-world, the involvement of understanding is an exis-
tential modification of project as a whole. In understanding the world,
being-in is always also understood. Understanding of existence as such
is always an understanding of world.

As factical, Dasein has always already transferred its potentiality
of being into a possibility of understanding.

In its character of project, understanding constitutes existentially
what we call the *sight* [*Sicht*] of Dasein. In accordance with the fundamental modes of its being which we characterized as the circumspection [Umsicht] of taking care, the considerateness [Rücksicht] of
concern, and as the sight geared toward being as such for the sake
of which Dasein is as it is, Dasein *is* equiprimordially sight existentially existing together with the disclosedness of the there. We shall
call the sight which is primarily and as a whole related to existence
transparency [*Durchsichtigkeit*]. We choose this term to designate correctly understood "self-knowledge" in order to indicate that it is not
a matter here of perceptually finding and gazing at a point which
is the self, but of grasping and understanding the full disclosedness
of being-in-the-world *throughout all* its essential constitutive factors.
Existing beings glimpse [sichtet] "themselves" only when they have
become transparent to themselves equiprimordially in their being with
the world, in being together with others as the constitutive factors of
their existence.

Conversely, the opacity [Undurchsichtigkeit] of Dasein is not
solely and primarily rooted in "egocentric" self-deceptions, but just
as much in ignorance [Unkenntnis] about the world.

147   We must, of course, guard against a misunderstanding of the
expression "sight." It corresponds to the clearedness [Gelichtetheit]
characterizing the disclosedness of the there. "Seeing" not only does
not mean perceiving with the bodily eyes, neither does it mean the
mere nonsensory perception of something objectively present in its
objective presence. *The* only peculiarity of seeing which we claim for
the existential meaning of sight is the fact that it lets beings accessible to it be encountered in themselves without being concealed. Of
course, every "sense" does this within its genuine realm of discovery.
But the tradition of philosophy has been primarily oriented from the
very beginning toward "seeing" as the mode of access of beings *and to*
being. To preserve this connection, one can formalize sight and seeing
to the point of establishing a universal term which characterizes every
access as access whatsoever to beings and to being.

By showing how all sight is primarily based on understanding—the circumspection of taking care is understanding as *comprehen-*
sion [*Verständigkeit*]—we have taken away from mere intuition [puren

Anschauen] its priority which noetically corresponds to the traditional ontological priority of objective presence. "Intuition" and "thought"* are both already remote derivatives of understanding. Even the phenomenological "intuition of essences" ["Wesensschau"] is based on existential understanding. We can decide about this kind of seeing only when we have obtained the explicit concepts of being and the structure of being, which only phenomena in the phenomenological sense can become.

The disclosedness of the there in understanding is itself a mode of the potentiality-of-being of Dasein. In the projectedness of its being upon the for-the-sake-of-which, together with the projectedness of its own being upon significance (world), lies the disclosedness of being in general.† An understanding of being is already anticipated in the projecting upon possibilities. Being is understood‡ in the project, but not ontologically grasped. Beings which have the kind of being of the essential project of being-in-the-world have as the constituent of their being the understanding of being. What we asserted earlier[8] dogmatically is now demonstrated in terms of the constitution of the being in which Dasein, as understanding, is its there. In accordance with the limits of this whole inquiry, a satisfactory clarification of the existential meaning of this understanding of being can only be attained on the basis of the temporal [temporalen] interpretation of being.

As existentials, attunement and understanding characterize the primordial disclosedness of being-in-the-world. In the mode of "being attuned" Dasein "sees" possibilities in terms of which it is. In the projective disclosure of such possibilities, it is always already attuned. The project of its ownmost potentiality of being is delivered over to the fact of thrownness into the there. With the explication of the existential constitution of the being of the there in the sense of thrown project does not the being of Dasein become still more mysterious? Indeed. We must first let the full mysteriousness [Rätselhaftigkeit] of this being [Sein] emerge, if only to be able to fail in a more genuine way in its "solution" and to raise the question anew of the being of thrown-projecting being-in-the-world.

In order to sufficiently bring even only the everyday mode of being of attuned understanding phenomenally to view, and if this is to be sufficient for the full disclosedness of the there, a concrete development of these existentials is necessary.

148

---

* To understand this as the "understanding" ["Verstand"], διάνοια, but not the "act of understanding" ["Verstehen"] from the understanding ["Verstand"].
† How does it "lie" there and what does beyng [Seyn] mean?
‡ This does not mean that being "is" ["sei"] due to the project.

8. Cf. § 4.

§ 32. *Understanding and Interpretation*

As understanding, Dasein projects its being upon possibilities. This *being toward possibilities* [*Sein zu Möglichkeiten*] that understands is itself a potentiality for being [Seinkönnen] because of the way these disclosed possibilities come back to Dasein. The project of understanding has its own possibility of development. We shall call the development of understanding *interpretation* [*Auslegung*]. In interpretation understanding appropriates what it has understood understandingly. In interpretation understanding does not become something different, but rather itself. Interpretation is existentially based in understanding, and not the other way around. Interpretation is not the acknowledgment of what has been understood, but rather the development of possibilities projected in understanding. In accordance with the train of these preparatory analyses of everyday Dasein, we shall pursue the phenomenon of interpretation in the understanding of the world, that is, in inauthentic understanding in the mode of its genuineness.

In terms of the significance of what is disclosed in understanding the world, the being of taking care of what is at hand learns to understand what the relevance can be with what is actually encountered. This means that circumspection discovers that the world which has already been understood is interpreted. What is at hand comes *explicitly* before 149   sight that understands. All preparing, arranging, setting right, improving, rounding out, occur in such a way that things at hand for circumspection are taken apart in their in-order-to and are taken care of according to what has become visible in this way. What has been circumspectly taken apart with regard to its in-order-to as such, what has been *explicitly* understood, has the structure of *something as something*. The circumspectly interpretive answer to the circumspect question of what this particular thing at hand is runs: it is for. . . . Saying what it is for is not simply naming something, but what is named is understood *as* that *as* which what is in question is to be taken. What is disclosed in understanding, what is understood, is always already accessible in such a way that in it its "as what" can be explicitly delineated. The "as" constitutes the structure of the explicitness of what is understood; it constitutes the interpretation. The circumspect, interpretive dealing with what is at hand in the surrounding world which "sees" this *as* a table, a door, a car, a bridge does not necessarily already have to analyze what is circumspectly interpreted in a particular *statement*. Any simple prepredicative seeing of what is at hand is in itself already understanding and interpretive. But does not the lack of this "as" constitute the simplicity of a mere perception of something? The seeing of this sight is always already understanding and interpreting. It contains in itself the explicitness of referential relations (of the in-order-to) which belong to the totality of relevance in terms of which what is simply encountered is understood. The articulation of what is understood in the interpreting approach to beings guided by the "something as something" lies *before* a thematic statement about it. The "as" does not first show up in the statement, but is only first stated, which is possible only because it is there as something to be stated. The fact that the explicitness of a statement can be lacking in simple looking, does not justify us in denying every articulate interpretation, and thus the as-structure, to this simple seeing. The simple seeing of things nearest to us in our having to do with . . . contains the structure of interpretation so primordially that a grasping of something which is, so to speak, *free of the as* requires a kind of reorientation. When we just stare at something, our just-having-it-before-us lies before us *as a failure to understand* it any more. This grasping which is free of the as is a privation of *simple* seeing, which understands; it is not more primordial than the latter, but derived from it. The ontic inexplicitness of the "as" must not mislead us into overlooking it as the *a priori* existential constitution of understanding.

But if any perception of useful things at hand always understands and interprets them, letting them be circumspectly encountered as something, does this not then mean that initially something merely objectively present is experienced which then is understood *as* a door, *as* a house? That would be a misunderstanding of the specific disclosive function of interpretation. Interpretation does not, so to speak, throw a "significance" over what is nakedly objectively present and does not stick a value on it, but what is encountered in the world is always already in a relevance which is disclosed in the understanding of world, a relevance which is made explicit by interpretation.

Things at hand are always already understood in terms of a totality of relevance. This totality need not be explicitly grasped by a thematic interpretation. Even if this totality of relevance has undergone such an interpretation, it recedes again into an undifferentiated understanding. It is precisely in this modality that it is the essential foundation of everyday, circumspect interpretation. This is always based on a *fore-having* [*Vorhabe*]. As the appropriation of understanding in being [Sein] that understands, the interpretation operates in being toward a totality of relevance which has already been understood. When something is understood but still veiled, it becomes unveiled [Enthüllung] by an act of appropriation that is always done under the guidance of a perspective which fixes that with regard to which what has been understood is to be interpreted. In each instance [jeweils], the interpretation is grounded in a *foresight* [*Vorsicht*] that "approaches" ["anschneidet"] what has been taken in fore-having with a definite interpretation in view. What is held in the fore-having and understood in a "fore-seeing" view becomes comprehensible through the interpretation. The interpretation can draw the conceptuality belonging to the beings to be interpreted from these themselves, or else the interpreta-

150

tion can force those beings into concepts to which they are opposed
in accordance with their kind of being. The interpretation has always
already decided, finally or provisionally, upon a definite conceptuality;
it is grounded in a *fore-conception* [*Vorgriff*].

The interpretation of something as something is essentially
grounded in fore-having, fore-sight, and fore-conception. Interpretation
is never a presuppositionless grasping of something previously given.
When a specific instance of interpretation (in the sense of a precise
textual interpretation) appeals to what "is there" ["dasteht"], then that
which initially "is there" is nothing other than the self-evident, undiscussed prejudice [Vormeinung] of the interpreter which necessarily lies
in every interpretive approach as that which is already "posited" with
interpretation in general, namely, that which is pre-given [vorgegeben]
in fore-having, fore-sight, and fore-conception.

How are we to conceive the character of this "fore"? Have we
done this when we formally say *"a priori"*? Why is this structure
appropriate to understanding, which we have characterized as a fundamental existential of Dasein? How is the structure of the "as" which
belongs to what is interpreted as such related to the fore-structure?
This phenomenon is obviously not to be dissolved "into pieces." But is
a primordial analytic to be ruled out? Should we accept such phenomena as "finalities"? Then the question would remain, why? Or do the
fore-structure of understanding and the as-structure of interpretation
show an existential-ontological connection with the phenomenon of
project? And does this phenomenon refer back to a primordial constitution of being of Dasein?

Before answering these questions, for which the preparation up
to this point is not at all sufficient, we must inquire whether what is
visible as the fore-structure of understanding and the as-structure of
interpretation does not itself already represent a unitary phenomenon
which has been used extensively in philosophical problematics, without what is used so universally measuring up to the primordiality of
ontological explication.

In the projecting of understanding, beings are disclosed in their
possibility. The character of possibility always corresponds to the kind
of being of the beings understood. Innerworldly beings in general are
projected upon the world, that is, upon a totality of significance in
whose referential relations taking care, as being-in-the-world, has rooted itself from the beginning. When innerworldly beings are discovered
along with the being of Dasein, that is, when they become intelligible,
we say that they have *meaning* [*Sinn*]. But strictly speaking, what is
understood is not the meaning, but beings [Seiende], or being [Sein].
Meaning is that wherein the intelligibility of something maintains
itself. That which can be articulated in disclosure that understands we

call meaning. The *concept of meaning* includes the formal framework of what necessarily belongs to what interpretation that understands articulates. *Meaning, structured by fore-having, fore-sight, and fore-conception, is the upon which of the project in terms of which something becomes intelligible as something.* Insofar as understanding and interpretation constitute the existential constitution of the being of the there, meaning must be conceived as the formal, existential framework of the disclosedness belonging to understanding. Meaning is an existential of Dasein, not a property that is attached to beings, which lies "behind" them or floats somewhere as a "realm between." Only Dasein "has" meaning in that the disclosedness of being-in-the-world can be "fulfilled" through the beings discoverable in it. *Thus only Dasein can be meaningful or meaningless.* This means: its own being and the beings disclosed with that being can be appropriated in an understanding [Verständnis] or they can be confined to incomprehensibility [Unverständnis].

This interpretation of the concept of "meaning" is fundamentally ontological-existential. If we adhere to it, then all beings whose mode of being is unlike Dasein must be understood as *unmeaningful* [*unsinniges*], as essentially bare of meaning as such. "Unmeaningful" does not mean here a value judgment, but expresses an ontological determination. *And only what is unmeaningful can be absurd* [*widersinnig*]. Objectively present things encountered in Dasein can, so to speak, assault its being [Sein]; for example, events of nature which break in on us and destroy us.

And when we ask about the meaning of being, our inquiry does not become profound and does not brood on anything which stands behind being, but questions being itself in so far as it stands within the intelligibility of Dasein. The meaning of being can never be contrasted with beings or with being as the supporting "ground" of beings, for "ground" is only accessible as meaning, even if that meaning itself is an abyss [Abgrund] of meaninglessness.

As the disclosedness of the there, understanding always concerns the whole of being-in-the-world. In every understanding of world, existence is also understood, and vice versa. Furthermore, every interpretation operates within the fore-structure which we characterized. Every interpretation which is to contribute some understanding must already have understood what is to be interpreted. This fact has always already been noticed, even if only in the realm of derivative ways of understanding and interpretation, in philological interpretation. The latter belongs to the scope of scientific cognition. Such cognition demands the rigor of demonstration giving reasons. Scientific proof must not already presuppose what its task is to found. But if interpretation always already has to operate within what is understood and nurture itself from this, how should it then produce scientific results without going in a circle,

152

especially when the presupposed understanding still operates in the common knowledge of human being and world? But according to the most elementary rules of logic, the *circle* is a *circulus vitiosus*. If this is so then the business of historical interpretation is thus banned *a priori* from the realm of rigorous knowledge. If the fact of the circle in understanding is not eliminated, historiography must be content with less rigorous possibilities of knowledge. It is permitted to compensate for this defect to some extent with the "spiritual significance" of its "objects." But even according to the opinion of historiographers themselves, it would be more ideal if the circle could be avoided and if there were the hope of finally creating a historiography which is as independent of the standpoint of the observer as the knowledge of nature is supposed to be.

153   *But to see a* vitiosum *in this circle and to look for ways to avoid it, even to "feel" that it is an inevitable imperfection, is to misunderstand understanding from the ground up.* It is not a matter of assimilating understanding and interpretation to a particular ideal of knowledge which is itself only a degeneration of understanding that has strayed into the legitimate task of grasping of what is objectively present in its essential unintelligibility. Rather, the fulfillment of the fundamental conditions of possible interpretation lies in not failing to recognize beforehand the essential conditions of the task. What is decisive is not to get out of the circle, but to get into it in the right way. This circle of understanding is not a circle in which any random kind of knowledge operates, but it is rather the expression of the existential *fore-structure* of Dasein itself. The circle must not be degraded to a *vitiosum*, not even to a tolerated one. A positive possibility of the most primordial knowledge is hidden in it which, however, is only grasped in a genuine way when interpretation has understood that its first, constant, and last task is not to let fore-having, fore-sight, and fore-conception be given to it by chance ideas and popular conceptions, but to secure the scientific theme by developing these in terms of the things themselves. Because, in accordance with its existential meaning, understanding is the potentiality for being of Dasein itself, the ontological presuppositions of historical [historischer] knowledge transcend in principle the idea of the rigor of the most exact sciences. Mathematics is not more exact than history, but only narrower with regard to the scope of the existential foundations relevant to it.

The "circle" in understanding belongs to the structure of meaning, and this phenomenon is rooted in the existential constitution of Dasein, that is, in interpretive understanding. Beings which, as being-in-the-world, are concerned about their being itself* have an ontologi-

---

* This "its being itself" is, however, intrinsically determined by the understanding of being, that is, by standing within the clearing of presence, where neither the clearing as such nor presence as such becomes thematic for representational thinking.

cal structure of the circle. However, if we note that the "circle" belongs ontologically to a kind of being of objective presence (subsistence), we shall in general have to avoid characterizing something like Dasein ontologically in terms of this phenomenon.

§ 33. *Statement [Aussage] as a Derivative Mode of Interpretation*

All interpretation is grounded in understanding. What is articulated as such in interpretation and is prefigured as articulable in understanding in general is meaning. Since the statement (the "judgment") is based on understanding and represents a derivative form of interpretation, it *also* "has" a meaning. Meaning, however, cannot be defined as what occurs "in" a judgment along with the act of judgment. The explicit analysis of the statement has several goals in our context.

154

On the one hand, we can demonstrate in the statement in what way the structure of the "as," which is constitutive for understanding and interpretation, can be modified. Understanding and interpretation thus come into sharper focus. Moreover, the analysis of the statement has a distinctive place in the fundamental-ontological problematic, because, in the decisive beginnings of ancient ontology, the λόγος functioned as the sole guide for the access to beings as they really are and for the determination of the being of beings. Finally, the statement has been regarded from ancient times as the primary and true "locus" of *truth*. This phenomenon is so intimately connected with the problem of being that our inquiry necessarily runs into the problem of truth as it proceeds; it already lies within the dimension of that problem, although not explicitly. The analysis of the statement should help prepare the way for this problematic.

In what follows we shall assign to the term *statement* three significations which are drawn from the phenomenon thus characterized. They are interconnected and delineate in their unity the full structure of the statement.

1. Primarily, statement means *pointing out [Aufzeigung]*. With this we adhere to the primordial meaning of λόγος as ἀπόφανσις: to let beings be seen from themselves. In the statement, "the hammer is too heavy," what is discovered for sight is not a "meaning," but a being in the mode of its being at hand. Even when this being is not near enough to be grasped and "seen," pointing out designates the being itself, not a mere representation of it, neither something "merely represented" nor even a psychical condition of the speaker, his representing of this being.

2. Statement is tantamount to *predication*. A "predicate" is "stated" about a "subject," the latter is *determined* by the former. What is stated in this signification of statement is not the predicate, but the "hammer

itself." What does this stating, that is, the determining, on the other hand, lies in the "too heavy." What is stated in the second signification of statement, what is determined as such, has been narrowed down in its content as opposed to what is stated in the first signification of this term. Every predication is what it is only as a pointing out. The second signification of statement has its foundation in the first. The elements which are articulated in predication, subject-predicate, originate within the pointing out. Determining does not first discover, but as a mode of pointing out initially *limits* seeing precisely to what shows itself—hammer—as such, in order to manifest *explicitly* what is manifest in its determinacy through the explicit *limitation* of looking. When confronted with what is already manifest, with the hammer which is too heavy, determining must first take a step back. "Positing the subject" dims beings down to focus on "the hammer there" in order to let what is manifest be seen *in* its determinable definite character through this dimming down. Positing the subject, positing the predicate, and positing them together are thoroughly "apophantic" in the strict sense of the word.

3. Statement means *communication* [*Mitteilung*], speaking forth. As such it has a direct relation to statement in the first and second meanings. It is letting someone see with us what has been pointed out in its definite character. Letting someone see with us shares with others the being pointed out in its definiteness. What is "shared" is the *being toward* [*Sein zum*] that which has been pointed out, it is a way of seeing something as in common. We must keep in mind that this being-toward is being-in-the-world, namely, in *the* world from which what is pointed out is encountered. Any statement, as a communication understood existentially, must have been expressed. As something communicated, what is spoken can be "shared" by others with the speaker even when they themselves do not have the beings pointed out and defined within reach or within sight. What is spoken can be "passed along" in further retelling. The orbit of communication which sees expands itself. But at the same time what is pointed out can become veiled again in this further retelling, although the knowledge and cognition growing in such hearsay always refers to beings themselves and does not "affirm" a "valid meaning" passed around. Even hearsay is a being-in-the-world and a being toward what is heard.

The theory of "judgment" prevalent today that is oriented toward the phenomenon of "validity" shall not be discussed at any length here. It is sufficient to refer to the very questionable character of this phenomenon of "validity" which, ever since Lotze, people have been fond of passing off as a "primal phenomenon" not to be traced further back. It owes this role only to its ontological lack of clarity. The "problematic" which has entrenched itself around this idolatry of the word is just as opaque. First, validity means the *"form" of the reality* which belongs to the content of the judgment insofar as it is unchangeable as opposed to the changeable "psychic" act of judgment. In light of the position of the question of being in general characterized in the introduction to this inquiry, we can hardly expect that "validity" as "ideal being" is going to be distinguished by any special ontological clarity. Second, validity means the validity of the meaning of the judgment which is valid for the "object" it has in view, and thus receives the significance of *"objective validity"* and objectivity in general. The meaning thus "valid" *for* beings, and which is valid "timelessly" in itself, is said to be "valid" also in the sense of being valid *for* every person who judges rationally. Third, validity means *bindingness*, "universal validity." If one then advocates a "critical" epistemology, according to which the subject does not "truly" "come out" to the object, then this valid character, as the validity of an object, objectivity, is based on the valid content of true (!) meaning. The three meanings of "validity" set forth, the way of being of the ideal, as objectivity, and as bindingness, are not only in themselves opaque, but constantly get confused with one another. Methodological caution requires that we do not choose such unstable concepts as the guide for our interpretation. We do not restrict the concept of meaning in advance to the signification of a "content of judgment," but we understand it as the existential phenomenon characterized in which the formal framework of what can be disclosed in understanding and articulated in interpretation becomes visible as such.

When we collect the three meanings of "statement" analyzed here in a unitary view of the complete phenomenon, the definition reads: *Statement is a pointing out which communicates and defines.* Now we must ask: what right do we have at all to conceive the statement as a mode of interpretation? If it is something of this sort, the essential structures of interpretation must be repeated in it. The statement's pointing out is accomplished on the basis of what is already disclosed in understanding, or what is circumspectly discovered. The statement is not an unattached kind of behavior which could of itself primarily disclose beings in general, but always already maintains itself on the basis of being-in-the-world. What we showed earlier[9] with regard to knowing the world is just as true of the statement. It needs a fore-having of something disclosed in general which it points out in the mode of determining. Furthermore, when one begins to determine something, one has a directed viewpoint with respect to what is to be stated. In the act of determining something [Bestimmungsvollzug], the function of determining is taken over by that with regard to which beings that have

157

---

9. Cf. § 13.

been presented are envisaged. Thus, the statement needs a fore-sight in which the predicate that is to be delineated and attributed is itself loosened, so to speak, from its implicit enclosure in the being itself. A significant articulation of what is pointed out always belongs to the statement as communication that defines; it operates within a definite set of concepts. The hammer is heavy, heaviness belongs to the hammer, the hammer has the property of heaviness. The fore-conception always also contained in the statement remains mostly inconspicuous because language always already contains a developed conceptuality. Like interpretation in general, the statement necessarily has its existential foundations in fore-having, fore-sight, and fore-conception.

But how does the statement become a *derivative* mode of interpretation? What has been modified in it? We can point out the modification by sticking with limiting cases of statements which function in logic as normal cases and as examples of the "simplest" phenomena of statement. That which logic makes thematic with the categorical statement, for example, "the hammer is heavy," it has always already understood "logically" before any analysis. As the "meaning" of the sentence, it has already presupposed, without noticing, the following: this thing, the hammer, has the property of heaviness. "Initially" there are no such statements in heedful circumspection. But it does have its specific ways of interpretation which can read as follows, as compared with the "theoretical judgment" just mentioned, and may take some such form as "the hammer is too heavy" or, even better, "too heavy, the other hammer!" The primordial act of interpretation does not lie in a theoretical sentence, but in circumspectly and heedfully putting away or changing the inappropriate tool "without wasting words." From the fact that words are absent, we may not conclude that the interpretation is absent. On the other hand, the circumspectly *spoken* interpretation is not already necessarily a statement in the sense defined. *Through what existential ontological modifications does the statement originate from circumspect interpretation?**

158

The being held in fore-having, for example, the hammer, is initially at hand as a useful thing. If this being is the "object" of a statement, as soon as we begin the statement, a transformation in the fore-having is already brought about beforehand. Something *at hand with which* we have to do or perform something, turns into something "about which" the statement that points it out is made. Fore-sight aims at something objectively present in what is at hand. Both *by* and *for* the way of looking, what is at hand is veiled as something at hand. Within this discovering of objective presence which covers over handiness, what is encountered as objectively present is determined in its being objectively present in such and such a way. Now the access is first available

---

* In what way can the statement be fulfilled by modifying the interpretation?

for something like *qualities*. That *as* which the statement determines what is objectively present is drawn *from* what is objectively present as such. The as-structure of interpretation has undergone a modification. The "as" no longer reaches out into a totality of relevance in its function of approaching what is understood. It is cut off with regard to its possibilities of the articulation of referential relations of significance which constitute the character of the surrounding world. The "as" gets pushed back to the uniform level of what is merely objectively present. It dwindles to the determination that belongs to the structure of just letting what is objectively present be seen. This leveling down of the primordial "as" of circumspect interpretation to the as of the determination of objective presence is the specialty of the statement. Only in this way does it gain the possibility of a pointing something out in a way that we merely look at it.

Thus the statement cannot deny its ontological provenance from an interpretation that understands. We call primordial the "as" of circumspect interpretation that understands (ἑρμηνεία) the existential-*hermeneutical* "as" in distinction from the *apophantical* "as" of the statement.

There are many interim stages between an interpretation which is completely wrapped up in heedful understanding and the extreme opposite case of a theoretical statement about objectively present things: statements about events in the surrounding world, descriptions of what is at hand, "reports on situations," noting and ascertaining a "factual situation," describing a state of affairs, telling about what has happened. These "sentences" cannot be reduced to theoretical propositional statements without essentially distorting their meaning. Like the latter, they have their "origin" in circumspect interpretation.

With the progress of knowledge about the structure of the λόγος, it was inevitable that this phenomenon of the apophantical "as" came to view in some form. The way in which it was initially seen is not a matter of chance, nor did it fail to have its influence on the history of logic to come.

When considered philosophically, the λόγος is itself a being [Seiendes] and, in accordance with the orientation of ancient ontology, something present. What is initially present, that is, what can be found like things, are words and the succession of words in which the λόγος is spoken. When we first look for the structure of the *logos* thus present, we find a *being present together* [Zusammenvorhandensein] of several words. What constitutes the unity of this together? As Plato knew, it consists in the fact that the λόγος is always λόγος τινός. With regard to the beings manifest in the λόγος, the words are combined to form one totality of words. Aristotle had a more radical view: every λόγος is σύνθεσις and διαίρεσις at the same time, neither the one—say, as a

159

"positive judgment"—nor the other—as a "negative judgment." Rather, every statement, whether affirmative or negative, whether false or true, is equiprimordially σύνθεσις and διαίρεσις. Pointing out is putting together and taking apart. However, Aristotle did not pursue this analytical question further to the problem: what phenomenon is it then within the structure of the λόγος that allows and requires us to characterize every statement as synthesis and diairesis?

What is to be got at phenomenally with the formal structures of "binding" and "separating," more precisely, with the unity of the two, is the phenomenon of "something as something." In accordance with this structure, something is understood with regard to something else, it is taken together with it, so that this confrontation that *understands, interprets*, and articulates, at the same time takes apart what has been put together. If the phenomenon of the "as" is covered over and, above all, veiled in its existential origin from the hermeneutical "as," then Aristotle's phenomenological point of departure disintegrates into the analysis of λόγος in a superficial "theory of judgment," according to which judgment is a binding or separating of representations and concepts.

Binding and separating can be further formalized to mean a "relating." The judgment is dissolved logistically into a system of "coordinations," it becomes the object of "calculation," but not a theme of ontological interpretation. The possibility and impossibility of the analytical understanding of σύνθεσις and διαίρεσις, or "relation" in the judgment in general, is closely bound up with the actual state of the fundamental ontological problematic.

To what extent this problematic has an effect on the interpretation of the λόγος and, on the other hand, to what extent the concept of "judgment" has, by a remarkable counter-movement, an effect on the ontological problematic, is shown by the phenomenon of the copula. It becomes evident in this "bond" that the structure of synthesis is initially posited as a matter of course and that it has also maintained the decisive interpretative function. But if the formal characteristics of "relation" and "binding" cannot contribute anything phenomenally to the factual structural analysis of the λόγος, the phenomenon intended with the term "copula" finally has nothing to do with bond and binding. Whether expressed explicitly in language or indicated in the verbal ending, the "is" and its interpretation are moved into the context of problems of the existential analytic if statements and an understanding of being are existential possibilities of being of Dasein itself. The development of the question of being (cf. Division Three of Part One) will then encounter again this peculiar phenomenon of being within the λόγος.

For the time being, we wanted to clarify with this demonstration of the derivation of the statement from interpretation and understanding the fact that the "logic" of λόγος is rooted in the existential analytic of Dasein. Recognizing the ontologically insufficient interpretation of the λόγος at the same time sharpens our insight into the lack of primordiality of the methodological basis on which ancient ontology developed. The λόγος is experienced as something objectively present and interpreted as such, and the beings which it points out have the meaning of objective presence as well. This meaning of being itself is left undifferentiated and uncontrasted with other possibilities of being so that being [Sein] in the sense of a formal being-something [Etwas-Sein] is at the same time fused with it and we are unable to obtain a clear-cut division between these two realms.*

§ 34. Da-sein and Discourse. Language

The fundamental existentials which constitute the being of the there, the disclosedness of being-in-the-world, are attunement and understanding. Understanding harbors in itself the possibility of interpretation, that is, the appropriation of what is understood. To the extent that attunement is equiprimordial with understanding, it maintains itself in a certain understanding. A certain possibility of interpretation also belongs to it. An extreme derivative of interpretation was made visible with the statement. The clarification of the third meaning of statement as communication (speaking forth) led us to the concept of saying and speaking, to which we purposely paid no attention up to now. The fact that language *only now* becomes thematic should indicate that this phenomenon has its roots in the existential constitution of the disclosedness of Dasein. *The existential-ontological foundation of language [Sprache] is discourse [Rede].*   161
In our previous interpretation of attunement, understanding, interpretation, and statement we have constantly made use of this phenomenon, but have, so to speak, suppressed it in the thematic analysis.

  Discourse is existentially equiprimordial with attunement and understanding. Intelligibility is also always already articulated before its appropriative interpretation. Discourse is the articulation of intelligibility. Thus it already lies at the basis of interpretation and statement. We called that which can be articulated in interpretation, and thus more primordially in speech, meaning [Sinn]. What is structured in discoursing articulation as such, we call the totality of significations [Bedeutungsganze]. This totality can be broken up into significations. As what is articulated of what can be articulated, significations are always bound up with meaning. If discourse, the articulation of the intelligibility of the there, is the primordial existential of disclosedness, and if disclosedness is primarily constituted by being-in-the-world,

---

* Husserl.

discourse must also essentially have a specifically *worldly* mode of being. The attuned intelligibility of being-in-the-world *expresses itself as discourse*. The totality of significations of intelligibility *is put into words* [*kommt zu Wort*]. Words accrue to significations. But word-things are not provided with significations.

The way in which discourse gets expressed is language. This totality of words in which discourse has its own "worldly" being [Sein] can thus be found as an innerworldly being [Seiendes] like something at hand. Language can be broken up into world-things objectively present. Discourse is existential language because the beings whose disclosedness it significantly articulates have the kind of being of being-in-the-world which is thrown and reliant upon the "world."*

As the existential constitution of the disclosedness of Dasein, discourse is constitutive for the existence of Dasein. *Listening* and *silence* are possibilities belonging to discoursing speech. The constitutive function of discourse for the existentiality of existence first becomes completely clear in these phenomena. First of all, we must develop the structure of discourse as such.

Discoursing is the "significant" structuring [Gliedern] of the intelligibility of being-in-the-world, to which being-with belongs, and which maintains itself in a particular way of heedful being-with-one-another. Being-with-one-another talks in assenting, refusing, inviting, warning, as talking things through, as getting back to someone, interceding, furthermore as "making statements" and as talking in "giving a talk." Discourse is discourse about. . . . That which discourse is *about* does not necessarily have the character of the theme of a definite statement; in fact, mostly it does not have it. Even a command is given about something; a wish is about something. And so is intercession. Discourse necessarily has this structural factor because it also constitutes the disclosedness of being-in-the-world and is prestructured in its own structure by this fundamental constitution of Dasein. What is talked about in discourse is always "addressed" in a particular view and within certain limits. In all discourse there is *what is spoken* as such, what is said as such when one actually wishes, asks, talks things over. . . . In this "something said," discourse communicates.

As the analysis has already indicated, the phenomenon of *communication* must be understood in an ontologically broad sense. "Communication" in which one makes statements, for example, giving information, is a special case of the communication that is grasped in principle existentially. Here the articulation of being-with-one-another understandingly is constituted. It brings about the "sharing" of being attuned together and of the understanding of being-with. Communication is never anything

162

---

* For language thrownness is essential.

like a conveying of experiences, for example, opinions and wishes, from the inside of one subject to the inside of another. Dasein-with is essentially already manifest in attunement-with and understanding-with. Being-with is "explicitly" *shared* in discourse, that is, it already *is*, only unshared as something not grasped and appropriated.

All discourse about . . . which communicates in what it says has at the same time the character of *expressing itself*. In talking, Dasein *ex*presses itself not because it has been initially cut off as "something internal" from something outside, but because as being-in-the-world it is already "outside" when it understands. What is expressed is precisely this being outside,* that is, the actual mode of attunement (of mood) which we showed to pertain to the full disclosedness of being-in. Being-in and its attunement are made known in discourse and indicated in language by intonation, modulation, in the tempo of talk, "in the way of speaking." The communication of the existential possibilities of attunement, that is, the disclosing of existence, can become the true aim of "poetic" speech.

Discourse is the structuring of the attuned intelligibility of being-in-the-world. Its constitutive factors are: what discourse is about (what is discussed), what is said as such, communication, and making known. These are not properties which can be just empirically snatched from language, but are existential characteristics rooted in the constitution of being of Dasein which first make something like language ontologically possible. Some of these factors can be lacking or remain unnoticed in the factical linguistic form of a particular discourse. The fact that they often are *not* "verbally" expressed is only an indication of a particular kind of discourse which, insofar as it is discourse, must always lie within the totality of these structures.

163

Attempts to grasp the "essence of language" have always taken their orientation toward a single one of these factors and have understood language guided by the idea of "expression," "symbolic forms," communication as "statement," "making known" experiences or the "form" of life. But nothing would be gained for a completely sufficient definition of language if we were to put these different fragmentary definitions together in a syncretistic way. What is decisive is to develop the ontological-existential totality of the structure of discourse beforehand on the basis of the analytic of Dasein.

The connection of discourse with understanding and intelligibility becomes clear through an existential possibility which belongs to discourse itself, listening. It is not a matter of chance that, when we have not heard "rightly," we say that we have not "understood." Listening is constitutive for discourse. And just as linguistic utterance is

---

* The there; being exposed as an open place.

based on discourse, acoustic perception is based on listening. Listening to . . . is the existential being-open of Dasein as being-with for the other. Listening even constitutes the primary and authentic openness of Dasein for its ownmost possibility of being, as in hearing the voice of the friend whom every Dasein carries with it. Dasein hears because it understands. As understandingly being-in-the-world with others, Dasein "listens to" ["hörig"] itself and to Dasein-with, and in this listening belongs [Hörigkeit zugehörig] to these. Listening to each other, in which being-with is developed, has the possible ways of following, going along with, and the privative modes of not hearing, opposition, defying, turning away.

On the basis of this existentially primary potentiality for hearing, something like *hearkening* becomes possible. Hearkening is itself phenomenally more primordial than what the psychologist "initially" defines as hearing, the sensing of tones and the perception of sounds. Hearkening, too, has the mode of being of a hearing that understands. "Initially" we never hear noises and complexes of sound, but the creaking wagon, the motorcycle. We hear the column on the march, the north wind, the woodpecker tapping, the crackling fire.

164   It requires a very artificial and complicated attitude in order to "hear" a "pure noise." The fact that we initially hear motorcycles and cars is, however, the phenomenal proof that Dasein, as being-in-the-world, always already maintains itself *together with* innerworldly things at hand and initially not at all with "sensations" whose chaos would first have to be formed to provide the springboard from which the subject jumps off finally to land in a "world." Essentially understanding, Dasein is initially together with what is understood.

Likewise, in the explicit listening to the discourse of the other we initially understand what is said; more precisely, we are already together with the other beforehand, with the being which the discourse is about. We do *not*, on the contrary, first hear what is expressed in the utterance. Even when speaking is unclear or the language is foreign, we initially hear *unintelligible* words, and not a multiplicity of tone data.

When what the discourse is about is heard "naturally," however, we can at the same time hear the way in which it is said, the "diction," but this, too, only by previously understanding what is spoken. Only thus is there a possibility of estimating whether the way in which it is said is appropriate to what the discourse that replies is about thematically.

Similarly, a counter-discourse that replies initially arises directly from understanding what the discourse is about, which has already been "shared" in being-with.

Only when the existential possibility of discourse and listening are given, can someone hearken [horchen]. He who "cannot hear" ["hören"] and "must feel" can perhaps hearken very well precisely for this reason. Just listening around is a privation of the hearing that understands. Discourse and listening are grounded in understanding. Understanding comes neither from a lot of talking, nor from busy listening around. Only one who already understands is able to listen.

Another essential possibility of discourse, keeping silent, has the same existential foundation. In talking with one another the person who is silent can "let something be understood," that is, one can develop an understanding more authentically than the person who never runs out of words. Speaking a lot about something does not in the least guarantee that understanding is thus furthered. On the contrary, talking at great length about something covers things over and brings what is understood into an illusory clarity, that is, the unintelligibility of the trivial. But to keep silent does not mean to be mute. On the contrary, one who is mute still has the tendency to "speak." Such a person has not only not proved that he can keep silent, he even lacks the possibility of proving this. And the person who is by nature accustomed to speak little is no better able to show that he can be silent and keep silent. One who never says anything is also unable to keep silent at a given moment. Authentic silence is possible only in genuine discourse. In order to be silent, Dasein must have something to say,* that is, must be in command of an authentic and rich disclosedness of itself. Then reticence makes manifest and puts down "idle talk." As a mode of discourse, reticence articulates the intelligibility of Dasein so primordially that it gives rise to a genuine potentiality for hearing and to a being-with-one-another that is transparent.

165

Since discourse is constitutive for the being of the there, that is, attunement and understanding, and since Dasein means being-in-the-world, Dasein as discoursing being-in has already expressed itself. Dasein has language. Is it a matter of chance that the Greeks, whose everyday existence lay predominantly in speaking with one another, and who at the same time "had eyes" to see, determined the essence of human being as ζῷον λόγον ἔχον† in the pre-philosophical as well as in the philosophical interpretation of Dasein? The later interpretation of this definition of human being in the sense of *animal rationale*, "rational living being," is not "false," but it covers over the phenomenal basis from which this definition of Dasein is taken. The human being shows itself as a being who speaks. This does not mean that the possibility of vocal utterance belongs to it, but that this being is in the mode of discovering world and Dasein itself. The Greeks do not

---

* and what calls for saying [das Zu-sagende]? (beyng) [Seyn].
† Human beings as the "gatherers," gathering toward beyng [Seyn]—presencing in the openness [Offenheit] of beings (but with the latter in the background).

have a word for language, they "initially" understood this phenomenon as discourse. However, since the λόγος came into their philosophical view predominantly as statement, the development of the fundamental structures of the forms and constituents of discourse was carried out following the guideline of *this logos*.* Grammar searched for its foundation in the "logic" of this logos. But this logic is based on the ontology of what is present. The basic stock of "categories of significance" which were passed over in subsequent linguistics, and are fundamentally still accepted as the criterion today, is oriented toward discourse as statement. If, however, we take this phenomenon in principle to have the fundamental primordiality and scope of an existential, the necessity arises of reestablishing linguistics on an ontologically more primordial foundation. The task of *freeing* grammar from logic requires *in advance* a *positive* understanding of the *a priori* fundamental structure of discourse in general as an existential, and cannot be carried out subsequently by improving and supplementing the tradition. Bearing this in mind, we must inquire into the basic forms in which it is possible to articulate what is intelligible in general, not only of the innerworldly beings that can be known in theoretical observation and expressed in propositions. A doctrine of significance will not emerge automatically from a comprehensive comparison of as many languages as possible, even including those that are most exotic. Nor is it sufficient to adopt the philosophical horizon within which W. von Humboldt took language as a problem. The doctrine of significance is rooted in the ontology of Dasein. Whether it flourishes or goes to waste depends upon the fate of this ontology.[10]

In the end, philosophical research must finally decide to ask what mode of being belongs to language in general. Is it an innerworldly useful thing at hand or does it have the mode of being of Dasein, or neither of the two? What kind of being does language have if there can be a "dead" language? What does it mean ontologically that a language grows or declines? We possess a linguistics, and the being of the beings which it has as its theme is obscure; even the horizon for any investigative question about it is veiled. Is it a matter of chance that initially and for the most part significations are "worldly," prefigured beforehand by the significance of the world, that they are indeed often predominantly "spatial"? Or is this "fact" existentially and ontologically necessary and why? Philosophical research will have to give up "philosophy of language" if it is to ask about the "things themselves" and attain the status of a problematic that has been clarified conceptually.

---

* Heidegger switches from Greek to Latin script here. In § 7 he also uses both Greek and Latin script. [TR]

10. Cf. Husserl's doctrine of signification in *Logische Untersuchungen*, vol. 2, Investigations 1 and 4–6. See also the more radical version of the problematic in *Ideen*, vol. 1, sect. 123 et sqq., p. 255ff.

166

The foregoing interpretation of language has the sole function of pointing out the ontological "place" for this phenomenon in the constitution of being of Dasein and above all of preparing the way for the following analysis, in which, taking as our guideline a fundamental kind of being belonging to discourse, in connection with other phenomena, we shall try to bring the everydayness of Dasein into view in a way that is ontologically more primordial.

B. The Everyday Being of the There and the Falling Prey of Dasein

In returning to the existential structures of the disclosedness of being-in-the-world, our interpretation has in a way lost sight of the everydayness of Dasein. The analysis must again regain this phenomenal horizon that was our thematic point of departure. Now the question arises: what are the existential characteristics of the disclosedness of being-in-the-world to the extent that the latter, as something everyday, maintains itself in the mode of being of the they? Is a specific attunement, a special understanding, discourse, and interpretation appropriate to the they? The answer to this question becomes all the more urgent when we remember that Dasein initially and for the most part is immersed in the they and mastered by it. Is not Dasein, as thrown being-in-the-world, initially thrown into the publicness of the they? And what else does this publicness mean than the specific disclosedness of the they?

167

If understanding must be conceived primarily as the potentiality-for-being of Dasein, we shall be able to gather from an analysis of the understanding and interpretation belonging to the they which possibilities of its being Dasein, as the they, has disclosed and appropriated to itself. These possibilities themselves, however, reveal an essential tendency of being of everydayness. And finally, when everydayness is explicated in an ontologically sufficient way, it unveils a primordial mode of being of Dasein in such a way that from it the phenomenon of thrownness, to which we have called attention, can be exhibited in its existential concreteness.

What is initially required is to make visible the disclosedness of the they, that is, the everyday mode of being of discourse, sight, and interpretation, in specific phenomena. With regard to these, the remark may not be superfluous that our interpretation has a purely ontological intention and is far removed from any moralizing critique of everyday Dasein and from the aspirations of a "philosophy of culture."

§ 35. *Idle Talk*

The expression "idle talk" is not to be used here in a disparaging sense. Terminologically, it means a positive phenomenon which constitutes the mode of being of the understanding and interpretation of everyday Dasein. For the most part, discourse expresses itself and has always already expressed itself. It is language. But then understanding and interpretation are always already contained in what is expressed [Ausgesprochenheit]. As expression, language harbors in itself an interpretedness of the understanding of Dasein. This interpretedness is no more merely objectively present than language is, but rather its being is itself of the character of Dasein. Dasein is initially, and in certain limits, constantly entrusted to this interpretedness that directs and apportions the possibilities of the average understanding and the attunement belonging to it. In the totality of its structured contexts of signification, expression preserves an understanding of the disclosed world and thus equiprimordially an understanding of the Dasein-with of others and of one's own being-in. The intelligibility already deposited in expression pertains to the discoveredness of beings actually obtained and handed down, as of the current intelligibility of being, and of the possibilities and horizons available for fresh interpretation and conceptual articulation. But above and beyond a mere reference to the fact of this interpretedness of Dasein, we must now ask about the existential mode of being of the discourse which is expressed and expressing itself. If it cannot be conceived as something objectively present, what is its being, and what does this being [Sein] say in principle about the everyday mode of being of Dasein?

Discourse expressing itself is communication. Its tendency of being aims at bringing the hearer to participate in disclosed being toward what is talked about in discourse.

In the language that is spoken when one expresses oneself, there already lies an average intelligibility; and, in accordance with this intelligibility, the discourse communicated can be understood, to a large extent, without the listener actually turning toward what is talked about in the discourse so as to have a primordial understanding of it. One understands not so much the beings talked about; rather, one already only listens to what is spoken about as such. This is understood, what is talked about is understood, only approximately and superficially. One means *the same thing* because it is in the *same* averageness that we have in a common understanding of what is said.

Hearing and understanding have attached themselves beforehand to what is spoken about as such. Communication does not "impart" the primary relation of being to the being spoken about, but being-with-one-another takes place in talking with one another and in heeding what is spoken about. What is important to it is that one speaks. The being-said, the saying, and the pronouncement provide a guarantee for the genuineness and appropriateness of the discourse and the understanding belonging to it. And since this discoursing has lost the primary relation of being [Sein] to the being [Seienden] talked about,

168

or else never achieved it, it does not communicate in the mode of a primordial appropriation of this being, but communicates by *gossiping* and *passing the word along*. What is spoken about as such spreads in wider circles and takes on an authoritative character. Things are so because one says so. Idle talk is constituted in this gossiping and passing the word along, a process by which its initial lack of grounds to stand on increases to complete groundlessness [Bodenlosigkeit]. And this is not limited to vocal gossip, but spreads to what is written, as "scribbling." In this latter case, gossiping is based not so much on hearsay. It feeds on sporadic superficial reading: the average understanding of the reader will *never be able* to decide what has been drawn from primordial sources with a struggle, and how much is just gossip. Moreover, the average understanding will not even want such a distinction, will not have need of it, since, after all, it understands everything.

169

The groundlessness of idle talk is no obstacle to its being public, but encourages it. Idle talk is the possibility of understanding everything without any previous appropriation of the matter. Idle talk already guards against the danger of getting stranded in such an appropriation. Idle talk, which everyone can snatch up, not only divests us of the task of genuine understanding, but develops an indifferent intelligibility for which nothing is closed off any longer.

Discourse, which belongs to the essential constitution of the being of Dasein, and also constitutes its disclosedness, has the possibility of becoming idle talk, and as such of not really keeping being-in-the-world open in an articulated understanding, but of closing it off and covering over innerworldly beings. To do this, one need not aim to deceive. Idle talk does not have the kind of being of *consciously passing off* something as something else. The fact that one has said something groundlessly and then passes it along in further retelling is sufficient to turn disclosing around into a closing off. For what is said is initially always understood as "telling," that is, as discovering. Thus, by its very nature, idle talk is a closing off since it *omits* going back to the foundation of what is being talked about.

This closing off is aggravated anew by the fact that idle talk, in which an understanding of what is being talked about is supposedly reached, holds any new questioning and discussion at a distance because it presumes it has understood, and in a peculiar way it suppresses them and holds them back.

This interpretedness of idle talk has always already settled itself down in Dasein. We get to know many things initially in this way, and some things never get beyond such an average understanding. Dasein can never escape the everyday way of being interpreted into which Dasein has grown initially. All genuine understanding, interpreting and communication, rediscovery and new appropriation come about in it, out of it, and against it. It is not the case that a Dasein, untouched

and unseduced by this way of interpreting, was ever confronted by the free land of a "world," so that it just looks at what it encounters. The
170   domination of the public way in which things have been interpreted has already decided upon even the possibilities of being attuned, that is, about the basic way in which Dasein lets itself be affected by the world. The they prescribes that attunement, it determines what and how one "sees."

Idle talk, which closes off in the way we described, is the mode of being of the uprooted understanding of Dasein. However, it does not occur as the objectively present condition of something objectively present, but it is existentially uprooted, and this uprooting is constant. Ontologically, this means that when Dasein maintains itself in idle talk, it is, as being-in-the-world, cut off from the primary and primordially genuine relations of being toward the world, toward Dasein-with, toward being-in itself. It keeps itself in suspension and yet in doing so it is still always together with the "world," with others, and toward itself. Only a being whose disclosedness is constituted by attuned and understanding discourse, that is, who in this ontological constitution is its there, who *is* "in-the-world," has the possibility of being of such uprooting which, far from constituting a nonbeing of Dasein, rather constitutes its most everyday and stubborn "reality."

However, it is in the nature of the obviousness and self-assurance of the average way of being interpreted that under its protection, the uncanniness of the suspension in which Dasein can drift toward an increasing groundlessness remains concealed to actual Dasein itself.

§ 36. *Curiosity*

In the analysis of understanding and the disclosedness of the there in general, we referred to the *lumen naturale* and called the disclosedness of being-in the *clearing* of Dasein in which something like sight first becomes possible. Sight was conceived with regard to the basic kind of disclosing characteristic of Dasein, namely, understanding, in the sense of the genuine appropriation of those beings to which Dasein can be related in accordance with its essential possibilities of being.

The basic constitution of the being of sight shows itself in a peculiar tendency of being which belongs to everydayness—the tendency toward "seeing." We designate it with the term *curiosity*, which is characteristically not limited to seeing and expresses the tendency toward a peculiar way of letting the world be encountered in perception. Our aim in interpreting this phenomenon is in principle existential and ontological. We do not restrict ourselves to an orientation toward cognition which even in the early stages of Greek philosophy, and not by accident, was conceived in terms of the "desire to see." The treatise
171   which stands first in the collection of Aristotle's treatises on ontol-

ogy begins with the sentence: πάντες ἄνθρωποι τοῦ εἰδέναι ὀρέγονται
φύσει.[11] The care for seeing is essential to the being of human being.
Thus an inquiry is introduced which attempts to discover the origin of
all scientific investigation of beings and their being by deriving it from
the kind of being of Dasein which we mentioned. This Greek interpretation of the existential genesis of science is not a matter of chance. It
brings to an explicit understanding what was prefigured in the statement of Parmenides: τὸ γὰρ αὐτὸ νοεῖν ἐστίν τε καὶ εἶναι. Being is
what shows itself in pure, intuitive perception, and only this seeing
discovers being. Primordial and genuine truth lies in pure intuition.
This thesis henceforth remains the foundation of Western philosophy.
The Hegelian dialectic has its motivation in it, and only on its basis
is that dialectic possible.

Above all, it was Augustine who noted the remarkable priority of
"seeing" in conjunction with his interpretation of *concupiscentia*.[12] *Ad ocu-
los enim videre proprie pertinet*, seeing properly belongs to the eyes. *Utimur
autem hoc verbo etiam in ceteris sensibus cum eos ad cognoscendum intendimus*.
But we use this word "to see" for the other senses, too, when we use
them in order to know. *Neque enim dicimus: audi quid rutilet; aut, olefac
quam niteat; aut, gusta quam splendeat; aut, palpa quam fulgeat: videri enim
dicuntur haec omnia*. For we do not say: hear how that glistens, or smell
how that shines, or taste how that glows, or feel how that gleams; but
we say of each: *see*, we say that all these things are seen. *Dicimus autem
non solum, vide quid luceat, quod soli oculi sentire possunt*, nor do we just
say: see how that glows when only the eyes can perceive it, *sed etiam, vide
quid sonet; vide quid oleat, vide quid sapiat, vide quid durum sit*. We also say:
see how that sounds, see how it smells, see how it tastes, see how hard
that is. *Ideoque generalis experientia sensuum concupiscentia sicut dictum est
oculorum vocatur, quia videndi officium in quo primatum oculi tenent, etiam
ceteri sensus sibi de similitudine usurpant, cum aliquid cognitionis explorant*.
That is why the experience of the senses in general is called "the pleasure
of the eyes," because even the other senses take upon themselves a certain resemblance to the function of seeing where cognition is concerned,
since there the function of seeing has a privilege.

What is this tendency to just-perceive about? Which existen-   *172*
tial constitution of Dasein becomes intelligible in the phenomenon
of curiosity?

Being-in-the-world is initially absorbed in the world taken care
of. Taking care is guided by circumspection which discovers things at
hand and preserves them in their discoveredness. Circumspection gives
to all our teaching and performing the route for moving forward, the
means of doing something, the right opportunity, the proper moment.

---

11. *Metaphysics* A i, 980a21.
12. *Confessions* X.35.

Taking care can come to rest in the sense of one's interrupting the performance and taking a rest or by finishing something. Taking care does not disappear in rest, but circumspection becomes free, it is no longer bound to the work-world. When it rests, care turns into circumspection which has become free. The circumspect discovery of the work-world has the character of being of de-distancing. Circumspection which has become free no longer has anything at hand which it has to bring near. Essentially de-distancing, it provides new possibilities of de-distancing for itself, that is, it tends to leave the things nearest at hand for a distant and strange world. Care turns into taking care of possibilities, resting and staying to see the "world" only in its *outward appearance*. Dasein seeks distance solely to bring it near in its outward appearance. Dasein lets itself be intrigued just by the outward appearance of the world, this then is a kind of being in which it makes sure that it gets rid of itself as being-in-the-world, and gets rid of being with the nearest everyday things at hand.

When curiosity has become free, it takes care to see not in order to understand what it sees, that is, to come to a being toward it, but *only* in order to see. It seeks novelty only to leap from it again to another novelty. The care of seeing is not concerned with comprehending and knowingly being in the truth, but with possibilities of abandoning itself to the world. Thus curiosity is characterized by a specific *not-staying* [Unverweilen] with what is nearest. Consequently, it also does not seek the leisure of reflective staying, rather it seeks restlessness and excitement from continual novelty and changing encounters. In not-staying, curiosity makes sure of the constant possibility of *distraction*. Curiosity has nothing to do with the contemplation that wonders at being, ϑαυμάζειν, it has no interest in wondering to the point of not understanding. Rather, it makes sure of knowing, but just in order to have known. The two factors constitutive for curiosity, *not-staying* in the surrounding world taken care of and *distraction* by new possibilities, are the basis of the third essential characteristic of this phenomenon, which we call *never dwelling anywhere*. Curiosity is everywhere and nowhere. This mode of being-in-the-world reveals a new kind of being of everyday Dasein, one in which it constantly uproots itself.

173

Idle talk also controls the ways in which one may be curious. It says what one is to have read and seen. The being everywhere and nowhere of curiosity is entrusted to idle talk. These two everyday modes of being of discourse and sight are not only objectively present side by side in their uprooting tendency, but *one* way of being drags the *other* with it. Curiosity, for which nothing is closed off, and idle talk, for which there is nothing that is not understood, provide themselves (that is, the Dasein existing in this way) with the guarantee of a supposedly genuine "lively life." But with this supposition a third phenomenon shows itself as characterizing the disclosedness of everyday Dasein.

§ 37. *Ambiguity*

When, in everyday being with one another, we encounter things that are accessible to everybody and about which everybody can say everything, we can soon no longer decide what is disclosed in genuine understanding and what is not. This ambiguity extends not only to the world, but likewise to being-with-one-another as such, even to the being of Dasein toward itself.

Everything looks as if it were genuinely understood, grasped, and spoken whereas basically it is not; or it does not look that way, yet basically is. Ambiguity not only affects the way we avail ourselves of what is accessible for use and enjoyment, and the way we manage it, but it has already established itself in understanding as a potentiality for being, and in the way Dasein projects itself and presents itself with possibilities. Not only does everyone know and talk about what is the case and what occurs, but everyone also already knows how to talk about what has to happen first, which is not yet the case, but "really" should be done. Everybody has always already guessed and felt beforehand what others also guess and feel. This being-on-the-track is based upon hearsay—whoever is "on the track" of something in a genuine way does not talk about it—and this is the most entangling way in which ambiguity presents possibilities of Dasein so that they will already be stifled in their power.

Even supposing that what *they* guessed and felt should one day be actually translated into deeds, ambiguity has already seen to it that the interest for what has been realized will immediately die away. This interest persists only, after all, in a kind of curiosity and idle talk, only as long as there is the possibility of a noncommittal just-guessing-with-someone. When one is on the track, and as long as one is on it, being "in on it" with someone precludes one's allegiance when what was guessed at is carried out. For then Dasein is actually forced back upon itself. Idle talk and curiosity lose their power. And they do take their revenge. In the light of the actualization of what they guessed, idle talk is quick to maintain that they could have done that, too, for, after all, they had guessed it, too. In the end, idle talk is indignant that what it guessed and constantly demanded now *actually* happens. After all, the opportunity to keep guessing is thus snatched away from it.

Since, however, the time span when Dasein becomes involved in the reticence of carrying something out, and even of genuinely getting stranded, is different from that of idle talk which "lives at a quicker

174

pace," so that viewed publicly it is essentially slower, idle talk will
have long since gone on to something else which is currently even
newer. That which had been surmised earlier, and has now been carried out, has come too late with regard to what is the very newest. In
their ambiguity, curiosity and idle talk make sure that what is done in
a genuine and new way is outdated as soon as it emerges before the
public. Only then can reticence become free in its positive possibilities,
when the idle talk covering it over has become ineffectual and the
"common" interest has died out.

The ambiguity of the way things have been interpreted publicly
passes off talking about things ahead of time and curious guessing as
what is really happening, and it stamps carrying things out and taking
action as something subsequent and of no importance. The understanding of Dasein in the they thus constantly *goes astray* in its projects with
regard to the genuine possibilities of being. Dasein is always ambigu-
ously "there," that is, in *the* public disclosedness of being-with-one-
another where the loudest idle talk and the most inventive curiosity
keep the "business" going, where everything happens in an everyday
way, and basically nothing happens at all.

Ambiguity is always tossing to curiosity what it seeks, and it
gives to idle talk the illusion of having everything decided in it.

This kind of being of the disclosedness of being-in-the-world,
however, also dominates being-with-one-another as such. The other is
initially "there" in terms of what they have heard about him, what
they say and know about him. Idle talk initially intrudes itself into the
midst of primordial being-with-one-another. Everyone keeps track of
the other, initially and first of all, watching how he will behave, what
he will say to something. Being-with-one-another in the they is not at
all a self-contained, indifferent side-by-sideness, but a tense, ambiguous
keeping track of each other, a secretive, reciprocal listening-in. Under
the mask of the for-one-another, the against-one-another is at play.

Here we must note that ambiguity does not first originate out of
an explicit intention to dissimulate [Verstellung] and distort [Verdre-
hung], that it is not called forth by the individual Dasein. It is already
implied in being-with-one-another, as *thrown* being-with-one-another
in a world. But publicly it is quite concealed, and *they* will always
protest the possibility that this interpretation of the kind of being of
interpreting the they could be correct. It would be a misunderstanding
if the explication of these phenomena were to seek to be confirmed by
the approval of the they.

The phenomena of idle talk, curiosity, and ambiguity were set
forth in such a way as to indicate that they are already interconnected in their being. The kind of being of this connection must now
be grasped existentially and ontologically. The basic kind of being of

everydayness is to be understood in the horizon of the structures of
the being of Dasein hitherto obtained.

§ 38. *Falling Prey [Verfallen] and Thrownness*

Idle talk, curiosity, and ambiguity characterize the way in which Dasein
is its "there," the disclosedness of being-in-the-world, in an everyday
way. As existential determinations, these characteristics are not objectively present in Dasein; they constitute its being. In them and in the
connectedness of their being, a basic kind of the being of everydayness
reveals itself, which we call the *entanglement* [*Verfallen*] of Dasein.

This term, which does not express any negative value judgment,
means that Dasein is initially and for the most part *together with* the
"world" that it takes care of. This absorption in . . . mostly has the
character of being lost in the publicness of the they. As an authentic potentiality for being a self, Dasein has initially always already
fallen away from itself and fallen prey to the "world." Falling prey to
the "world" means being absorbed in being-with-one-another as it is
guided by idle talk, curiosity, and ambiguity. What we called the inauthenticity of Dasein[13] may now be defined more precisely through the   *176*
interpretation of falling prey. But inauthentic and non-authentic by no
means signify "not really" ["eigentlich nicht"], as if Dasein utterly lost
its being in this kind of being. Inauthenticity does not mean anything
like no-longer-being-in-the-world, but rather it constitutes precisely a
distinctive kind of being-in-the-world which is completely taken in by
the world and the Dasein-with of the others in the they. Not-being-its-
self functions as a *positive* possibility of beings which are absorbed in
a world, essentially taking care of that world. This *non-being* must be
conceived as the kind of being of Dasein closest to it and in which it
mostly maintains itself.

Thus the falling prey of Dasein must not be interpreted as a "fall"
from a purer and higher "primordial condition." Not only do we not
have any experience of this ontically, but we also do not have possibilities and guidelines for such an ontological interpretation.

As factical being-in-the-world, Dasein, falling prey, has already
fallen *away from itself*; it has not fallen prey to some being which it does
or does not run into in the course of its being, but it has fallen prey
to the *world* which itself belongs to its being. Falling prey is an existential determination of Dasein itself, and says nothing about Dasein
as something objectively present, or about present relations to beings
from which it is "derived" or to beings with which it has subsequently
found itself in a *commercium*.

---

13. Cf. § 9.

The ontological-existential structure of falling prey would also be misunderstood if we wanted to attribute to it the meaning of a bad and deplorable ontic quality which could perhaps be removed in the advanced stages of human culture.

Neither in our first reference to being-in-the-world, as the fundamental constitution of Dasein, nor in our characterization of its constitutive structural factors, did we go beyond an analysis of the *constitution* of this kind of being and note its character as a phenomenon. It is true that the possible basic kinds of being-in, taking care and concern, were described. But we did not discuss the question of the everyday kind of being of these ways of being. It also became evident that being-in is quite different from a confrontation [Gegenüberstehen] which merely observes and acts, that is, the concurrent objective presence of a subject and an object. Nevertheless, it must have seemed that being-in-the-world functions as a rigid framework within which the possible relations of Dasein to its world occur, without the "framework" itself being touched upon in its kind of being. But this supposed "framework" itself belongs to the kind of being of Dasein. An *existential mode* of being-in-the-world is documented in the phenomenon of falling prey.

177

Idle talk discloses to Dasein a being toward its world, to others, and to itself—this being toward is such that these are understood, but in a mode of groundless floating. Curiosity discloses each and every thing, but in such a way that being-in is everywhere and nowhere. Ambiguity conceals nothing from the understanding of Dasein, but only in order to suppress being-in-the-world in this uprooted everywhere and nowhere.

With the ontological clarification of the kind of being of everyday being-in-the-world discernible in these phenomena, we first gain an existentially adequate determination of the fundamental constitution of Dasein. What structure does the "movement" of falling prey show?

Idle talk and the public interpretedness contained in it constitute themselves in being-with-one-another. Idle talk is not objectively present for itself within the world, as a product detached from being-with-one-another. Nor can it be volatilized to mean something "universal" which, since it essentially belongs to no one, "really" is nothing and "actually" only occurs in the individual Dasein that speaks. Idle talk is the kind of being of being-with-one-another itself, and does not first originate through certain conditions which influence Dasein "from the outside." But since, in idle talk and public interpretedness, Dasein itself presents itself with the possibility of losing itself in the they, of falling prey to groundlessness, that means that Dasein prepares for itself the constant temptation of falling prey. Being-in-the-world is in itself *tempting* [versucherisch].

Having already become a temptation for itself in this way, the way in which things have been publicly interpreted holds fast to Dasein in its falling prey. Idle talk and ambiguity, having-seen-everything and having-understood-everything, develop the supposition that the disclosedness of Dasein thus available and prevalent could guarantee to Dasein the certainty, genuineness, and fullness of all the possibilities of its being. The self-certainty and decisiveness of the they increasingly propagate the sense that there is no need of authentic, attuned understanding. The supposition of the they that one is leading and sustaining a full and genuine "life" brings a *reassurance* [*Beruhigung*] to Dasein, for which everything is in "the best order" and for whom all doors are open. Entangled being-in-the-world, tempting itself, is at the same time *tranquillizing* [*beruhigend*].

This tranquillization in inauthentic being, however, does not seduce one into stagnation and inactivity, but drives one to uninhibited "busyness." Being entangled in the "world" does not somehow come to rest. Tempting tranquillization *increases* entanglement. With special regard to the interpretation of Dasein, the opinion may now arise that understanding the most foreign cultures and "synthesizing" them with one's own may lead to the thorough and first genuine enlightenment of Dasein about itself. Versatile curiosity and restlessly knowing it all masquerade as a universal understanding of Dasein. But fundamentally it remains undetermined and unasked *what* is then really to be understood; nor has it been understood that understanding itself is a potentiality for being which must become free solely in one's *ownmost* Dasein. When Dasein, tranquillized and "understanding" everything, thus compares itself with everything, it drifts toward an alienation in which its ownmost potentiality for being-in-the-world is concealed. Entangled being-in-the-world is not only tempting and tranquillizing; it is at the same time *alienating*.

However, this alienation cannot mean that Dasein is factically torn away from itself. On the contrary, it drives Dasein into a kind of being intent upon the most exaggerated "self-dissection" which tries out all kinds of possibilities of interpretation, with the result that the "characterologies" and "typologies" which it points out are themselves too numerous to grasp. Yet this alienation, which *closes off* to Dasein its authenticity and possibility, even if only that of genuinely getting stranded, still does not surrender it to beings which it itself is not, but forces it into its inauthenticity, into a possible kind of being of *itself*. The tempting and tranquillizing alienation of falling prey has its own kind of movement with the consequence that Dasein gets *tangled up* [*verfängt*] in itself.

The phenomena pointed out of temptation, tranquillizing, alienation, and self-entangling (entanglement [Verfängnis]) characterize the specific kind of being of falling prey. We call this kind of "movement" of Dasein in its own being the *plunge* [*Absturz*]. Dasein plunges out of itself into itself, into the groundlessness and nothingness of inauthentic

178

everydayness. But this plunge remains concealed from it by the way things have been publicly interpreted so that it is interpreted as "getting ahead" and "living concretely."

The kind of movement of plunging into and within the groundlessness of inauthentic being in the they constantly tears understanding away from projecting authentic possibilities, and drags it into the tranquillized supposition of possessing or attaining everything. Since the understanding is thus constantly torn away from authenticity and dragged into the they (although always with a sham of authenticity), the movement of falling prey is characterized by *turbulence*.

179   Not only does falling prey determine being-in-the-world existentially; at the same time turbulence reveals the character of throwing and the movement of thrownness which can force itself upon Dasein in its attunement. Not only is thrownness not a "finished fact," it is also not a self-contained fact. The facticity of Dasein is such that Dasein, *as long as* it is what it is, remains tossed about [im Wurf] and sucked into the turbulence of the they's inauthenticity. Thrownness, in which facticity can be seen phenomenally, belongs to Dasein, which is concerned in its being about that being. Dasein exists factically.

But now that falling prey has been exhibited, have we not set forth a phenomenon which directly speaks *against* the definition in which the formal idea of existence was indicated? Can Dasein be conceived as a being whose being is concerned *with* potentiality for being if this being *has lost itself* precisely in its everydayness and "lives" *away from itself* in falling prey? Falling prey to the world is, however, phenomenal "evidence" *against* the existentiality of Dasein only if Dasein is posited as an isolated I-subject, as a self-point from which it moves away. Then the world is an object. Falling prey to the world is then reinterpreted ontologically as objective presence in the manner of innerworldly beings. However, if we hold on to the being of Dasein in the constitution indicated of *being-in-the-world*, it becomes evident that falling prey, *as the kind of being of this being-in*, represents the most elemental proof *for* the existentiality of Dasein. In falling prey, nothing other than our potentiality for being-in-the-world is the issue, even if in the mode of inauthenticity. Dasein *can* fall prey *only* because it is concerned with understanding, attuned being-the-world. On the other hand, *authentic* existence is nothing which hovers over entangled everydayness, but is existentially only a modified grasp of everydayness.

The phenomenon of falling prey does not give something like a "night view" of Dasein, a property occurring ontically which might serve to round out the harmless aspect of this being. Falling prey reveals an *essential*, ontological structure of Dasein itself. Far from determining its nocturnal side, it constitutes all of its days in their everydayness.

Our existential, ontological interpretation thus does not make any ontic statement about the "corruption of human nature," not because the necessary evidence is lacking but because its problematic is *prior to* any statement about corruption or incorruption. Falling prey is an ontological concept of motion. Ontically, we have not decided whether human being is "drowned in sin," in the *status corruptionis*, or whether it walks in the *status integritatis* or finds itself in an interim stage, the *status gratiae*. But faith and "worldview," when they state such and such a thing and when they speak about Dasein as being-in-the-world, must come back to the existential structures set forth, provided that their statements at the same time claim to be *conceptually* comprehensible.

The leading question of this chapter pursued the being of the there. Its theme was the ontological constitution of the disclosedness essentially belonging to Dasein. The being of disclosedness is constituted in attunement, understanding, and discourse. Its everyday mode of being is characterized by idle talk, curiosity, and ambiguity. These show the kind of movement of falling prey with the essential characteristics of temptation, tranquillization, alienation, and entanglement.

With this analysis, however, the totality of the existential constitution of Dasein has been laid bare in its main features and the phenomenal basis has been obtained for a "comprehensive" interpretation of the being of Dasein as care.

CHAPTER SIX
Care as the Being of Dasein

§ 39. *The Question of the Primordial Totality of the Structural Whole of Dasein*

Being-in-the-world is a structure that is primordially and constantly *whole*. In the previous chapters (Division One, Chapters Two–Five), this structure was clarified phenomenally as a whole and, always on this basis, in its constitutive moments. The preview given at the beginning[1] of the whole of the phenomenon has now lost the emptiness of its first general prefiguration. However, the phenomenal *manifoldness* of the constitution of the structural whole and its everyday kind of being can now easily distort the *unified* phenomenological view of the whole as such. But this view must be held in readiness more freely and more securely when we now ask the question toward which the preparatory fundamental analysis of Dasein was striving in general: *how is the totality of the structural whole that we pointed out to be determined existentially and ontologically?*

181

Dasein exists factically. We are asking about the ontological unity of existentiality and facticity, namely, whether facticity belongs essentially to existentiality. On the basis of the attunement essentially belonging to it, Dasein has a mode of being in which it is brought before itself and it is disclosed to itself in its thrownness. But thrownness is the mode of being of a being which always *is* itself its possibilities in such a way that it understands itself in them and in terms of them (projects itself upon them). Being-in-the-world, to which being together with things at hand belongs just as primordially as being-with others, is always for the sake of itself. But the self is initially and for the most part inauthentic, the they-self. Being-in-the-world is always already entangled. *The average everydayness of Dasein* can thus be determined as *entangled-disclosed, thrown-projecting being-in-the-world, which is concerned*

---

1. Cf. § 12.

with its ownmost potentiality in its being together with the "world" and in being-with with others.

Can we succeed in grasping this structural whole of the everydayness of Dasein in its totality? Can the being of Dasein be delineated in a unified way so that in terms of it the essential equiprimordiality of the structures pointed out becomes intelligible, together with the existential possibilities of modification which belong to it? Is there a way to attain this being phenomenally on the basis of the present point of departure of the existential analytic?

To put it negatively, it is beyond question that the totality of the structural whole is not to be reached phenomenally by means of cobbling together elements. This would require a blueprint. The being of Dasein, which ontologically supports the structural whole as such, becomes accessible by completely looking *through* this whole *at a* primordially unified phenomenon which already lies in the whole in such a way that it is the ontological basis for every structural moment in its structural possibility. Thus a "comprehensive" interpretation cannot consist of a process of piecing together what we have hitherto gained. The question of Dasein's existential character is essentially different from the question of the being of something objectively present. Everyday experience of the surrounding world, which is directed ontically and ontologically toward innerworldly beings, cannot present Dasein ontically and primordially for the ontological analysis. Similarly, our immanent perception of experiences is lacking an ontologically sufficient guideline. On the other hand, the being of Dasein is not to be deduced from an idea of human being. Can we gather from our previous interpretation of Dasein what ontic-ontological access to itself it requires, *from itself*, as the sole appropriate one?

An understanding of being belongs to the ontological structure of Dasein. In existing, it is disclosed to itself in its being. Attunement and understanding constitute the kind of being of this disclosedness. Is there an understanding attunement in Dasein in which it is disclosed to itself in a distinctive way?

If the existential analytic of Dasein is to maintain a fundamental clarity about its basic ontological function, then, in order to accomplish its preliminary task of setting forth the being of Dasein, it must search for the *most far-reaching* and *most primordial* possibilities of disclosure which lie in Dasein itself. The kind of disclosure in which Dasein brings itself before itself must be such that in it Dasein becomes accessible to itself as *simplified* in a certain way. Together with what has been disclosed to it, the structural whole of the being we seek must then come to light in an elemental way.

As a kind of attunement adequate for such methodological requirements, we shall take the phenomenon of anxiety as the basis of analysis. The elaboration of this fundamental kind of attunement and the ontological characteristics of what is disclosed in it as such take their point of departure from the phenomenon of entanglement, and distinguish anxiety from the related phenomenon of fear analyzed earlier. As a possibility of being of Dasein, together with the Dasein itself disclosed in it, anxiety provides the phenomenal basis for explicitly grasping the primordial totality of being of Dasein. Its being reveals itself as *care* [*Sorge*]. The ontological development of this fundamental existential phenomenon demands that we differentiate it from phenomena which at first might seem to be identified with care. Such phenomena are will, wish, predilection, and urge. Care cannot be derived from them because they are themselves founded upon it.

Like any ontological analysis, the ontological interpretation of Dasein as care, with whatever can be gained from the interpretation, is far removed from what is accessible to the pre-ontological understanding of being or even to our ontic acquaintance with beings. That the common understanding estranges what is known ontologically by referring it to that with which it is solely ontically acquainted is not surprising. Nonetheless, even the ontic approach with which we have tried to interpret Dasein ontologically as care might appear to be contrived in a far-fetched and theoretical way; to say nothing of the violence which one could see in our exclusion of the traditional and approved definition of human being. Thus we need a pre-ontological confirmation of the existential interpretation of Dasein as care. This lies in demonstrating that as soon as Dasein expressed anything about itself, it had already interpreted itself as *care* (*cura*), even though it did so only pre-ontologically.

183

The analytic of Dasein, which penetrates to the phenomenon of care, is to prepare the way for the fundamental, ontological problematic, *the question of the meaning of being in general.* In order to direct our view explicitly to this in the light of what we have gained, and go beyond the special task of an existential, *a priori* anthropology, we must look back at *the* phenomena which are most intimately connected with the leading question of being in order to get a more penetrating grasp of them. These phenomena are the modes of being explained hitherto: handiness and objective presence which determine innerworldly beings unlike Dasein. Because the ontological problematic has hitherto understood being primarily in the sense of objective presence ("reality," "world"-actuality), while the being of Dasein remained ontologically undetermined, we need to discuss the ontological connection of care, worldliness, handiness, and objective presence (reality). That leads to a more exact determination of the concept of *reality* in the context of a discussion of the epistemological questions oriented toward this idea which have been raised by realism and idealism.

Beings *are* independently of the experience, cognition, and comprehension through which they are disclosed, discovered, and determined. But being "is" only in the understanding* of that being to whose being something like an understanding of being belongs. Thus being [Sein] can be unconceptualized, but it is never completely uncomprehended. In ontological problematics, *being and truth* have been brought together since ancient times, if not even identified. This documents the necessary connection of being and understanding† [Verständnis], even if it is perhaps concealed in its primordial grounds. Thus for an adequate preparation of the question of being, we need an ontological clarification of the phenomenon of *truth*. This will be accomplished initially on the basis of that which our interpretation hitherto has gained with the phenomena of disclosedness and discoveredness, interpretation and statement.

184

The conclusion of the preparatory fundamental analysis of Dasein thus has as its theme the fundamental attunement of anxiety as a distinctive disclosedness of Dasein (§ 40), the being of Dasein as care (§ 41), the confirmation of the existential interpretation of Dasein as care in terms of the pre-ontological self-interpretation of Dasein (§ 42), Dasein, worldliness, and reality (§ 43), Dasein, disclosedness, and truth (§ 44).

§ 40. *The Fundamental Attunement of Anxiety as an Eminent Disclosedness of Dasein*

One possibility of being of Dasein is to give ontic "information" about itself as a being. Such information is possible only in the disclosedness belonging to Dasein which is based on attunement and understanding. To what extent is anxiety a distinctive attunement? How, in anxiety, is Dasein brought before itself in it through its own being, so that phenomenologically the being disclosed in anxiety is defined as such in its being, or adequate preparations can be made for doing so?

With the intention of penetrating to the being of the totality of the structural whole, we shall take our point of departure from the concrete analysis of entanglement carried out in the last chapter. The absorption of Dasein in the they and in the "world" taken care of reveals something like a *flight* of Dasein from itself as an authentic potentiality for being itself. This phenomenon of the flight of Dasein *from itself* and its authenticity seems, however, to be least appropriate to serve as a

---

* But this understanding as listening. But this never means that 'being' is only 'subjective,' but that being (qua the being of beings) qua difference "in" Dasein as what is thrown by the (throw).
† Thus: being and Dasein.

phenomenal foundation for the following inquiry. In this flight, Dasein precisely does *not* bring itself before itself. In accordance with its own-most trait of entanglement, this turning away leads *away from* Dasein. But in investigating such phenomena, our inquiry must guard against conflating ontic-existentiell characteristics with ontological-existential interpretation, and must not overlook the positive, phenomenal foundations provided for this interpretation by such a characterization.

It is true that existentielly the authenticity of being a self is closed off and pushed away in entanglement, but this closing off is only the *privation* of a disclosedness which reveals itself phenomenally in the fact that the flight of Dasein is a flight *from* itself. That from which Dasein flees is precisely what Dasein comes up "behind." Only because Dasein is ontologically and essentially brought before itself by the disclosedness belonging to it, *can* it flee *from* that from which it flees. Of course, in this entangled turning away, that from which it flees is *not grasped*, nor is it experienced in a turning toward it. But in turning away *from* it, it is "there," disclosed. On account of its character of being disclosed, this existentielly-ontic turning away makes it phenomenally possible to grasp existentially and ontologically what the flight is from. Within the ontic "away from," which lies in turning away, that from which Dasein flees can be understood and conceptualized by "turning toward" in a way which is phenomenologically interpretive.

Thus the orientation of our analysis toward the phenomenon of entanglement is not condemned in principle to be without any prospect of ontologically experiencing something about the Dasein disclosed in that phenomenon. On the contrary, it is precisely here that our interpretation is the least likely to be surrendered to an artificial self-conception of Dasein. It only carries out the explication of what Dasein itself discloses ontically. The possibility of penetrating to the being of Dasein by going along with it and pursuing it interpretatively in an attuned understanding increases, the more primordially that phenomenon is which functions methodologically as disclosive attunement. To say that anxiety accomplishes something like this is only an assertion for now.

We are not completely unprepared for the analysis of anxiety. It is true that we are still in the dark as to how it is ontologically connected with fear. Obviously they are kindred phenomena. What tells us this is the fact that both phenomena remain mostly undifferentiated, and we designate as anxiety what is really fear, and call fear what has the character of anxiety. We shall attempt to penetrate to the phenomenon of anxiety step by step.

The falling prey of Dasein to the they and the "world" taken care of, we called a "flight" from itself. But not every shrinking back from . . . , not every turning away from . . . is necessarily a flight.

185

Shrinking back from what fear discloses, from what is threatening, is founded upon fear and has the character of flight. Our interpretation of fear as attunement showed that what we fear is always a detrimental innerworldly being, approaching nearby from a definite region, which may remain absent. In falling prey, Dasein turns away from itself. What it shrinks back from must have a threatening character; yet this being has the same kind of being as the one which shrinks back from it—it is Dasein itself. What it shrinks back from cannot be grasped as something "fearsome"; because anything fearsome is always encountered as an innerworldly being. The only threat which can be "fearsome," and which is discovered in fear, always comes from innerworldly beings.

186

The turning away of falling prey is thus not a flight which is based on a fear of innerworldly beings. Any flight based on that kind of fear belongs still less to turning away, as turning away precisely *turns toward* innerworldly beings while absorbing itself in them. *The turning away of falling prey is rather based on anxiety, which in turn first makes fear possible.*

In order to understand this talk about the entangled flight of Dasein from itself, we must recall that being-in-the-world is the basic constitution of Dasein. *That about which one has anxiety is being-in-the-world as such.* How is what anxiety is anxious about phenomenally differentiated from what fear is afraid of? What anxiety is about is not an innerworldly being. Thus it essentially cannot have any relevance [Bewandtnis]. The threat does not have the character of a definite harmfulness which concerns what is threatened with a definite regard to a particular factical potentiality for being. What anxiety is about is completely indefinite. This indefiniteness not only leaves factically undecided which innerworldly being is threatening, it also means that innerworldly beings in general are not "relevant" ["relevant"]. Nothing which is at hand and present within the world functions as that which anxiety is anxious about. The totality of relevance discovered within the world of things at hand and objectively present is completely without importance [ohne Belang]. It collapses into itself. The world has the character of complete insignificance [Unbedeutsamkeit]. In anxiety we do not encounter this or that thing which, as threatening, could be relevant.

Thus neither does anxiety "see" a definite "there" and "over here" from which what is threatening approaches. The fact that what is threatening is *nowhere* characterizes what anxiety is about. Anxiety "does not know" what it is anxious about. But "nowhere" does not mean nothing; rather, region in general lies therein, and disclosedness of the world in general for essentially spatial being-in. Therefore, what is threatening cannot come closer from a definite direction within nearness, it is already "there"—and yet nowhere. It is so near that it is oppressive and takes away one's breath—and yet it is nowhere.

In what anxiety is about, the "it is nothing and nowhere" becomes manifest. The recalcitrance of the innerworldly nothing and nowhere means phenomenally that *what anxiety is about is the world as such.*   *187* The utter insignificance which makes itself known in the nothing and nowhere does not signify the absence of world, but means that innerworldly beings in themselves are so completely unimportant that, on the basis of this *insignificance* of what is innerworldly, the world in its worldliness is all that obtrudes itself.

What crowds in upon us is not this or that, nor is it everything objectively present together as a sum, but the *possibility* of things at hand in general, that is, the world itself. When anxiety has subsided, in our everyday way of talking we are accustomed to say "it was really nothing." This way of talking, indeed, gets at *what* it was ontically. Everyday discourse aims at taking care of things at hand and talking about them. That about which anxiety is anxious is not [nichts] innerworldly things at hand. But this not [Nichts] any thing at hand, which is all that everyday, circumspect discourse understands, is not completely nothing [Nichts]. The nothing of handiness is grounded in the primordial "something" ["Etwas"],* in the *world*. The world, however, ontologically belongs essentially to the being of Dasein as being-in-the-world. So if what anxiety is about exposes nothing, that is, the world as such, this means that *that about which anxiety is anxious is being-in-the-world itself.*†

Being anxious discloses, primordially and directly, the world as world. It is not the case that initially we deliberately look away from innerworldly beings and think only of the world about which anxiety arises, but anxiety, as a mode of attunement, first discloses the *world as world*. However, that does not mean that the worldliness of the world is grasped [begriffen] in anxiety.

Anxiety is not only anxiety about . . . , but is at the same time, as attunement, anxiety for. . . . That for which anxiety is anxious is not a *definite* kind of being and possibility of Dasein. The threat itself is, after all, indefinite and thus cannot penetrate threateningly to this or that factically concrete potentiality of being. What anxiety is anxious for is being-in-the-world itself. In anxiety, the things at hand in the surrounding world sink away, and so do innerworldly beings in general. The "world" can offer nothing more, nor can the Dasein-with of others. Thus anxiety takes away from Dasein the possibility of understanding itself, falling prey, in terms of the "world" and the public way of being interpreted. It throws Dasein back upon that for which it is anxious,

---

* Thus nothing to do with "nihilism."
† Determining being as such; what is absolutely unhoped for and not to be perdured [unaustragbare]—what estranges.

its authentic potentiality-for-being-in-the-world. Anxiety individuates Dasein to its ownmost being-in-the-world which, as understanding,

188 projects itself essentially upon possibilities. Thus along with that for which it is anxious, anxiety discloses Dasein as *being-possible*, and indeed as what, solely [einzig] from itself, can be individualized in individuation [Vereinzelung].

Anxiety reveals in Dasein its *being toward* its ownmost potentiality of being, that is, *being free for* the freedom of choosing and grasping itself. Anxiety brings Dasein before *its being free for . . . (propensio in)*, the authenticity of its being as possibility which it always already is. But, at the same time, it is this being to which Dasein as being-in-the-world is entrusted.

That *about which* anxiety is anxious reveals itself as that *for which* it is anxious: being-in-the-world. The sameness [Selbigkeit] of that about which and that for which one has anxiety extends even to anxiousness itself. For as attunement, anxiousness is a fundamental mode of being-in-the-world. *The existential sameness of disclosing and what is disclosed so that in what is disclosed the world is disclosed as world, as being-in, individualized, pure, thrown potentiality for being, makes it clear that with the phenomenon of anxiety a distinctive kind of attunement has become the theme of our interpretation.* Anxiety individualizes and thus discloses Dasein as "solus ipse." This existential "solipsism," however, is so far from transposing an isolated subject-thing into the harmless vacuum of a worldless occurrence that it brings Dasein in an extreme sense precisely before its world as world, and thus itself before itself as being-in-the-world.

Again, everyday discourse and the everyday interpretation of Dasein furnish the most unbiased evidence that anxiety as a basic attunement is disclosive in this way. We said earlier that attunement reveals "how one is." In anxiety one has an *"uncanny"* [*"unheimlich"*] feeling. Here, with anxiety, the peculiar indefiniteness of that which Dasein finds itself involved in anxiety initially finds expression: the nothing and nowhere. But uncanniness means at the same time not-being-at-home. In our first phenomenal indication of the fundamental constitution of Dasein, and the clarification of the existential meaning of being-in in contradistinction to the categorial signification of "insideness," being-in was defined as dwelling with . . . , being familiar with. . . .[2] This characteristic of being-in was then made more concretely visible through the everyday publicness of the they which brings tran-

189 quillized self-assurance—"being-at-home" ["Zuhause-sein"] with all its obviousness, into the average everydayness of Dasein.[3] Anxiety, on the other hand, fetches Dasein back out of its entangled absorption in the "world." Everyday familiarity collapses. Dasein is individualized

---

2. Cf. § 12.
3. Cf. § 27.

[vereinzelt], but *as* being-in-the-world. Being-in enters the existential "mode" of *not-being-at-home* [*Un-zuhause*]. The talk about "uncanniness" ["Unheimlichkeit"] means nothing other than this.

Now, however, what falling prey, as flight, is fleeing from becomes phenomenally visible. It is not a flight *from* innerworldly beings, but precisely *toward* them as the beings among which taking care of things, lost in the they, can linger in tranquillized familiarity. Entangled flight *into* the being-at-home of publicness is flight *from* not-being-at-home, that is, from the uncanniness which lies in Dasein as thrown, as being-in-the-world entrusted to itself in its being. This uncanniness constantly pursues Dasein and threatens its everyday lostness in the they, although not explicitly. This threat can factically go along with the complete security and self-sufficiency of the everyday way of taking care. Anxiety can arise in the most harmless situations. Nor does it have any need for darkness, in which things usually become uncanny to us more easily. In the dark there is emphatically "nothing" to see, although the world is *still* "there" *more obtrusively*.

If we interpret the uncanniness of Dasein existentially and ontologically as a threat which concerns Dasein itself and which comes from Dasein itself, we are not asserting that uncanniness has always already been understood in factical anxiety in this sense. The everyday way in which Dasein understands uncanniness is the entangled turning away which "dims down" not-being-at-home. The everydayness of this fleeing, however, shows phenomenally that anxiety as a fundamental kind of attunement belongs to Dasein's essential constitution as being-in-the-world, which, as an existential, is never objectively present, but *is* itself always in the mode of factical Dasein, that is, in the mode of an attunement. Tranquillized, familiar being-in-the-world is a mode of the uncanniness of Dasein, not the other way around. *Not-being-at-home** *must be conceived existentially and ontologically as the more primordial phenomenon*.

And only because anxiety always already latently determines being-in-the-world, can being-in-the-world, as being together with the "world" taking care of things and attuned, be afraid. Fear is anxiety which has fallen prey to the "world." It is inauthentic and concealed from itself as such.

Factically, the mood of uncanniness remains for the most part   *190*
existentielly uncomprehended. Moreover, with the dominance of falling prey and publicness, "real" ["eigentliche"] anxiety is rare. Often, anxiety is "physiologically" conditioned. This fact is an *ontological* problem in its facticity, not only with regard to its ontic causes and course of development. The physiological triggering of anxiety is possible only because Dasein is anxious in the very ground of its being.

---

* (Ex-propriation [Enteignis]).

Still more rare than the existentiell fact of real anxiety are the attempts to interpret this phenomenon in its fundamental, existential-ontological constitution and function. The reasons for this lie partly in the general neglect of the existential analytic of Dasein, particularly in the failure to recognize the phenomenon of attunement.[4] The factical rarity of the phenomenon of anxiety, however, cannot deprive it of its suitability for taking over a methodical function *in principle* for the existential analytic. On the contrary, the rarity of the phenomenon is an indication of the fact that Dasein, which mostly remains concealed from itself in its authenticity because of the way things get publically interpreted by the they, can be disclosed in a primordial sense in its fundamental attunement.

191   It is true that it is the nature of every kind of attunement to disclose complete being-in-the-world in all its constitutive factors (world, being-in, self). However, in anxiety there lies the possibility of a distinctive disclosure, since anxiety individualizes. This individuality fetches Dasein back from its falling prey and reveals to it authenticity and inauthenticity as possibilities of its being. The fundamental possibilities of Dasein, which are always my own,* show themselves in anxiety as they are, undisguised [unverstellt], by innerworldly beings to which Dasein, initially and for the most part, clings.

To what extent has this existential interpretation of anxiety gained a phenomenal basis for answering the leading question of the being of the totality of the structural whole of Dasein?

§ 41. *The Being of Dasein as Care*

With the intention of grasping the totality of the structural whole ontologically, we must first ask whether the phenomenon of anxiety and

---

* Not egotistical, but to be taken over as thrown.

4. It is not a matter of chance that the phenomena of anxiety and fear, which have never been distinguished in a thoroughgoing way, were constitutive for the scope of Christian theology ontically and also ontologically, although in very narrow limits. This always happened when the anthropological problem of the being of human being toward God gained priority, and phenomena such as faith, sin, love, and repentance guided the questions. Cf. Augustine's doctrine of *timor castus* and *servilis*, which is often discussed in his exegetical writings and letters. On fear in general, cf. "De diversis quaestionibus octoginta tribus," qu. 33; "De metu," qu. 34; "Utram non aliud amandum sit, quam metu carere," qu. 35; "Quid amandum sit" (Migne, ed., P. L., vol. 7, 23 sqq.).
   Apart from the traditional context of an interpretation of *poenitentia* and *contritio*, Luther treated the problem of fear in his commentary on Genesis, here, of course, least of all conceptually, but all the more penetratingly by way of edification. Cf. *Enarrationes in genesin*, chap. 3, WW. (Erl. edition), *Exegetica opera latina*, vol. 1, 177 et sqq.
   S. Kierkegaard got furthest of all in the analysis of the phenomenon of anxiety, again in the theological context of a "psychological" exposition of the problem of original sin. Cf. *Der Begriff der Angst*, 1844, *Ges. Werke* (Diedrichs), vol. 5.

what is disclosed in it are able to give us the whole of Dasein in a way that is phenomenally equiprimordial, so that our search for totality can be fulfilled in this givenness. The complete inventory of what lies in it can be enumerated: as attunement, being anxious is a way of being-in-the-world; that about which we have anxiety is thrown being-in-the-world; that for which we have anxiety is our potentiality-for-being-in-the-world. The complete phenomenon of anxiety thus shows Dasein as factical, existing being-in-the-world. The fundamental, ontological characteristics of this being are existentiality, facticity, and falling prey. These existential determinations are not pieces belonging to something composite, one of which might sometimes be missing, but a primordial coherence [Zusammenhang] is woven in them which constitutes the totality of the structural whole that we are seeking. In the unity of the determinations of the being of Dasein that we have mentioned, its being becomes ontologically comprehensible as such. How is this unity itself to be characterized?

Dasein is a being which is concerned in its being about that being. The "is concerned about . . ." has become clearer in the constitution of being of understanding as self-projective being toward its ownmost potentiality-for-being. This potentiality is that for the sake of which any Dasein is as it is. Dasein has always already compared itself, in its being, with a possibility of itself. Being free *for* its ownmost potentiality-for-being, and thus for the possibility of authenticity and inauthenticity, shows itself in a primordial, elemental concretion in anxiety. But ontologically, being toward one's ownmost potentiality-for-being means that Dasein is always already *ahead* of itself in its being. Dasein is always already "beyond itself," not as a way of behaving toward beings which it is *not*, but as being toward the potentiality-for-being which it itself is. This structure of being of the essential "being concerned about" we formulate as the *being-ahead-of-itself* of Dasein.

But this structure concerns the whole of the constitution of Dasein. Being-ahead-of-itself does not mean anything like an isolated tendency in a worldless "subject," but characterizes being-in-the-world. But to being-in-the-world belongs the fact that it is entrusted [überantwortet] to itself, that it is always already thrown *into a world*. The fact that Dasein is left to itself [Überlassenheit] shows itself primordially and concretely in anxiety. More completely formulated, being-ahead-of-itself means *being-ahead-of-itself-in-already-being-in-a-world*. As soon as this essentially unitary structure is seen phenomenally, what we worked out earlier in the analysis of worldliness also becomes clearer. There we found that the referential totality of significance, which is constitutive for worldliness, is "anchored" in a for-the-sake-of-which. The fact that this referential totality of the manifold relations of the in-order-to is bound up with that which Dasein is concerned about does

192

not signify that an objectively present "world" of objects is welded together with a subject. Rather, it is the phenomenal expression of the fact that the constitution of Dasein, whose wholeness is now delineated explicitly as being-ahead-of-itself-in-already-being-in . . . , is primordially a whole. Expressed differently: existing is always factical. Existentiality is essentially determined by facticity.

Furthermore, the factical existing of Dasein is not only in general and indifferently a thrown potentiality-for-being-in-the-world, but is always already also absorbed in the world taken care of. In this entangled being-together-with, fleeing from uncanniness (which mostly remains covered over by latent anxiety because the publicness of the they suppresses everything unfamiliar) announces itself, whether it does so explicitly or not, and whether it is understood or not. Entangled *being-together*-with innerworldly things at hand taken care of is essentially included in being-ahead-of-oneself-already-being-in-the-world.

The formal existential totality of the ontological structural whole of Dasein must thus be formulated in the following structure: the being of Dasein means being-ahead-of-oneself-already-in (the world) as being-together-with (innerworldly beings encountered). This being fills in the significance of the term *care*, which is used in a purely ontological and existential way. Any ontically intended tendency of being, such as worry or carefreeness, is ruled out.

193   Since being-in-the-world is essentially care, being-together-with things at hand could be taken in our previous analyses as *taking care* of them, while being with the Dasein-with of others encountered within the world could be taken as *concern*. Being-together-with is taking care, because as a mode of being-in it is determined by its fundamental structure, care. Care not only characterizes existentiality, abstracted from facticity and falling prey, but encompasses the unity of these determinations of being. Nor does care mean primarily and exclusively an isolated attitude of the ego toward itself. The expression "care for oneself," following the analogy of taking care and concern, would be a tautology. Care cannot mean a special attitude toward the self, because the self is already characterized ontologically as being-ahead-of-itself; but in this determination the other two structural moments of care, already-being-in . . . and being-together-with, are *co-posited* [*mitgesetzt*].

In being-ahead-of-oneself as the being toward one's ownmost potentiality-of-being lies the existential and ontological condition of the possibility of *being free for* authentic existentiell possibilities. It is the potentiality-for-being for the sake of which Dasein always is as it factically is. But since this being toward the potentiality-for-being is itself determined by freedom, Dasein *can* also be related to its possibilities *unwillingly*, it *can* be inauthentic and factically it is this initially and for the most part. The authentic for-the-sake-of-which remains ungrasped, the project of one's potentiality-of-being is left to the disposal of the they. Thus in being-ahead-of-itself, the "self" actually means the self in the sense of the they-self. Even in inauthenticity, Dasein remains essentially ahead-of-itself, just as the entangled fleeing of Dasein from itself still shows *the* constitution of being of a being that *is concerned about its being*.

As a primordial structural totality, care lies "before" every factical "attitude" and "position" of Dasein, that is, it is always already *in* them as an existential *a priori*. Thus this phenomenon by no means expresses a priority of "practical" over theoretical behavior. When we determine something objectively present by merely looking at it, this has the character of care just as much as a "political action," or resting and having a good time. "Theory" and "praxis" are possibilities of being for a being whose being must be defined as care.

The phenomenon of care in its totality is essentially something that cannot be split up; thus any attempts to derive it from special acts or drives, such as willing and wishing or urge and predilection, or of constructing it out of them, will be unsuccessful.   194

Willing and wishing are necessarily rooted ontologically in Dasein as care and are not simply ontologically undifferentiated experiences, which occur in a "stream" that is completely indeterminate as to the meaning of its being. This is no less true for predilection and urge. They, too, are based upon care insofar as they are purely demonstrable in Dasein in general. This does not exclude the fact that urge and predilection are ontologically constitutive even for beings which are only "alive." The basic ontological constitution of "life," however, is a problem in its own right, and can be developed only reductively and privatively in terms of the ontology of Dasein.

Care is ontologically "prior" to the phenomena we mentioned, which can, of course, always be adequately "described" within certain limits without the complete ontological horizon needing to be visible or even known as such. For the present fundamental ontological study, which neither aspires to a thematically complete ontology of Dasein nor even to a concrete anthropology, it must suffice to suggest how these phenomena are existentially grounded in care.

The potentiality-for-being for the sake of which Dasein is, has itself the mode of being of being-in-the-world. Accordingly, the relation to innerworldly beings lies in it ontologically. Even if only privatively, care is always taking care and concern. In willing, a being that is understood, that is, projected upon its possibility, is grasped as something to be taken care of or to be brought to its being through concern. *For this reason,* to any willing there always belongs something willed and this thing willed has already been determined in terms of a for the-sake-of-which. If willing is to be possible ontologically, the following factors are constitutive for it: the previous disclosedness of the for-the-sake-of-which in general (being-ahead-of-oneself), the disclosedness of what can be taken care of (world as the wherein of already-being), and the understanding self-projection of Dasein upon a potentiality-for-being toward a possibility of the being "willed." The underlying wholeness of care shows through in the phenomenon of willing.

As something factical, the understanding self-projection of Dasein is always already together with a discovered world. From this world it takes its possibilities, initially in accordance with the interpretedness of the they. This interpretation has from the outset restricted the possible options of choice to the scope of what is familiar, attainable, feasible, to what is correct and proper. The leveling down of the possibilities of *195* Dasein to what is initially available in an everyday way at the same time results in a phasing out of the possible as such. The average every-dayness of taking care of things becomes blind to possibility and gets tranquillized with what is merely "real." This tranquillization not only does not rule out a high degree of busyness in taking care of things, rather, it arouses it. It is not the case that positive, new possibilities are then willed, but what is available is "tactically" changed in such a way that there is an illusion of something happening.

All the same, under the leadership of the they, this tranquillized "willing" does not signify that being toward one's potentiality-for-being has been extinguished, but only that it has been modified. Being toward possibilities then shows itself for the most part as mere *wishing*. In the wish, Dasein projects its being toward possibilities which not only remain ungrasped in taking care of things, but whose fulfillment is not even thought about and expected. On the contrary, the predominance of being-ahead-of-itself in the mode of mere wishing brings with it a lack of understanding of factical possibilities. Being-in-the-world whose world is primarily projected as a wish-world has lost itself utterly in what is available, but in such a way that in the light of what is wished for, what is available (all the things at hand) is never enough. Wishing is an existential modification of understanding self-projection which, having fallen prey to thrownness, simply indulges in possibilities. Such indul-gence *closes off* possibilities; what is "there" in such wishful indulgence becomes the "real world." Ontologically, wishing presupposes care.

In indulgence, being-in-the-world-already-among . . . has prior-ity. Being-ahead-of-itself-in-already-being-in is modified accordingly. Entangled indulgence reveals the *inclination* of Dasein to be "lived" by the world in which it is. This inclination shows the character of being out for something. Being-ahead-of-itself has gotten lost in a "just-always-already-among." That "toward which" one is inclined is a let-ting oneself be attracted by the sort of thing in which the inclination loses itself. When Dasein, so to speak, sinks down into an inclination, then it is not just an inclination that is present, rather, the complete structure of care is modified. Blinded, it puts all possibilities in the service of the inclination.

On the other hand, the *urge* "to live" is a "toward" which brings its own drive along with it. It is "toward at any cost." Urge seeks to crowd out other possibilities. Here, too, being-ahead-of-oneself is inauthentic if one is invaded by an urge coming from the very thing that is urging one on. The urge can outrun one's actual attunement and understanding. But Dasein is not then—and never is—a "mere urge" to which other relations of dominating and leading are sometimes added, but as a modification of complete being-in-the-world, it is always already care.

In pure urge, care has not yet become free, although it first makes it ontologically possible for Dasein to be urged on by itself. On the other hand, in an inclination care is always already bound. Inclination and urge are possibilities rooted in the thrownness of Dasein. The urge "to live" is not to be destroyed; the inclination to be "lived" by the world is not to be eradicated. But because and only because they are ontologically based in care, both are to be modified ontically and existentielly by care as something authentic.

The expression "care" means an existential and basic ontological phenomenon which nevertheless is *not simple* in its structure. This ontologically elemental totality of the care structure cannot be reduced to an ontic "primal element," just as being certainly cannot be "explained" in terms of beings. Finally, we shall see that the idea of being in general is no more "simple" than the being of Dasein. The characterization of care as being-ahead-of-itself-in-already-being-in—as being-together-with—makes clear that this phenomenon, too, is still structurally *articulated* in itself. But is that not a phenomenal indication that the ontological question must be pursued still further until we can set forth a *still more primordial* phenomenon which ontologically supports the unity and totality of the structural manifold of care? Before we follow up this question, we need to look back and appropriate more precisely what has been interpreted up to now with the intention of seeing the fundamental ontological question of the meaning of being in general. But first we must show that what is ontologically "new" in this interpretation is ontically rather old. The explication of the being of Dasein as care does not force Dasein under a contrived idea, but brings us existentially to a concept which has already been disclosed ontically and existentielly.

§ 42. *Confirmation of the Existential Interpretation of Dasein as Care in Terms of the Pre-ontological Self-interpretation of Dasein*

In the foregoing interpretations, which finally led to exposing care as the being of Dasein, the most important thing was to arrive at the appropriate *ontological* foundations of the being which we ourselves actually are and which we call "human being." For this purpose, it was necessary from the outset to change the direction of our analysis from the approach presented by the traditional definition of human being, which is an approach that is ontologically unclarified and fundamentally questionable. In comparison with this definition, the existential and ontological interpretation might seem strange, especially if "care" is understood just ontically as "worry" and "troubles." Accordingly, we shall cite a document that is pre-ontological in character, even though its demonstrative power is "only historical."

Let us bear in mind, however, that in this document Dasein expresses itself about itself "primordially," unaffected by any theoretical interpretation and without aiming to propose any. Furthermore, let us observe that the being of Dasein is characterized by historicality, though this must be demonstrated ontologically. *If* Dasein is "historical" in the basis of its being, then a statement that comes from its history and goes back to it, and that is *prior* to any scientific knowledge, takes on a special importance which, however, is never purely ontological. The understanding of being which lies in Dasein itself expresses itself pre-ontologically. What is cited in the following document is to make clear the fact that our existential interpretation is not a mere fabrication, but as an ontological "construction" it is well grounded and has been sketched out beforehand in elemental ways.

The following self-interpretation of Dasein as "care" is preserved in an old fable:[5]

198

> Cura cum fluvium transiret, videt cretosum lutum sustulitque cogitabunda atque coepit fingere. dum deliberat quid iam fecisset. Jovis intervenit. rogat eum Cura ut det illi spiritum, et facile impetrat. cui cum vellet Cura nomen ex sese ipsa imponere, Jovis prohibuit suumque nomen ei dandum esse dictitat. dum Cura et Jovis disceptant, Tellus surrexit simul suumque nomen esse volt cui corpus praebuerit suum. sumpserunt Saturnum iudicem, is sic aecus iudicat: 'tu Jovis quia spiritum dedisti, in morte spiritum, tuque Tellus, quia dedisti corpus, corpus recipito, Cura enim quia prima finxit, teneat quamdiu vixerit. sed quae nunc de nomine eius vobis controversia est, homo vocetur, quia videtur esse factus ex humo.'

5. The author found the following pre-ontological evidence for the existential and ontological interpretation of Dasein as care in the essay of K. Burdach, "Faust und die Sorge," *Deutsche Vierteljahrsschrift für Literaturwissenschaft und Geistesgeschichte* I (1923): 1ff. Burdach shows that Goethe took the care fable, handed down as no. 220 of the fables of Hyginus, and reworked it for the second part of his *Faust*. Cf. particularly pp. 40ff. The above text is cited according to F. Bucheler, *Rheinisches Museum* 41 (1886): 5, translation Burdach, ibid., pp. 41ff.

Once when "Care" was crossing a river, she saw some clay; she thoughtfully took a piece and began to shape it. While she was thinking about what she had made, Jupiter came by. "Care" asked him to give it spirit, and this he gladly granted. But when she wanted her name to be bestowed upon it, Jupiter forbade this and demanded that it be given his name instead. While "Care" and Jupiter were arguing, Earth (Tellus) arose, and desired that her name be conferred upon the creature, since she had offered it part of her body. They asked Saturn to be the judge. And Saturn gave them the following decision, which seemed to be just: "Since you, Jupiter, have given its spirit, you should receive that spirit at death; and since you, Earth, have given its body, you shall receive its body. But since 'Care' first shaped this creature, she shall possess it as long as it lives. And because there is a dispute among you as to its name, let it be called 'homo,' for it is made out of humus (earth)."

This pre-ontological document becomes especially significant not only in that "care" is here seen as that to which human Dasein belongs "for its lifetime," but also because this priority of "care" emerges in connection with the familiar interpretation of human being as a compound of body (earth) and spirit. *Cura prima finxit.* This being has the "origin" of its being in care. *Cura teneat, quamdiu vixerit*: this being is not released from its origin, but is held fast and dominated by it, as long as this being "is in the world." "Being-in-the-world" has the character of being of "care." It does not get its name (*homo*) with regard to its being, but in relation to that of which it consists (*humus*). The decision as to wherein the "primordial" being of this creature is to be seen is left to Saturn, "time."[6] The pre-ontological characterization of the essence of human being expressed in this fable thus has envisaged from the very beginning *the* mode of being which rules its *temporal sojourn in the world.*

The history of the signification of the ontic concept of "*cura*" permits us to see still further fundamental structures of Dasein. Burdach[7] calls our attention to an ambiguity of the term "*cura*," according to which it means not only "anxious effort," but also "carefulness,"

199

---

6. Cf. Herder's poem "Das Kind der Sorge," *Suphan* 29: 75.

7. "Faust und die Sorge," p. 49. Already with the Stoics, μέριμνα was a stable term, and it comes back in the New Testament, in the Vulgate, as *sollicitudo.* The direction followed in our existential analytic of Dasein toward "care" occurred to the author in connection with attempts at an interpretation of Augustinian, that is, Greek and Christian, anthropology with regard to the basic foundations attained in the ontology of Aristotle.

"dedication." Thus Seneca writes in his last letter (*Ep.* 124): "Of the four existing natures (tree, animal, human being, God), the last two, which alone are endowed with reason, are distinguished in that God is immortal, human being mortal. The good of the One, namely of God, is fulfilled by its nature; but that of the other, human being, is fulfilled by *care* (*cura*): *unius bonum natura perficit, dei scilicet, alterius cura, hominis.*"

The *perfectio* of human being—becoming what one can be in being free for one's ownmost possibilities (project)—is an "accomplishment" of "care." But, equiprimordially, care determines the fundamental mode of this being according to which it is delivered over (thrownness) to the world taken care of. The "ambiguity" of "care" refers to a *single* basic constitution in its essentially twofold structure of thrown project.

As compared with the ontic interpretation, the existential and ontological interpretation is not only a theoretical and ontic generalization. That would only signify that ontically all the human being's behavior is "full of care" and guided by his "dedication" to something. The "generalization" is an *a priori-ontological* one. It does not refer to ontic qualities that constantly keep emerging, but to a constitution of being which always already is taken as a basis. This constitution first makes it ontologically possible that this being can be addressed ontically as *cura*. The existential condition of the possibility of "the cares of life" and "dedication" must be conceived in a primordial, that is, ontological sense as care.

The transcendental "universality" of the phenomenon of care and all fundamental existentials has, on the other hand, that broad scope through which the basis is given on which *every* ontic interpretation of Dasein with a worldview moves, whether it understands Dasein as "the cares of life" and need, or in an opposite manner.

The "emptiness" and "generality" of the existential structures which obtrude themselves ontically have their *own* ontological definiteness and fullness. The whole of the constitution of Dasein itself is not simple in its unity, but shows a structural articulation which is expressed in the existential concept of care.

Our ontological interpretation of Dasein has brought the pre-ontological self-interpretation of this being as "care" to the *existential concept* of care. The analytic of Dasein does not aim, however, at an ontological basis for anthropology; it has a fundamental, ontological goal. This goal has tacitly determined the course of our considerations, our choice of phenomena, and the limits to which our analysis may penetrate. With regard to our leading question of the meaning of being and its development, our inquiry must now, however, *explicitly* secure what has been gained so far. But something like this cannot be attained by an external synopsis of what has been discussed. Rather, what could only be roughly indicated at the beginning of the existential analytic must be sharpened to a more penetrating understanding of the problem with the help of what we have gained.

§ 43. *Dasein, Worldliness, and Reality*

The question of the meaning of being is possible at all only if something like an understanding of being *is*. An understanding of being belongs to the kind of being of that being which we call Dasein. The more appropriately and primordially we have succeeded in explicating this being, the surer we are to attain our goal in the further course of working out the problem of fundamental ontology.

While pursuing the tasks of a preparatory existential analytic of Dasein, we developed an interpretation of understanding, meaning, and interpretation. Our analysis of the disclosedness of Dasein showed furthermore that, with that disclosedness, Dasein is revealed equiprimordially in accordance with its fundamental constitution of being-in-the-world with regard to the world, being-in, and the self. Furthermore, in the factical disclosedness of world, innerworldly beings are also discovered. This means that the being of these beings is always already understood in a certain way, although not appropriately conceived ontologically. The pre-ontological understanding of being comprehends all beings which are essentially disclosed in Dasein, but the understanding of being itself has not yet articulated itself according to the various modes of being.

201

At the same time, our interpretation of understanding showed that, in accordance with its entangled kind of being, it has initially and for the most part transposed itself into an understanding of "world." Even when it is not only a matter of ontic experience, but of ontological understanding, the interpretation of being initially orients itself toward the being of innerworldly beings.* Here the being of things initially at hand is passed over, and beings are first conceived as a context of things (*res*) objectively present. *Being* acquires the meaning of *reality*.[8] Substantiality becomes the basic characteristic of being. Corresponding to this diversion in the understanding of being, even the ontological understanding of Dasein moves into the horizon of this concept of being. Like other beings, *Dasein* too is *objectively present as real*. Thus *being in general* acquires the meaning of *reality*.† Accordingly,

---

* Here we must differentiate: φύσις, ἰδέα, οὐσία, *substantia, res,* objectivity, objective presence.
† Reality as 'actuality' and *realitas* as 'whatness' ['Sachheit']. The middle position of Kant's concept of "objective reality."

8. Cf. above, pp. 89ff. and 99f.

the concept of reality has a peculiar priority in the ontological problematic. This priority diverts the path to a genuine existential analytic of Dasein; it also diverts our view of the being of innerworldly things initially at hand. Finally, it forces the problematic of being in general into a direction which lies off course. The other modes of being are defined negatively and privatively with regard to reality.

Therefore, not only the analytic of Dasein, but the development of the question of the meaning of being in general must be wrested from a one-sided orientation toward being in the sense of reality. We must demonstrate that reality is not only *one* kind of being *among* others, but stands ontologically in a definite foundational context with Dasein, world, and handiness. To demonstrate this, we must discuss in principle the *problem of reality*, its conditions and limitations.

Under the heading "the problem of reality" various questions are clustered: (1) whether the beings which are supposedly "transcendent to consciousness" *are* at all; (2) whether this reality of the "external world" can be sufficiently *proven*; (3) to what extent this being, if it is real, is to be known in its being-in-itself; (4) what the meaning of this being, reality, signifies in general. The following discussion of the problem of reality treats three things with regard to the question of fundamental ontology: (*a*) reality as a problem of being and the demonstrability of the "external world," (*b*) reality as an ontological problem, (*c*) reality and care.

202

(a) Reality as a Problem of Being and the Demonstrability of the "External World"

Of these questions enumerated about reality, the one which comes first is the ontological question of what reality signifies in general. However, as long as a pure ontological problematic and methodology was lacking, this question (if it was asked explicitly at all) was necessarily confounded with a discussion of the "problem of the external world"; for the analysis of reality is possible only on the basis of an appropriate access to what is real. But intuitive cognition has always been viewed as the way to grasp what is real. Intuitive cognition [anschauende Erkennen] "is" as a kind of behavior of the soul, of consciousness. Since the character of the in-itself and independence belongs to reality, the question of the possible independence "from consciousness" of what is real, or of the possible transcendence of consciousness in the "sphere" of what is real, is coupled with the question of the meaning of reality. The possibility of an adequate ontological analysis of reality depends on how far that *from which* there is independence, *what* is to be transcended, is *itself* clarified with regard to its *being*. Only in this way can the kind of being that belongs to transcendence be ontologically grasped. And, finally, the primary kind of access to what is real must be secured by deciding the question whether cognition can take over this function at all.

These inquiries, which *take precedence over* any possible ontological question about reality, have been carried out in the foregoing existential analytic. From this analytic we saw that cognition is a *founded* mode of access to what is real. The real is essentially accessible only as innerworldly being [Seiendes]. Every access to such beings is ontologically based on the fundamental constitution of Dasein, on being-in-the-world. This has the primordial constitution of being of care (being ahead of itself—already being in a world—as being together with innerworldly beings).

The question of whether there is a world at all and whether its being can be demonstrated, makes no sense at all if it is raised by Dasein as being-in-the-world—and who else should ask it? Moreover, it is encumbered with an ambiguity. World as the wherein of being-in, and "world" as innerworldly beings, that in which one is absorbed in taking care, are confused or else not distinguished at all. But world is essentially disclosed *with the being* of Dasein; "world" is always already also discovered with the disclosedness of world. Of course, innerworldly beings in the sense of what is real, as merely objectively present, can still remain covered over. But even what is real is discoverable only on the basis of a world already disclosed. And only on this basis can what is real still remain *concealed*. One asks the question about the "reality" of the "external world" without previously clarifying the *phenomenon of world* as such. Factically, the problem of the external *world* is constantly oriented toward innerworldly beings (things and objects). Thus these discussions drift into a problematic which ontologically can hardly be disentangled.

The complexity of these questions, and the confusion of what one would like to demonstrate with what is demonstrated and with what guides the demonstration, is shown in Kant's "Refutation of Idealism."[9] Kant calls it "a scandal of philosophy and human reason in general"[10] that there is still no cogent proof for "the existence of things outside us" which will do away with any skepticism. He himself proposes such a proof as the foundation of his "theorem" that "the mere, but empirically determined, consciousness of my own existence proves the existence of objects in space outside of me."[11]

First we must explicitly note that Kant uses the term "existence" ["Dasein"] to designate the kind of being which we have called

203

---

9. Cf. *Kritik der reinen Vernunft*, B 274ff., and the amended additions in the Preface to the second edition, p. xxxix; note, also, "On the Paralogisms of Pure Reason," pp. 399ff., especially p. 412.
10. Ibid., Preface, note.
11. Ibid., B 275.

"objective presence" in our present inquiry. "Consciousness of my existence" means for Kant consciousness of my being present in the sense of Descartes. The term "existence" means both the presence of consciousness and the presence of things.

The proof for the "existence of things outside of me" is supported by the fact that change and persistence belong equiprimordially to the nature of time. My presence, that is, the presence given in the inner sense of a manifold of representations, is change that is present. But the definiteness of time presupposes something present which persists. This, however, cannot be "in us," "because precisely my existence in time can first be determined by this persisting thing."[12] With the change present "in me," which is posited empirically, a present thing which persists "outside of me" is also posited. This persisting thing is the condition of the possibility of the presence of change "in me." The experience of the being-in-time of representations equiprimordially posits changing things "in me" and persisting things "outside of me."

Of course, this proof is not a causal inference and, accordingly, not burdened with the prejudices of such proof. Kant gives, so to speak, an "ontological proof" in terms of the idea of temporal beings. At first, it appears as if Kant has abandoned the Cartesian position of the prediscovered isolated subject. But that is only illusion. The fact that Kant requires any proof at all for the "existence of things outside of me" already shows that he takes the subject, the "in me," as the starting point for this problematic. The proof itself is then carried out by departing from the empirically given change *"in me."* For only "in me" is "time" experienced, and time carries the burden of the proof. It provides the foundation for leaping into the "outside of me" in the course of the proof. Moreover, Kant emphasizes the fact that "the problematic kind of [Idealism] which . . . only alleges our inability to prove an existence outside of our own by immediate experience is reasonable and in accordance with a fundamental, philosophical way of thinking, namely, before a sufficient proof has been found, never to permit a decisive judgment."[13]

But even if the ontic priority of the isolated subject and of inner experience were given up, ontologically the position of Descartes would, after all, be retained. What Kant proves—if we admit that his proof and its basis are correct at all—is that beings that are changing and beings that are permanent are necessarily present together. But

204

---

12. Ibid., B 275.
13. Ibid., B 274/275.

ordering two objectively present things on the same level does not as yet mean that subject and object are objectively present together. And even if this were proven, what is ontologically decisive would still remain covered over: the fundamental constitution of the "subject," of Dasein, as being-in-the-world. *The being present together of the physical and the psychical is ontically and ontologically completely different from the phenomenon of being-in-the-world.*

Kant presupposes the difference *and the connection* of the "in me" and "outside of me"—factically he is justified, but from the sense of his proof without justification. It has not been proven that whatever is decided about the being present together of what changes and what persists when one takes time as a guideline also applies to the connection between the "in me" and the "outside of me." But if the whole of the difference and connection of the "inside" and "outside" presupposed in the proof were seen, if what is presupposed with this presupposition were ontologically understood, then the possibility of believing that a proof for the "existence of things outside of me" was still lacking and necessarily would collapse.

205

The "scandal of philosophy" does not consist in the fact that this proof is still lacking up to now, but *in the fact that such proofs are expected and attempted again and again.* Such expectations, intentions, and demands grow out of an ontologically insufficient way of positing *that from which,* independently and "outside" of which, a "world" is to be proven as objectively present. It is not that the proofs are insufficient, but the kind of being of the being that does the proving and requests proofs is *not defined enough.* For this reason the illusion can arise that, with this demonstration of the necessary objective presence together of two objectively present things, something is proved or even able to be proved about Dasein as being-in-the-world. Correctly understood, Dasein defies such proofs, because it always already *is* in its being what the later proofs first deem necessary to demonstrate for it.

If one wanted to conclude from the impossibility of the proofs for the objective presence of things outside of us that this is thus "merely to be accepted on *faith*,"[14] this distortion [Verkehrung] of the problem would not be overcome. The preconceived opinion would persist that basically and ideally a proof must be possible. The inappropriate way of approaching the problem is still endorsed when one confines oneself to a "faith in the reality of the external world," even if this faith is explicitly "acknowledged." Although one is not offering a stringent

---

14. Ibid., Preface, note.

proof, one is still in principle demanding a proof and trying to satisfy that demand.[15]

206   Even if one wanted to fall back on the fact that the subject must presuppose, and indeed always already does unconsciously presuppose, the fact that the "external world" is objectively present, one would still be starting with the construct of an isolated subject. The phenomenon of being-in-the-world would no more be met with than it would be by demonstrating that the physical and the psychical are objectively present together. With such presuppositions, Dasein always already comes "too late"; for insofar as it carries out [vollzieht] this presupposing as a being [Seiendes] (and this would not be possible otherwise), it is, *as a being* always already in a world. "Earlier" than any presupposition that Dasein makes, or any of its ways of behavior, there is the *"a priori"* of its constitution of being in the mode of being of care.

Faith in the reality of the "external world," whether justified or not; *proving* this reality, whether sufficiently or insufficiently; *presupposing* it, whether explicitly or not—such attempts that have not mastered their own basis with complete transparency presuppose a subject which is initially *worldless*; they presuppose a subject which is not certain of its world, and which basically must first make itself certain of a world. Thus, being-in-the-world is from the very beginning geared to interpreting, opining, being certain, and having faith, a kind of behavior which is in itself always already a founded mode of being-in-the-world.

The "problem of reality" in the sense of the question whether an external world is objectively present or demonstrable, turns out to be an impossible one, not because its consequences lead to inextricable impasses, but because the very being which serves as its theme repudiates such a line of questioning, so to speak. It is not a matter of proving that and how an "external world" is objectively present, but of demonstrating why Dasein as being-in-the-world has the tendency of "initially" burying the "external world" in nullity "epistemologically" in order to then resurrect it through proofs. The reason for this lies in the falling prey of Dasein and in the diversion motivated therein of the primary understanding of being to the being of objective presence. If the line of questioning in this ontological orientation is "critical," it finds something merely "inner" as what is objectively present and alone certain. After the primordial phenomenon of being-in-the-

---

15. Cf. W. Dilthey, "Beiträge zur Lösung der Frage vom Ursprung unseres Glaubens an die Realität der Außenwelt und seinem Recht" (1890), *Ges. Schr.* 5, 1, pp. 90ff. Right at the beginning of this treatise, Dilthey says unmistakably: "For if there is to be a universal truth for man, thinking must, following the method first given by Descartes, clear a path from the facts of consciousness toward outer reality," ibid., p. 90.

world has been shattered, the isolated subject is all that remains and it becomes the basis that is then joined together with a "world."

The multiplicity of attempts at a solution of the "problem of reality" developed through the various kinds of realism and idealism, and in the positions which mediate between them, cannot be discussed in this inquiry at any great length. Certainly there is a core of genuine understanding to be found in each of these solutions, but it would be wrong if one wanted to achieve a tenable solution to the problem by counting up how much is correct in each case. Rather, what is needed is the basic insight that the various epistemological directions do not so much go off the track epistemologically, but that, because they neglect the existential analytic of Dasein in general, they do not even attain the basis for a phenomenally secured problematic. Nor is this *basis* to be attained by subsequent phenomenological improvements of the concept of the subject and consciousness.* Such a procedure would not guarantee that the inappropriate *line of questioning* would not, after all, remain.

207

With Dasein as being-in-the-world, innerworldly beings have already been disclosed. This existential and ontological statement seems to agree with the thesis of *realism* that the external world is objectively present in a real way. Since the objective presence of innerworldly beings is not denied in this existential statement, it agrees in its result, so to speak, doxographically, with the thesis of realism. But it is distinguished in principle from all realism in that realism believes that the reality of the "world" needs proof, and at the same time is capable of proof. Both views are directly negated in the existential statement. But what completely separates it from realism is the lack of ontological comprehension in realism. After all, it tries to explain reality ontically by real connections of interaction between real things.

As opposed to realism, *idealism*, no matter how contrary† and untenable it might be, has a fundamental priority, if it does not misunderstand itself as "psychological" idealism. If idealism emphasizes the fact that being and reality are only "in consciousness," this expresses the understanding that being cannot be explained by beings. But to the extent that it remains unclarified *that* an understanding of being occurs here and *what* this understanding of being means ontologically, how it is possible, and that it belongs to the constitution of being of Dasein,‡ idealism constructs the interpretation of reality in a vacuum. The fact that being cannot be explained by beings, and that reality is only possible in the understanding of being, does not absolve us from

---

* Leap into Da-sein.
† Namely, to existential and ontological experience.
‡ And Dasein belongs to the essence of being as such.

asking about the being of consciousness, of the *res cogitans* itself. If the idealist thesis is to be followed consistently, the ontological analysis of consciousness is prescribed as an inevitable prior task. Only because being is "in consciousness," that is, intelligible in Dasein, can Dasein also understand and conceptualize characteristics of being such as independence, "in itself," reality in general. Only for that reason are "independent" beings accessible to circumspection as encountered in the world.

208

If the term idealism amounts to an understanding of the fact that being is never explicable* by beings, but is always already the "transcendental" for every being, then the sole correct possibility of a philosophical problematic lies in idealism. Then Aristotle was no less of an idealist than Kant. If idealism means the reduction of all beings to a subject or a consciousness which are only distinguished by the fact that they remain *undetermined* in their being and are characterized at best negatively as "unthinglike," then this idealism is methodologically no less naïve than the grossest realism.

It is still possible that one may give the problematic of reality *priority* over any orientation in terms of "standpoints" by maintaining the thesis that every subject is what it is only for an object and vice versa. But with this formal approach the terms of the correlation, like the correlation itself, remain undetermined ontologically. But at bottom the whole correlation is necessarily thought as "somehow" *existent* [*seiend*] and thus that it must be thought with regard to a definite idea of being. Of course, if the existential and ontological basis is secured beforehand with the evidence of being-in-the-world, this correlation can be known subsequently as a formalized, ontologically indifferent relation.

Our discussion of the unexpressed presuppositions of efforts to solve the problem of reality in ways which are merely "epistemological" shows that this problem must be taken back into the existential analytic of Dasein as an ontological problem.[16]

---

* Ontological difference.

16. Nicolai Hartmann, following the procedure of Scheler, recently used the thesis of knowledge as a "relation of being" as the foundation of his ontologically oriented epistemology. Cf. *Grundzüge einer Metaphysik der Erkenntnis*, second enlarged edition, 1925. But Scheler and Hartmann both fail in the same way, in spite of all the differences between their phenomenological point of departure, to recognize that "ontology" fails in its traditional, basic orientation with regard to Dasein and that precisely the "relation of being" (cf. above, pp. 79ff.) contained in knowledge forces us to its *fundamental* revision, not just to a critical improvement. Because Hartmann underestimates the inexplicit scope of influence of an ontologically unclarified positing of the relation of being, he is forced into a "critical realism" which is basically completely foreign to the level of problematic put forth by him. For Hartmann's interpretation of ontology, cf. "Wie ist kritische Ontologie überhaupt möglich?" in the *Festschrift für Paul Natorp*, 1924, p. 124ff.

(b) Reality as an Ontological Problem

If the term reality* refers to the being of innerworldly beings (*res*) objectively present (and nothing else is understood by this), that means for the analysis of this mode of being that *innerworldly* beings are ontologically to be comprehended only when the phenomenon of innerworldliness has been clarified. But innerworldliness is based on the phenomenon of *world*, which in turn belongs to the fundamental constitution of Dasein as an essential structural factor of being-in-the-world. Again, being-in-the-world is ontologically bound up with the structural totality of the being of Dasein which we characterized as care. But thus we have characterized the foundations and the horizons that must be clarified if an analysis of reality is to be possible. In this connection the character of the in-itself first becomes ontologically intelligible. By taking our orientation toward this context of problems, we have interpreted the being of innerworldly beings in our earlier analyses.[17]

To be sure, within certain limits, a phenomenological characterization of the reality of what is real can already be given without an explicit existential and ontological basis. Dilthey tried this in the treatise which we mentioned above. What is real is experienced in impulse and will. Reality is *resistance*, more precisely, the character of resisting. The analytic elaboration of the phenomenon of resistance is what is positive in Dilthey's treatise, and is the best concrete substantiation of his idea of a "descriptive and analytic psychology." But he is kept from correctly working out the analysis of the phenomenon of resistance by the epistemological problematic of reality. The "principle of phenomenality" does not let Dilthey arrive at an ontological interpretation of the being of consciousness. "The will and its inhibition emerge within the same consciousness."[18] What kind of being belongs to this "emerging"? What is the meaning of the being of the "within"? What relation of being does consciousness bear to what is real itself? All this needs an ontological determination. That this was not done can be finally explained by the fact that Dilthey left "life"—"behind" which one of course cannot go—standing in such a way that it is ontologically undifferentiated. However, an ontological interpretation of Dasein does not mean that we must go back ontically to some other being [Seiendes].

---

* Not reality as factuality [Sachheit].

17. Cf., above all, § 16: The Worldliness of the Surrounding World Announcing Itself in Innerworldly Beings; § 18: Relevance and Significance: The Worldliness of the World; § 29: Dasein as Attunement—On the Being-in-itself of Innerworldly Beings, cf. 101ff.
18. Cf. Dilthey, "Beiträge," p. 134.

210   The fact that Dilthey was epistemologically refuted cannot prevent us from making fruitful use of what is positive in his analyses, which is precisely what has not been understood in these refutations.

Thus recently Scheler took up Dilthey's interpretation of reality.[19] He champions a "voluntative theory of Dasein." Here Dasein is understood in the Kantian sense as objective presence. The "being of objects is given immediately only in relation to drive and will." Scheler, like Dilthey, not only emphasizes the fact that reality is never primarily given in thinking and grasping, above all he also refers to the fact that knowledge itself is, again, not judgment and that knowing is a "relation of being."

Fundamentally, what we have already said about the ontological indefiniteness of Dilthey's foundations is valid for this theory too. Nor can the fundamental ontological analysis of "life" be inserted afterwards as a support. It bears and conditions the analysis of reality, the full explication of resistance and its phenomenal presuppositions. We encounter resistance in not-getting-through, as an obstacle to wanting-to-get-through. But with this willing, something must already have been disclosed, something which drive and will are *out to get*. The ontic indefiniteness of what they are out to get must not, however, be overlooked ontologically or, for that matter, be understood as if it were nothing. Being out to get . . . , which comes up against resistance and must "come up against it," is itself already *together with* a totality of relevance. But the discoveredness of that totality is grounded in the disclosedness of the referential totality of significance. *The experience of resistance, that is, the discovery of resistance in striving, is ontologically possible only on the basis of the disclosedness of world.* Resistance characterizes the being of innerworldly beings. Experiences of resistance factically determine only the extent and direction in which beings encountered within the world are discovered. Their sum does not first introduce the disclosure of world, but presupposes it. The "against" and the "counter to" are supported in their ontological possibility by disclosed being-in-the-world.

211   Nor is resistance experienced in a drive or a will "emerging" in its own right. These turn out to be modifications of care. Only beings with this kind of being are able to run up against something resistant in the world. Thus, if reality is defined by resistance, we must consider two things. On the one hand, we have only gotten at one characteristic

---

19. Cf. *Die Formen des Wissens und die Bildung*, lecture, 1925, notes 24 and 25. A remark made when correcting the galleys: Scheler has now published his inquiry, long since announced, on "Erkenntnis und Arbeit" in the collection of treatises just published, *Die Wissensformen und die Gesellschaft*, 1926. Section VI of this treatise (p. 455) gives a more detailed presentation of the "voluntative theory of Dasein" in connection with an appreciation and critique of Dilthey.

of reality among others, and on the other hand the already disclosed world is necessarily presupposed for resistance. Resistance characterizes the "external world" in the sense of innerworldly beings, but never in the sense of world. *"Consciousness of reality" is itself a way of being-in-the-world.* Every "problematic of the external world" necessarily goes back to this basic existential phenomenon.

If the *"cogito sum"* is to serve as the point of departure for the existential analytic, we not only need to turn it around, but we need a new ontological and phenomenal confirmation of its content. Then the first statement is *"sum,"* in the sense of I-am-in-a-world. As such a being, "I am" ["bin ich"] in the possibility of being toward various modes of behavior (*cogitationes*) as ways of being together with innerworldly beings. In contrast, Descartes says that *cogitationes* are indeed objectively present and an ego is also objectively present as a worldless *res cogitans.*

(c) Reality and Care

As an ontological term, reality is related to innerworldly beings. If it serves to designate this kind of being in general, then handiness and objective presence function as modes of reality. But if one lets this world keep its traditional* meaning, it means being in the sense of the sheer objective presence of things. But not all objective presence is the objective presence of things. "Nature," which "surrounds" us, is indeed an innerworldly being, but it shows neither the kind of being of handiness, nor of objective presence as "natural things." In whatever way one interprets this being of "nature," *all* modes of being of innerworldly beings are ontologically founded in the worldliness of the world, and thus in the phenomenon of being-in-the-world. From this there arises the insight that neither does reality have priority within the modes of being of innerworldly beings, nor can this mode of being even characterize something like world and Dasein in an ontologically adequate way.

Reality is *referred back to the phenomenon of care* in the order of ontological foundational contexts and possible categorial and existential demonstration. The fact that reality is ontologically grounded in the being of Dasein cannot mean that something real can only be what it is in itself when and as long as Dasein exists.   212

However, only as long as Dasein *is,* that is, as long as there is the ontic possibility of an understanding of being, "is there" [*gibt es*] being [*Sein*]. If Dasein does not exist, then there "is" no "independence" either, nor "is" there an "in itself." Such matters are then neither

---

* Prevalent today.

comprehensible nor incomprehensible. If Dasein does not exist then innerworldly beings, too, can neither be discovered, nor can they lie in concealment. *Then* it can neither be said that beings are, nor that they are not. It can *now* indeed be said that as long as there is an understanding of being and thus an understanding of objective presence, that *then* beings will still continue to be.

As we have noted, being (not beings) is dependent upon the understanding of being; that is, reality (not the real) is dependent upon care. This dependency protects our further analytic of Dasein from an uncritical interpretation of Dasein constantly intruding itself—an interpretation that follows the guideline of the idea of reality. Only the orientation toward existentiality which is interpreted in an ontologically *positive* way can guarantee that, in the factical course of the analysis of "consciousness," of "life," some meaning of reality, even if it is undifferentiated, is not taken as a ground [zugrundegelegt].

The fact that beings having the kind of being of Dasein cannot be comprehended in terms of reality and substantiality has been expressed by the thesis that the *substance of human being* [*Substanz des Menschen*] *is existence* [*Existenz*]. Interpreting existentiality as care and distinguishing it from reality do not, however, signal the end of the existential analytic, but only lets the maze of problems in the question of being and its possible modes, and the meaning of such modifications, emerge more sharply. Only if an understanding of being *is*, are beings accessible as beings; only if beings of the kind of being of Dasein are, is an understanding of being as beings [Seinsverständnis als Seiendes] possible.

§ 44. *Dasein, Disclosedness, and Truth*

From time immemorial, philosophy has associated* truth with being. The first discovery of the being of beings by Parmenides "identifies" being [Sein] with the perceptive understanding of being [Sein]: τὸ γὰρ αὐτὸ νοεῖν ἐστίν τε καὶ εἶναι.[20] In his sketch of the history of the discovery of ἀρχαί,[21] Aristotle emphasizes the fact that the philosophers before him were led by "the things themselves" to question further: αὐτὸ τὸ πρᾶγμα ὡδοποίησεν αὐτοῖς καὶ συνηνάγκασε ζητεῖν.[22] He also characterizes the same fact with the words: ἀναγκαζόμενος δ' ἀκολουθεῖν τοῖς φαινομένοις;[23] he (Parmenides) was compelled to

213

---

* φύσις is intrinsically ἀλήθεια, since κρύπτεσθαι φιλεῖ.

20. Diels, *Fragmente der Vorsokratiker*, fragment 3.
21. *Metaphysics* I.
22. Ibid., 984a18ff.
23. Ibid., 986b31.

follow what showed itself in itself. In another passage he says: ὑπ᾽ αὐτῆς τῆς ἀληθείας ἀναγκαζόμενοι,[24] compelled by "truth" itself, they carried out their investigations. Aristotle designates this inquiry as φιλοσοφεῖν περὶ τῆς ἀληθείας,[25] "philosophizing" about the "truth" or even as ἀποφαίνεσθαι περὶ τῆς ἀληθείας,[26] as demonstrating something and letting it be seen with regard to the "truth" and in the scope of "truth." Philosophy itself is defined as ἐπιστήμη τις τῆς ἀληθείας,[27] the science of "truth." But at the same time it is characterized as an ἐπιστήμη, ἣ θεωρεῖ τὸ ὄν ἧ ὄν,[28] as the science that considers beings as beings, that is, with regard to their being [Sein].

What does it mean to speak of "inquiring into 'truth,' " or the science of "truth"? Is "truth" made thematic in this inquiry in the sense of a theory of knowledge or of judgment? Obviously not, for "truth" means the same thing as the "matter" ["Sache"], "what shows itself." But then what does the expression "truth" mean if it can be used as a term for "beings" ["Seiendes"] and "being" ["Sein"]?

But if *truth* rightfully has a primordial connection with *being*, then the phenomenon of truth moves into the orbit* of the problematic of fundamental ontology. But must not this phenomenon have been encountered already within our preparatory fundamental analysis, the analytic of Dasein? What ontic-ontological connection does "truth" have with Dasein and with its ontic characteristic which we call the understanding of being? Can the reason why being necessarily goes together with truth and vice versa be pointed out in terms of this understanding?

These questions cannot be avoided. Because being in fact does "go together" with truth, the phenomenon of truth has already been one of the themes of our earlier analysis, although not explicitly under this name. Now we must explicitly delimit the phenomenon of truth, giving precision to the problem of being and pinpointing the problems contained therein. In doing this, we shall not simply summarize what we have said previously. The investigation takes a new point of departure.†

214

Our analysis starts from (a) the *traditional concept of truth* and attempts to lay bare its ontological foundations. In terms of these foundations the *primordial* phenomenon of truth becomes visible. On the basis of this, (b) the *derivative character* [*Abkünftigkeit*] of the traditional

---

* Not only, but into the *middle*.
† This is the real place to begin the leap into Da-sein.

24. Ibid., 984b10.
25. Ibid., A, 983b2, cf. 988a20.
26. Ibid., 993b17.
27. Ibid., 993b20.
28. Ibid., 1003a21.

concept of truth can be indicated. Our investigation makes clear that the question of the *kind of being* of truth also necessarily belongs to the question of the "essence" of truth. Together with this we must (c) clarify the ontological meaning of saying that "there is truth," and also clarify the kind of necessity with which "we must presuppose" that there "is" ["gibt"] truth.

(a) The Traditional Concept of Truth and Its Ontological Foundations

Three theses characterize the traditional interpretation of the essence of truth and the way it is supposed to have been first defined:

1. The "locus" of truth is the proposition [Aussage] (judgment).
2. The essence of truth lies in the "agreement" of the judgment with its object.
3. Aristotle, the father of logic, attributed truth to judgment as its primordial locus; he also initiated the definition of truth as "agreement."

A history of the concept of truth, which could only be presented on the basis of a history of ontology, is not intended here. A few characteristic references to familiar matters may serve to introduce the analytical discussions.

Aristotle says: παθήματα τῆς ψυχῆς τῶν πραγμάτων ὁμοιώματα,[29] the "experiences" of the soul, the νοήματα ("representations"), are correspondences [Angleichungen] to things. This assertion, which is by no means presented as an explicit definition of the essence of truth, also became the occasion for the development of the later formulation of the essence of truth as *adaequatio intellectus et rei*. Thomas Aquinas,[30] who refers this definition to Avicenna (who, in turn, adopted it from Isaac Israeli's *Book of Definitions* [tenth century]), also uses the terms *correspondentia* (correspondence) and *convenientia* (coming together) for *adaequatio* (agreement).

215   The neo-Kantian epistemology of the nineteenth century frequently characterized this definition of truth as an expression of a methodologically undeveloped naïve realism, and declared it to be incommensurate with any formulation of the question of truth which had gone through Kant's "Copernican revolution." But Kant too held onto this concept of truth, so much so that he did not even bring it up as a matter for discussion. This has been overlooked, even though Brentano already called our attention to it. Thus, Kant says: "The old

29. *De interpretatione* 1.16a6.
30. Cf. *Quaest. disp. de veritate*, qu. I, art. 1.

and celebrated question with which it was supposed that one might drive the logicians into a corner is this: *'what is truth?'* The explanation of the name of truth—namely, that it is the agreement of knowledge with its object—will be here granted and presupposed."[31]

"If truth consists in the agreement of knowledge with its object, then this object must be distinguished from others; for knowledge is false if it does not agree with the object to which it is related, even if it should contain something which might well be valid for other objects."[32] And in the introduction to the transcendental dialectic Kant says: "Truth and illusion are not in the object so far as it is intuited, but in the judgment about it so far as it is thought."[33]

Of course the characterization of truth as "agreement," *adaequatio,* ὁμοίωσις, is very general and empty. But it will still have some justification if it can hold its own irrespective of the interpretation that this distinctive predicate "knowledge" will support. We now ask about the foundations of this "relation." *What is tacitly co-posited in the relational totality*—adaequatio intellectus et rei? *What ontological character does what is co-posited itself have?*

What does the term "agreement" ["Übereinstimmung"] mean in general? The agreement of something with something has the formal character of the relation [Beziehung] of something to something. Every agreement, and thus "truth" as well, is a relation. But not every relation is an agreement. A sign points to what is shown. Showing is a relation, but not an agreement between the sign and what is shown. But obviously every agreement does not mean something like the *convenientia* laid down in the definition of truth. The number 6 agrees with 16 minus 10. These numbers agree; they are equal with regard to the question of how much. Equivalence [Gleichheit] is *one* kind of agreement. Something like a "with respect to" belongs to it structurally. What is that with respect to which what is related in the *adaequatio* agrees? In clarifying the "truth relation" we must also take into account what is peculiar to the terms of this relation. With respect to what do *intellectus* and *res* agree? In their kind of being and essential content do they supply anything at all with respect to which they can agree? If it is impossible for *intellectus* and *res* to be equivalent because they are not of the same species, are they then perhaps similar? But knowledge is supposed to "give" the matter *just as* it is. "Agreement" has the relational character of "just as." In what way is this relation possible, as a relation between *intellectus* and *res*? From these questions it becomes clear that it is not sufficient for the clarification of the structure of truth simply to presuppose this

216

---

31. *Kritik der reinen Vernunft,* B82.
32. Ibid., p. 83.
33. Ibid., p. 250.

relational totality; rather we must go back and ask about the context of being which supports this totality as such.

Do we need for this purpose to unfold the "epistemological" problematic with regard to the subject-object relation, or can the analysis limit itself to an interpretation of the "immanent consciousness of truth," thus remaining "within the sphere" of the subject? According to general opinion, what is true is knowledge. But knowledge is judging. In judging, one must distinguish between judging as a *real* psychical procedure and what is judged as an *ideal* content. It is of the latter that we say it is "true." In contrast, the real psychical procedure is either objectively present or not. Accordingly, the ideal content of judgment stands in a relation of agreement. Thus this relation pertains to a connection between an ideal content of judgment and the real thing as that *about* which one judges. Is agreement real or ideal in its kind of being, or neither of the two? *How should the relation between an ideal being and a real thing objectively present be grasped ontologically?* This relation does, after all, subsist [besteht] and it subsists in factical judging not only between the content of judgment and the real object, but rather at the same time between the ideal content and the real act of judgment. Does it subsist still more inwardly here?

Or are we not allowed to ask about the ontological meaning of the relation between the real and the ideal (μέθεξις)? The relation is supposed to *subsist*. What does subsistence mean ontologically?

217　Why should this not be a legitimate question? Is it a matter of chance that this problem has not made any headway for more than two thousand years? Does the distortion [Verkehrung] of the question already lie in the beginning, in the ontologically unclarified separation of the real and the ideal?

And is not the separation of the real act [Vollzug] and the ideal content thoroughly illegitimate with regard to the "actual" judging of what is judged? Is not the reality of knowing and judging sundered into two kinds of being, two "levels," that can never be pieced together so as to get at the kind of being of knowing? Is not psychologism correct in rejecting this separation, even if it neither clarifies ontologically the kind of being that belongs to the thinking of what is thought, nor even recognizes it as a problem?

If we go back to the separation between the act of judgment and its content, we shall not further our discussion of the kind of being that belongs to the *adaequatio*, but only make plain the indispensability of clarifying the kind of being of knowing itself. The analysis necessary for this must attempt to bring to view the phenomenon of the truth that characterizes knowledge. When does truth become phenomenally explicit in knowing itself? When knowing shows itself [sich ausweist] *as true.* This self-demonstration [Selbstausweisung] assures it of its truth. Thus the relation of agreement must become visible in the phenomenal context of demonstration.

Let someone make the true statement with his back to the wall: "The picture on the wall is hanging crookedly." This statement demonstrates itself when the speaker turns around and perceives the picture hanging crookedly on the wall. What is shown in this demonstration? What is the meaning of confirming [Bewährung] this statement? Do we perhaps establish an agreement between "knowledge" or "what is known" with the thing on the wall? Yes and no, depending on whether our interpretation of the expression "what is known" is phenomenally adequate. To what is the speaker related when he judges without perceiving the picture, but "only representing" it? Possibly to "representations"? Certainly not, if representation is supposed to mean here representing as a psychical event. Nor is he related to representations in the sense of what is represented, if we mean by that a "picture" of the real thing on the wall. Rather, the statement that is "only representing" is, in accordance with its ownmost meaning, related to the real picture on the wall. What one has in mind is the real picture, and nothing else. Any interpretation that inserts something else here as what one has in mind in a statement that merely represents falsifies the phenomenal state of affairs about which a statement is made. Making statements is a being toward the existent thing itself. And what is demonstrated by perception? Nothing else than *that* this being *is* the very being [Seiende] that was meant in the statement. What comes to be demonstrated is that the expressive being [aussagende Sein] toward that which has been spoken about is a pointing out of the being; *that* it *reveals* [*entdeckt*] the being toward which it is. What gets demonstrated is the being-revealing [Entdeckend-sein] of the statement. What is to be confirmed is *that* it *discovers* the being toward which it is. What is demonstrated is the discovering-being of the statement. Here knowing remains related solely to the being [Seiende] itself in the act of demonstration. It is in this being that the confirmation takes place. The being that one has in mind shows itself *as* it is in itself, that is, it shows that *it*, in its selfsameness, is just as *it* is discovered or pointed out in the statement. Representations are not compared, neither among themselves nor in *relation* to the real thing. What is to be demonstrated is not an agreement of knowing with its object, still less something psychical with something physical, but neither is it an agreement between the "contents of consciousness" among themselves. What is to be demonstrated is solely the being-discovered of the being itself, *that being* in the "How" of its discoveredness. This is confirmed by the fact that what is stated (that is, the being itself) shows itself *as the very same thing. Confirmation* means *the being's showing itself in its self-sameness.*[34] Confirmation is accomplished on the basis of the being's showing itself. That is possible only in that the knowing that asserts and is confirmed is itself a *discovering being toward* real beings in its ontological meaning.

To say that a statement *is* true means that it discovers the being in itself. It asserts, it shows, it lets beings "be seen" (ἀπόφανσις) in their discoveredness. The *being-true* [*Wahrsein*] (*truth*) of the statement must be understood as *discovering* [*entdeckend-sein*]. Thus, truth by no means has the structure of an agreement between knowing and the object in the sense of a correspondence of one being (subject) to another (object).

219

Being-true as discovering is, in turn, ontologically possible only on the basis of being-in-the-world. This phenomenon, in which we recognized a basic constitution of Dasein, is the *foundation* [*Fundament*] of the primordial phenomenon of truth. This is now to be followed up in a more penetrating manner.

(b) The Primordial Phenomenon of Truth and the Derivative Character of the Traditional Concept of Truth

Being-true (truth) means to-be-discovering [entdeckend-sein]. But is this not a highly arbitrary definition of truth? With such violent definitions of the concept we might succeed in eliminating the idea of agreement from the concept of truth. Must we not pay for this dubious gain by letting the "good" old tradition fall into nothingness? However, this seemingly *arbitrary* definition contains only the *necessary* interpretation of what the oldest tradition of ancient philosophy primordially anticipated [ahnte] and even understood in a pre-phenomenological way. The being-true of the λόγος as ἀπόφανσις is the ἀληθεύειν in the manner of ἀποφαίνεσθαι: to let beings be seen in their unconcealment (discoveredness), taking them out of their concealment. The ἀλήθεια which is equated by Aristotle with πρᾶγμα and φαινόμενα in the passages cited above refers to the "things themselves," that which shows itself, *beings in the how of their discoveredness.* And is it a coincidence

---

34. For the idea of demonstration as "identification," cf. Husserl's *Logische Untersuchungen,* vol. 2, part 2, Investigation 6. On "Evidence and Truth," see ibid., § 36–39, p. 115ff. The usual treatments of the *phenomenological* theory of truth are limited to what is said in the *critical* prolegomena (vol. 1) and they note the connection with Bolzano's theory of the proposition. In contrast, the *positive* phenomenological interpretations, which are quite different from Bolzano's theory, are left alone. The only person who took up these investigations in a positive sense was E. Lask whose *Logik der Philosophie* (1911) is influenced just as strongly by the Sixth Investigation "Über sinnliche und kategoriale Anschauungen," p. 128ff.) as his *Lehre vom Urteil* (1912) was by the sections cited on evidence and truth.

that in one of the fragments of Heraclitus[35]—the *oldest* fragments of philosophical doctrine which *explicitly* treat the λόγος—the phenomenon of truth in the sense of discoveredness (unconcealment), as we have set it forth, shows through? Those who do not understand are contrasted with the λόγος and with one who speaks the λόγος and understands it. The λόγος is φράζων ὅκως ἔχει, it tells how beings comport themselves. In contrast to those who do not understand, what they do remains in concealment, λανθάνει; they forget (ἐπιλανθάνονται), that is, for them it sinks back into concealment. Thus unconcealment, ά-λήθεια, belongs to the λόγος. To translate this word as "truth," and especially to define this expression conceptually in theoretical ways, is to cover over the meaning of what the Greeks posited at the basis—as "self-evident" and as pre-philosophical—of the terminological use of ἀλήθεια.

In citing such evidence we must guard against uninhibited word-mysticism. Nevertheless, in the end, it is the business of philosophy to preserve the *power of the most elemental words* in which Dasein expresses itself and to protect them from being flattened by the common understanding to the point of unintelligibility, which in its turn functions as a source of illusory problems.

What we stated earlier,[36] in a more or less dogmatic interpretation, about λόγος and ἀλήθεια has now received its phenomenal demonstration. The "definition" of truth that we have proposed does not *shake off* the tradition, but is rather its primordial *appropriation*. This will be even more the case if we succeed in demonstrating whether and how theory had to arrive at the idea of agreement on the basis of the primordial phenomenon of truth.

Nor is the "definition" of truth as disclosedness and disclosing a mere explanation of words; rather, it grows out of the analysis of the relations of Dasein which we are initially accustomed to call "true."

Being-true as discovering is a manner of being of Dasein. What makes this discovering itself possible must necessarily be called "true" in a still more primordial sense. *The existential and ontological foundations of discovering itself first show the most primordial phenomenon of truth.*

Discovering is a way of being of being-in-the-world. Taking care, whether in circumspection or in looking in a leisurely way, discovers innerworldly beings. The latter become what is discovered. They are "true" in a secondary sense. Primarily "true," that is, discovering, is Dasein. Truth in the secondary sense does not mean to be discovering [Entdeckend-sein] (discovery), but to be discovered [Endeckt-sein] (discoveredness).

---

35. Cf. Diels, *Fragmente der Vorsokratiker*, Heraclitus, fragment 1.
36. Cf. pp. H 32ff.

But we showed in our earlier analysis of the worldliness of the world and of innerworldly beings that the discoveredness of inner-worldly beings is *grounded* in the disclosedness of the world. How-ever, disclosedness is the basic character of Dasein in accordance with which it *is* its there. Disclosedness is constituted by attunement, understanding, and discourse, and pertains equiprimordially to the world, being-in, and the self. The structure of care as *being-ahead-of-itself*—already-being-in-a-world—as being together with innerworldly beings holds within itself the disclosedness of Dasein. *With* and *through* it is discoveredness; thus only with the *disclosedness* of Dasein is the *most primordial* phenomenon of truth attained. What is shown earlier with regard to the existential constitution of the there[37] and in relation to the everyday being of the there[38] pertains to nothing less than the most primordial phenomenon of truth. Insofar as Dasein essentially *is* its disclosedness, and, as disclosed, it discloses and discovers, it is essentially "true." *Dasein is "in the truth."* This statement has an ontological meaning. It does not mean that Dasein is ontically always, or even only at times, inducted [eingeführt] "into every truth," but that the disclosedness of its ownmost being belongs to its existential constitution.

By considering what we have gained so far, the full existential meaning of the statement, "Dasein is in the truth," can be summarized by the following considerations:

1. *Disclosedness in general* belongs essentially to the constitution of being of Dasein. It encompasses the totality of the structure of being that has become explicit through the phenomenon of care. To care there belongs not only being-in-the-world, but being together with inner-worldly beings. The discoveredness of innerworldly beings is equiprimordial with the being of Dasein and its disclosedness.

2. *Thrownness* belongs to the constitution of being of Dasein as a constituent of its disclosedness. In thrownness it is revealed that Dasein is in each instance always mine and that this Dasein is always already in a definite world and together with a definite range of defi-nite innerworldly beings. Disclosedness is essentially factical.

3. *Project* belongs to the constitution of being of Dasein: disclosive being toward its own potentiality-of-being. Dasein *can*, as an under-standing being, understand *itself* in terms of the "world" and others, or else in terms of its ownmost potentiality-of-being. This possibility means that Dasein discloses itself to itself in and as its ownmost poten-tiality-of-being. This *authentic* disclosedness shows the phenomenon of the most primordial truth in the mode of authenticity. The most

---

37. Cf. § 29.
38. Cf. § 34b.

221

primordial and authentic disclosedness in which Dasein can be as a potentiality-of-being is the *truth of existence*. Only in the context of an analysis of the authenticity of Dasein does it receive its existential, ontological definiteness.

4. *Falling prey* belongs to the constitution of being of Dasein. Initially and for the most part, Dasein is lost in its "world." Understanding, as a project upon possibilities of being, has shifted itself into its world. Absorbing oneself in the they signifies that one is dominated by the public way of interpreting. What is discovered and disclosed stays in the mode in which it has been disguised and closed off by idle talk, curiosity, and ambiguity. Being toward beings has not been extinguished, but uprooted. Beings are not completely concealed, rather they are what is discovered, and at the same time distorted. They show themselves, but in the mode of semblance [Schein]. Similarly, what was previously discovered sinks back again into dissemblance and concealment. *Because it essentially falls prey to the world, Dasein is in "untruth" in accordance with its constitution of being.* This term is used here ontologically, as is the expression "falling prey." Any ontically negative "value judgment" is to be avoided in its existential and analytic use. Being closed off and covered over belong to the *facticity* of Dasein. The full existential and ontological meaning of the statement, "Dasein is in the truth," also says equiprimordially that "Dasein is in untruth." But only insofar as Dasein is disclosed is it also closed off; and insofar as innerworldly beings are always already discovered with Dasein, are such beings covered over (hidden) or disguised as possible innerworldly beings to be encountered.

Thus, Dasein must explicitly and essentially appropriate what has also already been discovered, defend it *against* semblance [Schein] and dissemblance [Verstellung], and ensure itself of its discoveredness again and again. All new discovery takes place not on the basis of complete concealment, but takes its point of departure from discoveredness in the mode of semblance. Beings look like . . . , that is, they are in a way already discovered, and yet they are still disguised [verstellt].

Truth (discoveredness) must always first be wrested [abgerungen] from beings. Beings are torn from concealment. Each and every factical discoveredness is, so to speak, always a kind of *robbery*. Is it a matter of chance that the Greeks express themselves about the essence of truth with a *privative* expression (ἀ-λήθεια)? When Dasein expresses itself in this way, does not a primordial understanding of its own being make itself known—namely, the understanding (even if it is only pre-ontological) that being-in-untruth constitutes an essential determination of being-in-the-world?

The fact that the goddess of truth who leads Parmenides places him before two paths, that of discovering and that of concealment,

means nothing other than Dasein is always already in the truth and
223   untruth. The path of discovering is gained only in κρίνειν λόγῳ, in
distinguishing between them understandingly and in deciding for the
one rather than the other.[39]

The existential and ontological condition for the fact that being-in-
the-world is determined by "truth" and "untruth" lies in *the* constitution of being of Dasein* which we characterized as *thrown project*. It is
a constituent of the structure of care.

The existential and ontological interpretation of the phenomenon
of truth has shown: (1) Truth in the most primordial sense is the disclosedness of Dasein, to which belongs the discoveredness of innerworldly beings; (2) Dasein is equiprimordially in truth and untruth.

These statements will only be fully comprehensible within the
horizon of the traditional interpretation of the phenomenon of truth if
it can be shown: (1) Truth, understood as an agreement, has its origin
in disclosedness by way of a definite modification; (2) The kind of
being of disclosedness itself leads to the fact that initially its derivative
modification comes into view and guides the theoretical explication of
the structure of truth.

Statement and its structure, the apophantical "as," are based on
interpretation and its structure, the hermeneutical "as," and furthermore on understanding, on the disclosedness of Dasein. But here truth
is regarded as a distinctive determination of statements thus derived.
Accordingly, the roots of the truth of statement reach back to the disclosedness of understanding.[40] But now the phenomenon of *agreement*
must be shown *explicitly* in its derivative character above and beyond
this indication of the origin of the truth of statements.

Being together with innerworldly beings, and taking care of them,
discovers. But to the disclosedness of Dasein discourse essentially
belongs.[41] Dasein expresses itself; *itself*—as a being [Sein] toward beings
224   that discovers. And in statements it expresses itself as such about beings
that have been discovered. Statements communicate beings in the How
[Wie] of their discoveredness. Dasein, perceiving the communication,
brings itself to a discovering being toward the beings discussed. The
statements made are made about something, and in what they are about
they contain the discoveredness of beings. This discoveredness is pre-

---

* Of Da-sein and thus of standing-in [Inständigkeit].

39. K. Reinhardt first grasped and solved the much mistreated problem of the connection
of the two parts of Parmenides' poem (cf. *Parmenides und die Geschichte der griechischen
Philosophie*, 1916), although he does not explicitly point out the ontological foundation
for the connection between ἀλήθεια and δόξα.
40. Cf. above, § 33: "Statement as a Derivative Mode of Interpretation."
41. Cf. § 34.

served in what is expressed. What is expressed becomes, so to speak, an innerworldly thing at hand that can be taken up and spoken about further. Because the discoveredness has been preserved, what is expressed (what is thus at hand) has in itself a relation to any beings about which it is a statement. Discoveredness is always a discoveredness of. . . . Even when Dasein repeats what has been said, it comes into a being toward the very beings that have been discussed. But it is and believes itself exempt from a primordial repetition of the act of discovering.

Dasein does not need to bring itself to beings in "original" experience, but it nevertheless remains in a being toward these beings. Discoveredness is appropriated to a large extent not by one's own discovering, but by hearsay of what has been said. Absorption in what has been said belongs to the kind of being of the they. What is expressed as such takes over the being toward those beings discovered in the statement. But if they are to be explicitly appropriated with regard to their discoveredness, this means that the statement should be shown to be one that discovers. But the statement expressed is something at hand, in such a way that, as preserving discoveredness, it has in itself a relation to the beings discovered. To demonstrate that it is something that discovers, means to demonstrate how the statement in which discoveredness is preserved is related *to* these beings. The statement is something at hand. The beings to which it has a discovering relation are innerworldly things at hand or objectively present. Thus the relation presents itself as something objectively present. But this relation lies in the fact that the discoveredness preserved in the statement is always a discoveredness of. . . . The judgment "contains something valid for the objects" (Kant). But the relation itself now acquires the character of objective presence by getting switched over to a relation between objectively present things. Discoveredness of . . . becomes the objectively present conformity of something objectively present, of the statement expressed, *to* something objectively present, the being spoken about. And if this conformity is then viewed only as the relation between objectively present things, that is, if the kind of being of the terms of the relation is understood without differentiation as merely objectively present things, then the relation shows itself as the objectively present conformity of two objectively present things.

When the statement has been expressed [Ausgesprochenheit], *the dis-*   225
coveredness of beings moves into the kind of being of innerworldly things
*at hand. But to the extent that in this discoveredness, as a **discoveredness***
***of** . . . , a relation to things objectively present persists, discoveredness (truth)*
itself in its turn becomes an objectively present relation between objectively
present things (intellectus and res).

The existential phenomenon of discoveredness, which is based on the disclosedness of Dasein, becomes an objectively present property (though one which still contains the character of a relation in itself) and, as such a property, is split up into an objectively present relation. Truth as disclosedness and as being toward discovered beings—a being that itself discovers—has become truth as the agreement between innerworldly things objectively present. Thus we have shown the ontological derivation of the traditional concept of truth.

However, what comes last in the order of the existential and ontological foundational context is regarded ontically and factically as what is first and nearest. But the necessity of this fact is again based in the kind of being of Dasein itself. Absorbed in taking care of things, Dasein understands itself in terms of what it encounters within the world. The discoveredness belonging to discovering is initially found within the world in what has been *ex*pressed. But not only is truth encountered as something objectively present, rather the understanding of being in general initially understands all beings as objectively present. If the "truth" that we encounter initially in an ontic way is understood ontologically in the way closest to us, then the λόγος (statement) gets understood as λόγος τινός (statement about . . . , discoveredness of . . .), but the phenomenon gets interpreted as objectively present with regard to its possible objective presence. But because objective presence is equated with the meaning of being in general, the question whether this kind of being of truth, and its initially encountered structure, are primordial or not *can* not come alive at all. *The understanding of being of Dasein which was initially dominant, and has still not been overcome today in a **fundamental** and **explicit** way, itself covers over [verdeckt] the primordial phenomenon of truth.*

At the same time, we must not overlook the fact that for the Greeks, who were the first to develop this initial understanding of being systematically [wissenschaftlich] and to bring it to dominance, this primordial understanding of truth was, at the same time, alive, even if pre-ontologically, and it even held its own against the concealment implicit in their ontology—at least in Aristotle.[42]

226   Aristotle never defended the thesis that the primordial "locus" of truth is judgment. Rather, he says that the λόγος is the kind of being of Dasein which can either discover *or* cover over. This *double possibility* is what is distinctive about the being-true of the λόγος; it is the comportment which *can also cover over*. And since Aristotle never asserted this thesis, he was never in the position of "expanding" the concept of truth of λόγος to pure νοεῖν. The "truth" of αἴσθησις and of the seeing of the "Ideas," is the primordial discovering. And only because νόησις primarily discovers, can the λόγος, too, have the function of discovering as διανοεῖν.

---

42. Cf. *Nichomachean Ethics* Z, and *Metaphysics* θ, 10.

The thesis that the genuine "locus" of truth is judgment not only invokes Aristotle unjustly, it also fails with regard to its content to recognize the structure of truth. The statement is not the primary "locus" of truth, but the *other way around*; the statement as a mode of appropriation of discoveredness and as a way of being-in-the-world is based on discovering, or rather the *disclosedness* of Dasein. The most primordial "truth" is the "locus" of the statement, and this primordial truth is the ontological condition of the possibility that statements can be true or false (discovering or covering over).

Understood in its most primordial sense, truth belongs to the fundamental constitution of Dasein. The term signifies an existential. But thus we have already sketched out our answer to the question of the kind of being of truth and the meaning of the necessity of the presupposition that "there is truth."

(c) The Kind of Being of Truth and the Presupposition of Truth

Constituted by disclosedness, Dasein is essentially in the truth. Disclosedness is an essential kind of being of Dasein. *"There is"* [*"gibt es"*] *truth only insofar as Dasein is and as long as it is.* Beings are discovered only *when* Dasein *is*, and only *as long as* Dasein *is* are they disclosed. Newton's laws, the law of contradiction, and any truth whatsoever, are true only as long as Dasein *is*. Before there was any Dasein, there was no truth; nor will there be any after Dasein is no more. For in such a case truth as disclosedness, discovering, and discoveredness *cannot* be. Before Newton's laws were discovered, they were not "true." From this it does not follow that they were false or even that they would become false if ontically no discoveredness were possible any longer. Just as little does this "restriction" imply a diminution of the being true of "truths."

227

The fact that before Newton his laws were neither true nor false cannot mean that the beings which they point out in a discovering way did not previously exist. The laws became true through Newton, through them beings in themselves became accessible for Dasein. With the discoveredness of beings, they show themselves precisely as the beings that previously already were. To discover in this way is the kind of being of "truth."

That there are "eternal truths" will not be adequately proven until it is successfully demonstrated that Dasein has been and will be for all eternity. As long as this proof is lacking, the statement remains a fanciful assertion which does not gain in legitimacy by being generally "believed" by philosophers.

In accordance with the essential kind of being appropriate to Dasein, all truth is relative to the being of Dasein. Is this relativity tantamount to saying that all truth is "subjective"? If one interprets "subjective" to mean "left to the arbitrariness of the subject," then certainly not. For in accordance with its very meaning, discovering exempts statements from the province of "subjective" arbitrariness and brings discovering Dasein before beings themselves. And only *because* "truth," as discovering, *is a kind of being of Dasein*, can it be removed from the arbitrariness of *Dasein's* discretion [Belieben]. Even the "universal validity" of truth is rooted solely in the fact that Dasein can discover and free beings in themselves. Only thus can this being in itself [Seiende an ihm selbst] be binding for every possible statement, that is, for every possible way of pointing them out. If truth has been correctly understood, is it not in the least jeopardized by the fact that it is ontically possible only in the "subject," and stands or falls with the being [Sein] of that "subject"?

The meaning of the presupposition of truth, too, becomes intelligible in terms of the existentially conceived kind of being of truth. *Why must we presuppose that there is truth?* What does "presuppose" mean? What do "must" and "we" mean? What does it mean, "there is truth"? "We" presuppose truth because, "we," existing in the kind of being of Dasein, *are* "in the truth." We do not presuppose it as something "outside" and "above" us to which we are related along with other "values" too. We do not presuppose "truth," rather* *truth* makes it ontologically possible that we can *be* in such a way that we "presuppose" something at all. Truth first *makes possible* something like presupposition.

228

What does "presupposing" mean? To understand something as the ground of the being of [Sein] another being [Seienden]. Such understanding of a being in its context of being is possible only on the basis of disclosedness, that is, the discovering of Dasein. To presuppose "truth" then means to understand it as something for the sake of which Dasein is. But Dasein is always already ahead of itself; that lies in its constitution of being as care. It is a being that is concerned in its being about its ownmost potentiality-of-being. Disclosedness and discovering belong essentially to the being [Sein] and potentiality-for-being of Dasein as being-in-the-world, and this includes circumspectly discovering and taking care of innerworldly beings. In the constitution of being of Dasein as care, in being ahead of itself, lies the most primordial "presupposing." *Because this presupposing itself belongs to the being of Dasein, "we" have to also presuppose "ourselves" as determined by disclosedness.* This "presupposing" that lies in the being of Dasein is not related to beings unlike Dasein, which are there in addition to Dasein, but solely to Dasein itself. The truth which is presupposed, or which "is there," by which its being [Sein] is to be defined, has the kind of being, or meaning of being, of Dasein itself. We have to "make" the

---

* but the essence of truth places us in the "prior" of what is spoken to us!

presupposition of truth because it *is* already "made" with the being of the "we."

We *must* presuppose truth, it *must be* as the disclosedness of Dasein, just as Dasein itself *must* always be as my own and this particular Dasein. This belongs to the essential thrownness of Dasein into the world. *Has Dasein as itself ever freely decided, and will it ever be able to decide, whether it wants to come into "Dasein" or not?* "In itself" we cannot see why beings should be *discovered*, why *truth* and *Dasein* must be. The usual refutation of skepticism, of the denial of either the being or the knowability of "truth," gets stuck halfway. What it shows by means of formal argumentation is simply the fact that when someone judges, truth has been presupposed. This suggests that "truth" belongs to statements, that pointing out, according to meaning, is a discovering. *Why* that must be never gets *clarified*; it has never been clarified wherein the ontological ground for this necessary connection of the being of statement and truth lies. Similarly, the kind of being of truth and the meaning of presupposing with its ontological foundation in Dasein itself remain completely obscure. Moreover, one fails to recognize the fact that truth is already presupposed even when no one *judges* insofar as Dasein is at all.

229

A skeptic can no more be refuted than the being of truth can ever be "proved." If the skeptic, who denies the truth, factically *is*, he does *not* even *need* to be refuted. Insofar as he *is,* and has understood himself in this being, he has extinguished Dasein, and thus truth, in the despair of suicide. The necessity of truth cannot be proven because Dasein cannot first be subjected to proof for its own part. It has no more been demonstrated that there has ever "been" a "real" skeptic (although that is what has, at bottom, been believed in the refutations of skepticism, in spite of what these undertake to do) than it has been demonstrated that there are any "eternal truths." Perhaps such skeptics have been more frequent than one would innocently like to believe when one tries to overturn "skepticism" by formal dialectics.

Thus, with the question of the being of truth and the necessity of its presupposition, as well as that of the essence of knowledge, an "ideal subject" has generally been posited. The motive for this, whether it is made explicit or is tacit, lies in the requirement that philosophy should have the "*a priori*" as its theme, rather than "empirical facts" as such. There is some justification for this requirement, though it still needs to be grounded ontologically. But is this requirement satisfied by positing an "ideal subject"? Is it not a *fantastically idealized* subject? Is not precisely the *a priori* character of that merely "factual" subject, of Dasein, missed with the concept of such a subject? Is it not an attribute of the *a priori* character of the factical subject (that is, of the facticity of Dasein) that it is equiprimordially in truth and untruth?

The ideas of a "pure ego" and a "consciousness in general" are so far from including the *a priori* character of "real" subjectivity that they pass over the ontological character of facticity of Dasein and its constitution of being, or rather they do not see it at all. Rejection of a "consciousness in general" does not mean the negation of the *a priori*, any more than the positing of an idealized subject guarantees a factually based *a priori* character of Dasein.

The claim that there are "eternal truths," as well as the confusion of the phenomenally based "ideality" of Dasein with an idealized absolute subject, belong to the remnants of Christian theology within the philosophical problematic that have not yet been radically eliminated.

230   The being of truth stands in a primordial connection with Dasein. And only because Dasein exists as constituted by disclosedness—that is, by understanding—can something like being be understood, only thus is an understanding of being possible at all.

"There is" [Es gibt] being—not beings—only insofar as truth is. And truth *is* only insofar as, and as long as, Dasein is. Being and truth "are" equiprimordially. What does it mean that being "is," when being should be distinguished from all beings?* One can inquire into this concretely only if the meaning of being and the scope of the understanding of being in general have been clarified. Only then can one also analyze primordially what belongs to the concept of a science *of being as such*, its possibilities and transformations. And in delimiting this inquiry and its truth, inquiry as the discovery *of beings* and their truth will have to be ontologically defined.

The answer to the question of the meaning of being is still lacking. What has the fundamental analysis of Dasein that we have carried out thus far provided for working out this question? By freeing the phenomenon of care, we clarified the constitution of being of that being to whose being [Sein] something like an understanding of being belongs. The being of Dasein was thus at the same time distinguished from the modes of being (handiness, objective presence, reality) that characterize beings unlike Dasein. We clarified understanding itself on this basis; and thus, at the same time, the methodological transparency of the procedure of interpreting being by understanding and interpreting it has been guaranteed.

If the primordial constitution of being of Dasein has been attained in care, then the understanding of being contained in care must also be grasped on this basis; that is, it must be possible to delineate the meaning of being. But *is* the most primordial, existential, and ontological constitution of Dasein disclosed with the phenomenon of care? Does the structural manifoldness in the phenomenon of care give the most primordial totality of the being of factical Dasein? Has the inquiry up to now brought Dasein *as a whole* into view at all?

---

* Ontological difference.

Dasein and Temporality

§ 45. *The Result of the Preparatory Fundamental Analysis of Dasein and the Task of a Primordial, Existential Interpretation of this Being*

What was gained by our preparatory analysis of Dasein, and what are we looking for? We have *found* the fundamental constitution of the being in question, being-in-the-world, whose essential structures are centered in disclosedness. The totality of this structural whole revealed itself as care. The being of Dasein is contained in care. The analysis of this being took as its guideline existence,[1] which was defined by way of anticipation as the essence of Dasein. The term existence formally indicates that Dasein *is* as an understanding potentiality-of-being which is concerned in its being about its being. I myself am in each instance the being [Seiende] existing [seiend] in this way. The development of the phenomenon of care provided an insight into the concrete constitution of existence, that is, into its equiprimordial connection with facticity and with the entanglement of Dasein.

We are *searching* for the answer to the question of the meaning of being in general, and above all the possibility of radically developing this basic question* of all ontology. But freeing the horizon in which something like being in general becomes intelligible amounts to clarifying the possibility of the understanding of being [Sein], in general, an understanding which itself belongs to the constitution of that being [Seienden], which we call Dasein.[2] However, the understanding

---

* By which, however, "onto-logic" is transformed at the same time (cf. *Kant and the Problem of Metaphysics*, section IV).

1. Cf. § 9.
2. Cf. §§ 6, 21, and 43.

of being can only be *radically* clarified as an essential factor in the being of Dasein, if the being [Seiende] to whose being [Sein] it belongs has been *primordially* interpreted in itself with regard to its being [Sein].

Are we entitled to the claim that in characterizing Dasein ontologically as care we have given a *primordial* interpretation of this being? By what standard is the existential analytic of Dasein to be measured with regard to its primordiality or nonprimordiality? What then do we mean by the *primordiality* of an ontological interpretation?

Ontological inquiry is a possible way of interpretation which we characterized as the working-out and appropriation of an understanding.[3] Every interpretation has its fore-having, its fore-sight, and its fore-conception. If such an interpretation becomes an explicit task of an inquiry, the totality of these "presuppositions" (which we call the *hermeneutical situation*) needs to be clarified and made secure beforehand, both in a fundamental experience of the "object" to be disclosed and in terms of that experience. In ontological interpretation, beings are to be freed with regard to their own constitution of being. Such an interpretation obliges us first to give a phenomenal characterization of the being [Seiende] we have taken as our theme and thus bring it into the scope of our fore-having with which all the subsequent steps of our analysis are to conform. But at the same time these steps need to be guided by the possible fore-sight of the kind of being of the being [Seienden] in question. Fore-having and fore-sight then prefigure (fore-conception) at the same time the conceptuality to which all the structures of being are to be brought.

But a *primordial* ontological interpretation requires not only in general that the hermeneutical situation be secured in conformity with the phenomena, but also the explicit assurance that *the totality* of the beings taken as its theme have been brought to a fore-having. Similarly, it is not sufficient just to make a first sketch [Vorzeichnung] of the being of these beings, even if it is phenomenally based. If we are to have a fore-sight of being, we must see it with respect to the *unity* of the possible structural factors belonging to it. Only then can the question of the meaning of the unity that belongs to the totality of being of all beings be asked and answered with phenomenal certainty.

Did the existential analysis of Dasein, which we have carried out, arise from such a hermeneutical situation that will guarantee the primordiality which fundamental ontology requires? Can we proceed from the results attained—that the being of Dasein is care—to the question of the primordial unity of this structural totality?

What is the status of the fore-sight which has been guiding our ontological procedure up to now? We defined the idea of existence as a potentiality-of-being, a potentiality that understands and is con-

232

---

3. Cf. § 32.

cerned about its own being. But this *potentiality-of-being* that is always *mine* is free for authenticity or inauthenticity, or for a mode in which neither of these has been differentiated.[4] Our previous interpretation, starting out with average everydayness, confined itself to the analysis of indifferent or inauthentic existing. Of course, it was possible and necessary to reach a concrete definition of the existentiality of existence even in this way. Still, our ontological characterization of the constitution of existence was flawed by an essential lack. Existence means potentiality-of-being, but also authentic potentiality-of-being. As long as the existential structure of authentic potentiality-of-being is not incorporated in the idea of existence, the fore-sight guiding an *existential* interpretation lacks primordiality.

233

And what is the situation with the fore-having of the hermeneutical situation up to now? When and how did our existential analytic make sure that by starting with everydayness it forced the *whole* of Dasein—this being from its "beginning" to its "end"—into the phenomenological view that gives us our theme? We did assert that care is the totality of the structural whole of the constitution of being of Dasein.[5] But have we not at the very beginning of our interpretation renounced the possibility of bringing Dasein as a whole to view? Everydayness is, after all, precisely the being [Sein] "between" birth and death. And if existence determines the being of Dasein, and if its essence is co-constituted* by potentiality-of-being, then, as long as Dasein exists, it must always, as such a potentiality, *not yet be* something? A being whose essence [Essenz] is made up of existence essentially opposes itself to the possibility of being comprehended as a whole being. Not only has the hermeneutical situation given us no assurance of "having" the whole being up to now; it is even questionable whether the whole being is attainable at all, and whether a primordial, ontological interpretation of Dasein must not get stranded—on the kind of being [Sein] of the thematic being [Seienden] itself.

One thing has become unmistakable: *The existential analytic of Dasein up to now cannot lay claim to primordiality.* Its fore-having never included more than the *inauthentic* being of Dasein, of Dasein as *less than whole* [unganzes]. If the interpretation of the being of Dasein is to become primordial as a foundation for the development of the fundamental question of ontology, it will have to bring the being of Dasein in its possible *authenticity* and *wholeness* [Ganzheit] existentially to light beforehand.

Thus the task arises of placing Dasein as a whole in our fore-having. However, that means that we must first unpack the question

---

* at the same time: *already*-being.

4. Cf. § 9.
5. Cf. § 41.

234   of this being's potentiality-for-being-whole. As long as Dasein is, something is always still outstanding: what it can and will be. But the "end"
itself belongs to what is outstanding. The "end" of being-in-the-world
is death. This end, belonging to the potentiality-of-being, that is, to
existence, limits and defines the possible totality of Dasein. The being-
at-an-end* of Dasein in death, and thus its being a whole, can, however,
be included in our discussion of the possible *being* whole in a phenomenally appropriate way only if an ontologically adequate, that is,
an *existential* concept of death has been attained. But as far as Dasein†
goes, death *is* only in an existentiell *being toward death* [*Sein zum Tode*].‡
The existential structure of this being [Sein] turns out to be the ontological constitution of the potentiality-for-being-whole of Dasein. Thus,
the whole existing Dasein can be brought into our existential fore-
having. But can Dasein also exist as a whole *authentically*? How is the
authenticity of existence to be defined at all if not with reference to
authentic existing? Where do we get our criterion for this? Obviously
Dasein itself in its being must present the possibility and manner of
its authentic existence, if such existence is neither imposed upon it
ontically, nor ontologically fabricated. But an authentic potentiality-of-
being is attested by conscience. Like death, this phenomenon of Dasein
requires a genuinely existential interpretation. It leads to the insight
that an authentic potentiality-of-being of Dasein lies in wanting-to-
have-a-conscience. This existentiell possibility, however, tends, from
the meaning of its being, to be made definite in an existentiell way
by being toward death.

   With the demonstration of an *authentic potentiality-for-being-whole*
of Dasein our existential analytic secures the constitution of the *primordial* being of Dasein. But the authentic *potentiality-for-being-whole*
becomes visible as a mode of care. With this the phenomenally adequate basis for a primordial interpretation of the meaning of being of
Dasein is also secured.

   The primordial ontological ground of the existentiality of Dasein,
however, is *temporality* [*Zeitlichkeit*]. The articulated structural totality
of the being of Dasein as care first becomes existentially intelligible
in terms of temporality. The interpretation of the meaning of being
of Dasein cannot stop with this fact. The existential-temporal analysis of this being [Seienden] needs concrete confirmation. We must
go back and free the ontological structures of Dasein already gained
with regard to their temporal meaning. Everydayness reveals itself as
a mode of temporality. But by thus repeating our preparatory funda-

---

* 'being'-toward-the-end [Zum-Ende-'sein'].
† Thought in accordance with the essence of Dasein.
‡ being of nonbeing.

mental analysis of Dasein, the phenomenon of temporality itself will, at the same time, become more transparent. In terms of temporality, it *235* becomes intelligible why Dasein is and can be historical in the ground of its being and, *as historical [geschichtliches]*, it can develop historiography [Historie].

If temporality constitutes the primordial meaning of being of Dasein, and if this being [Seienden] is concerned *about its being* in its very being, then care must need "time" and thus reckon with "time." The temporality of Dasein develops a "time calculation." The "time" experienced in such calculation is the proximate phenomenal aspect of temporality. From it originates the everyday, vulgar understanding of time. And this understanding develops into the traditional concept of time.

The clarification of the origin of the "time" "in which" inner-worldly beings are encountered, of time as within-timeness, reveals an essential possibility of the temporalization of temporality. With this clarification the understanding prepared itself for a still more primordial temporalization of temporality. In it is based the understanding of being that is constitutive for the being of Dasein. The projection [Entwurf] of a meaning of being in general can be accomplished in the horizon of time.*

Thus the inquiry comprised in this Division will traverse the following stages: The possible being whole of Dasein and being toward death (Chapter One); the attestation of Dasein of an authentic potentiality-of-being and resolution (Chapter Two); the authentic potentiality-for-being-a-whole of Dasein and temporality as the ontological meaning of care (Chapter Three); temporality and everydayness (Chapter Four); temporality and historicity (Chapter Five); temporality and within-timeness as the origin of the vulgar concept of time (Chapter Six).[6]

---

* Presencing [An-wesenheit] (Arrival and Event).

6. In the nineteenth century S. Kierkegaard explicitly grasped and thought through the problem of existence as existentiell in a penetrating way. But the existential* problematic is so foreign to him that in an ontological regard he is completely under the influence of Hegel and his view of ancient philosophy. Thus more is to be learned philosophically from his "edifying" writings than from his theoretical work—with the exception of the treatise on the concept of anxiety.
   * and, to be sure, the fundamental ontological one, i.e. aiming at the question of being as such in general.

CHAPTER ONE

The Possible Being-a-Whole of Dasein and Being-toward-Death

§ 46. *The Seeming Impossibility of Ontologically Grasping and Determining Dasein as a Whole*

The inadequacy of the hermeneutical situation from which the fore-going analysis originated must be overcome. With regard to the fore-having of the whole of Dasein, which must necessarily be obtained, we must ask whether this being, as something existing, can become accessible at all in its being a whole. There seem to be important reasons that speak against the possibility of our required task, reasons that lie in the constitution of Dasein itself.

Care, which forms the totality of the structural whole of Dasein, obviously contradicts a possible being whole of this being according to its ontological sense. The primary factor of care, "being ahead of itself," however, means that Dasein always exists for the sake of itself. "As long as it is," up until its end, it is related to its potentiality-of-being. Even when it, still existing, has nothing further "ahead of it," and has "settled its accounts," its being is still determined by "being ahead of itself." Hopelessness, for example, does not tear Dasein away from its possibilities, but is only one of its own modes of *being toward* these possibilities. Even when one is without illusions and "is ready *for* any-thing," the "ahead of itself" is there. This structural factor of care tells us unambiguously that something is always still *outstanding* [*aussteht*] in Dasein which has not yet become "real" as a potentiality-of-its-being. A *constant unfinished quality* [*Unabgeschlossenheit*] thus lies in the essence of the basic constitution of Dasein. This lack of wholeness means that there is still something outstanding in one's potentiality-for-being.

However, if Dasein "exists" in such a way that there is absolutely nothing more outstanding for it, it has also already thus become no-longer-being-there. Eliminating what is outstanding in its being [Sein] is equivalent to annihilating its being. As long as Dasein *is* as a being, it has never attained its "wholeness." But if it does, this attainment becomes the absolute loss of being-in-the-world. It is then never again to be experienced *as a being*.

The reason for the impossibility of experiencing Dasein ontically as an existing whole, and thus of defining it ontologically in its wholeness, does not lie in any imperfection of our *cognitive faculties*. The hindrance lies on the side of the *being* of this being. Whatever cannot even *be an* experience that would claim to grasp Dasein withdraws itself fundamentally from any possibility of being experienced. But is it not then a hopeless undertaking to try to discern the ontological wholeness of being of Dasein?

As an essential structural factor of care, "being ahead of itself" cannot be eliminated. But is what we concluded from this tenable? Did we not conclude in a merely formal argumentation that it is impossible to grasp the whole of Dasein? Or did we not at bottom inadvertently posit Dasein as something objectively present ahead of which something not yet objectively present constantly moves along? Did our argumentation grasp not-yet-being and the "ahead-of-itself" in a genuinely *existential* sense? Did we speak about "end" and "totality" in a way phenomenally appropriate to Dasein? Did the expression "death" have a biological significance or one that is existential and ontological, or indeed was it sufficiently and securely defined at all? And have we actually exhausted all the possibilities of making Dasein accessible in its totality?

We have to answer these questions before the problem of the wholeness of Dasein can be dismissed as nothing. The question of the wholeness of Dasein, both the existentiell question about a possible potentiality-for-being-a-whole, as well as the existential question about the constitution of being of "end" and "wholeness," contain the task of a positive analysis of the phenomena of existence set aside up to now. In the center of these considerations we have the task of characterizing ontologically the being-toward-the-end of Dasein and of achieving an existential concept of death. Our inquiry related to these topics is structured in the following way: the possibility of experiencing the death of others, and the possibility of grasping the whole of Dasein (§ 47); what is outstanding, end, and wholeness (§ 48); how the existential analysis of death is distinguished from other possible interpretations of this phenomenon (§ 49); preliminary sketch of the existential and ontological structure of death (§ 50); being toward death and the everydayness of Dasein (§ 51); everyday being toward death and the complete existential concept of death (§ 52); the existential project of an authentic being toward death (§ 53).

§ 47. The Possibility of Experiencing the Death of Others and the Possibility of Grasping Dasein as a Whole

When Dasein reaches its wholeness in death, it simultaneously loses the being of the there. The transition to no-longer-Dasein lifts Dasein right out of the possibility of experiencing this transition and of understanding it as something experienced. This kind of thing is denied to each and every Dasein in relation to itself. The death of others, then, is all the more penetrating. In this way, an end of Dasein becomes "objectively" accessible. Dasein can attain an experience of death all the more because it is essentially being-with with others. This "objective" givenness of death must then make possible an ontological delimitation of the wholeness of Dasein.

Thus from the kind of being that Dasein possesses as being-with-one-another, we might glean the fairly obvious information that when the Dasein of others has come to an end, it might be chosen as a substitute theme for our analysis of the wholeness of Dasein. But does this lead us to our intended goal?   238

Even the Dasein of others, when it has reached it wholeness in death, is a no-longer-Dasein in the sense of no-longer-being-in-the-world. Does not dying mean going-out-of-the-world and losing being-in-the-world? Yet, the no-longer-being-in-the-world of the deceased (understood in an extreme sense) is still a being [ein Sein] in the sense of the mere objective presence [Nur-noch-vorhandensein] of a corporeal thing encountered. In the dying of others that remarkable phenomenon of being can be experienced that can be defined as the sudden transition [Umschlag] of a being [Seienden] from the kind of being of Dasein (or of life) to no-longer-Dasein. The *end* of the being qua Dasein is the *beginning* of this being [Seienden] qua something merely [bloßen] present.

This interpretation of the transition from Dasein to something only just present, however, misses the phenomenal content in that the being still remaining does not represent a mere [pures] corporeal thing. Even the objectively present corpse is, viewed theoretically, still a possible object for pathological anatomy whose understanding is oriented toward the idea of life. This something which is only-just-present is "more" than a *lifeless*, material thing. In it we encounter something *unliving* which has lost its life.

But even this way of characterizing what still remains does not exhaust the complete phenomenal findings with regard to Dasein.

The "deceased," as distinct from the dead body, has been torn away from "those remaining behind" and is the object of "being taken care of" ["Besorgens"] in funeral rites, burial, and the cult of graves.

And that is so because the kind of being of the deceased is "still more" than a thing at hand in the surrounding world to be taken care of. In lingering together with him in mourning and commemorating, those remaining behind *are with* him, in a mode of concern which honors him. Thus the relation of being to the dead must not be grasped as a being together with something at hand which *takes care of it*.

In such being-with with the dead, the deceased *himself* is no longer factically "there." However, being-with always means being-with-one-another in the same world. The deceased has abandoned our *"world"* and left it behind. Nonetheless, it is *in terms of this world* that those remaining can still *be with him*.

The more appropriately the no-longer-being-there of the deceased is grasped phenomenally, the more clearly it can be seen that in such being-with with the dead, the real having-come-to-an-end of the deceased is precisely *not* experienced. Death does reveal itself as a loss, but as a loss experienced by those remaining behind. However, in suffering this loss, the loss of being as such, which the dying person "suffers," does not become accessible. We do not experience the dying of others in a genuine sense; we are at best always just "near by" ["dabei"].

Even if it were possible and feasible to clarify "psychologically" the dying of others in this being nearby, this would by no means let us grasp the way of being we have in mind, namely, coming-to-an-end. We are asking about the ontological meaning of the dying of the person who dies, as a potentiality-of-being of *his* being [Sein], and not about the way of being-with and the still-being-there of the deceased with those left behind. Taking death as experienced in others as the theme of our analysis of the end of Dasein and its wholeness cannot give us what it is supposed to give, either ontically or ontologically.

After all, taking the dying of others as a substitute theme for the ontological analysis of the finished character of Dasein and its wholeness rests on an assumption that demonstrably fails altogether to recognize the kind of being of Dasein. That is what one presupposes when one is of the opinion that any Dasein could arbitrarily be replaced by another, so that what cannot be experienced in one's own Dasein is accessible in another Dasein. But is this assumption really so groundless?

The fact that one Dasein *can be represented* [*Vertretbarkeit*] by another belongs indisputably to the possibilities-of-being of being-with-one-another in the world. In the everydayness of taking care of things, constant use of such representability is made in many ways. Every going somewhere, every imparting of something, is representable in the scope of the "surrounding world" taken care of. The broad multiplicity of ways of being-in-the-world in which one person can be represented by another extends not only to the well polished modes of publicly being with one another, but concerns as well the possibilities of taking care of things limited to definite circles, tailored to professions, social classes, and stages of life. But the very meaning of such representation is such that it is always a representation "in" and "together with" something, that is, in taking care. Everyday Dasein understands itself initially and for the most part, however, in terms of *what* it is accustomed to take care of. "One *is*" what one does. With regard to this being (the everyday being-absorbed-with-one-another in the "world" taken care of), representability is not only possible in general, but is even constitutive for being-with-one-another. *Here* one Dasein can and must, within certain limits, "*be*" another Dasein.

240

However, the possibility of representation gets completely stranded when it is a matter of representing the possibility of being that constitutes the coming-to-an-end of Dasein and gives it its wholeness as such. *No one can take the other's dying away from him.* Someone can go "to his death for an other." However, that always means to sacrifice oneself for the other "*in a definite cause.*" Such dying for . . . can never mean that the other has thus had his death in the least taken away. Every Dasein itself must take dying upon itself in every instance. Insofar as it "is," death is always essentially my own. And it indeed signifies a peculiar possibility of being in which it is absolutely a matter of the being of my own Dasein. In dying, it becomes evident that death* is ontologically constituted by mineness and existence.[1] Dying is not some given occurrence, but a phenomenon to be understood existentially in an eminent sense still to be delineated more closely.

But if "ending," as dying, constitutes the wholeness of Dasein, then the being of the wholeness itself must be conceived as an existential phenomenon of my own Dasein. In "ending," and in the being a whole of Dasein which is thus constituted, there is, according to its essence, no representation. The way out suggested fails to recognize this existential fact when it proposes the dying of others as a substitute theme for the analysis of wholeness.

Thus the attempt to make the being a whole of Dasein phenomenally accessible in an appropriate way gets stranded again. But the result of these considerations is not just negative. They were oriented toward the phenomenon, even if rather crudely. We have indicated that death is an existential phenomenon. Our inquiry is thus forced into a purely existential orientation toward Dasein which is in each case one's own. For the analysis of death as dying, there remains only the

---

* The relation of Dasein to death; death itself = its arrival—entrance, dying.

1. Cf. § 9.

possibility of bringing this phenomenon either to a purely *existential* concept or, on the other hand, of renouncing any ontological understanding of it.

Furthermore, it was evident in our characterization of the transition from Dasein to no-longer-being-there as no-longer-being-in-the-world that the going-out-of-the-world of *Dasein* in the sense of dying must be distinguished from a going-out-of-the-world of what is only living. The ending of what is only alive we formulate terminologically as perishing [Verenden]. The distinction can become visible only by distinguishing the ending characteristic of Dasein from the ending of a life.[2] Dying can, of course, also be conceived physiologically and biologically. But the medical concept of *"exitus"* does not coincide with that of perishing.

From the previous discussion of the ontological possibility of conceiving of death, it becomes clear at the same time that substructures of beings of a different kind of being (objective presence or life) thrust themselves to the fore unnoticeably and threaten to confuse the interpretation of the phenomenon, even the *first* appropriate *presentation* [*Vorgabe*] of it. We can cope with this problem only by looking for an ontologically adequate way of defining constitutive phenomena for our further analysis, such as end and totality.

§ 48. *What is Outstanding, End, and Wholeness*

Our ontological characterization of end and wholeness can only be preliminary in the scope of this inquiry. To perform this task adequately we must not only set forth the *formal* structure of end in general and wholeness in general. At the same time, we must disentangle the structural variations possible for them in different realms, that is, deformalized variations which are related to beings with a defined content and which are structurally determined in terms of their being [Sein]. This task again presupposes a sufficiently unequivocal and positive interpretation of the kinds of being that require a regional separation of the whole of beings. The understanding of these ways of being, however, requires a clarified idea of being in general. The task of adequately carrying out the ontological analysis of end and wholeness gets stranded not only because the theme is so far-reaching, but because there is a difficulty in principle: in order to master this task, we must presuppose that precisely what we are seeking in this inquiry (the meaning of being in general) is something that we have found already and with which we are quite familiar.

---

2. Cf. § 10.

241

In the following considerations, the "variations" in which we are chiefly interested are those of end and wholeness; these are ontological determinations of Dasein which are to lead to a primordial interpretation of this being. With constant reference to the existential constitution of Dasein already developed, we must try to decide how ontologically inappropriate to Dasein are the concepts of end and wholeness which initially press themselves upon us, no matter how categorically indefinite they remain. The rejection of such concepts must be further developed into a positive *assignment* to their specific realms. In this way our understanding of end and wholeness in their variant forms as existentials will be strengthened, and this guarantees the possibility of an ontological interpretation of death.

242

But even if the analysis of the end and wholeness of Dasein assumes such a wide-ranging orientation, this cannot mean that the existential concepts of end and wholeness are to be obtained by way of a deduction. On the contrary, it is a matter of taking the existential meaning of the coming-to-an-end of Dasein from Dasein itself and of showing how this "ending" can constitute a *being whole* of that being that *exists*.

What has been discussed up to now about death can be formulated in three theses:

1. As long as Dasein is, a not-yet belongs to it, which it will be—what is constantly outstanding.
2. The coming-to-its-end of what is not-yet-at-an-end (in which what is outstanding is, according to its being, removed) has the character of no-longer-Dasein.
3. Coming-to-an-end implies a mode of being in which each and every actual Dasein simply cannot be represented by someone else.

In Dasein there is inevitably [undurchstreichbar] a constant "lack of wholeness" which finds its end in death. But may we interpret the phenomenal fact that this not-yet "belongs" to Dasein as long as it is to mean that it is something *outstanding* [*Ausstand*]? With regard to what kind of beings do we speak of something outstanding? The expression means indeed what "belongs" to a being, but is still lacking. Outstanding, as lacking, is based on a belongingness. For example, the remainder of a debt still to be paid is outstanding. What is outstanding is not yet available. Liquidating the "debt" as paying off what is outstanding means that the money "comes in" and the remainder is paid off, so that the not-yet is, as it were, filled out until the sum owed is "all together." Thus, to be outstanding means that what belongs together is not yet together. Ontologically, this implies the unhandiness of those pieces to be still to be supplied which have the same kind of being as those already at hand. The latter in their turn do not have their kind of being modified by having the remainder come in. The persisting untogetherness is liquidated by a cumulative placing together. *The being [Seiende] in which something is still outstanding has the kind of being of something at hand.* We characterize the together, or the untogether based on it, as a *sum*.

243   The untogether belonging to such a mode of the together, lacking as something outstanding, can, however, by no means ontologically define the not-yet that belongs to Dasein as its possible death. Dasein does not have the kind of being of a thing at hand in the world at all. The together of the being that Dasein is "in running its course" until it has completed "its course" is not constituted by a "progressive" piecing-on of beings that, somehow and somewhere, are already at hand in their own right. That Dasein should *be* together only when its not-yet has been filled out is so far from being the case that precisely then it no longer is. Dasein always already exists in such a way that its not-yet *belongs* to it. But are there not beings which are as they are and to which a not-yet can belong, without these beings necessarily having the kind of being of Dasein?

For example, one can say that the last quarter of the moon is outstanding until it is full. The not-yet decreases with the disappearance of the shadow covering it. And yet the moon is, after all, always already objectively present as a whole. Apart from the fact that the moon is never *wholly* to be grasped even when it is full, the not-yet by no means signifies a not-yet-*being*-together of parts belonging together, but rather pertains only to the way we *grasp* it perceptually. The not-yet that belongs to Dasein, however, not only remains preliminarily and at times inaccessible to one's own or to others' experience, it "is" not yet "real" ["wirklich"] at all. The problem is not a matter of our *grasp* of the not-yet of the character of Dasein, but rather the possible *being* or *nonbeing* of this not-yet. Dasein, as itself, has to *become*, that is, *be*, what it is not yet. In order to thus be able, by comparison, to define the *being of the not-yet of the character of Dasein*, we must take into consideration these beings to whose kind of being becoming belongs.

For example, the unripe fruit moves toward its ripeness. In ripening, what it not yet is is by no means pieced together as something not-yet-objectively-present. The fruit ripens itself, and this ripening characterizes its being as fruit. Nothing we can think of which could be added on could remove the unripeness of the fruit, if this being did not ripen *of itself*. The not-yet of unripeness does not mean something other which is outstanding that could be objectively present in and with it in a way indifferent to the fruit. It means the fruit itself in its specific kind of being. The sum that is not yet complete is, as something at hand, "indifferent" to the unhandy remainder that is lacking. Strictly speaking, it can be neither indifferent to it nor not indifferent.

The ripening fruit, however, is not only not indifferent to its unripeness   *244*
as an other to itself, but, ripening, it *is* the unripeness. The not-yet is
already included in its own being, by no means as an arbitrary determination, but as a constituent. Correspondingly, Dasein, too, *is always
already its not-yet*[3] as long as it is.

What constitutes the "unwholeness" in Dasein, the constant being-
ahead-of-itself, is neither a summative together which is outstanding,
nor even a not-yet-having-become-accessible, but rather a not-yet that
every Dasein, as the being that it is, has to be. Still, the comparison
with the unripeness of the fruit does show essential differences despite
some similarities. To reflect on these differences means that we should
recognize how indefinite our previous discussion of end and ending
has hitherto been.

Ripening is the specific being of the fruit. It is also a kind of
being of the not-yet (unripeness), and is formally analogous to Dasein
in that the latter, as well as the former, always already *is* its not-yet
in a sense yet to be defined. But even then, this does not mean that
ripeness as "end" and death as "end" coincide with regard to their
ontological structure as ends. With ripeness, the fruit *fulfills* [*vollendet*]
itself. But is the death at which Dasein arrives a fulfillment in this
sense? It is true that Dasein has "completed its course" with its death.
Has it thus necessarily exhausted its specific possibilities? Rather, are
these possibilities not precisely what get taken from it? Even "unful-
filled" Dasein ends. On the other hand, Dasein so little needs to ripen
only with its death that it can already have gone beyond its ripeness
before the end. For the most part, it ends in unfulfillment, or else
disintegrated and used up.

Ending does not necessarily mean fulfilling oneself. It thus
becomes more urgent to ask *in what sense, if any, death must be grasped
as the ending of Dasein.*

In the first instance, ending means *stopping* [*Aufhören*], and it
means this in senses that are ontologically different. The rain stops.
It is no longer objectively present. The road stops. This ending does
not cause the road to disappear, but this stopping rather determines
the road as this objectively present one. Hence ending, as stopping,   *245*
can mean either to change into the absence of objective presence or,
however, to be objectively present only when the end comes. The latter kind of ending can again be determinative for an *unfinished* thing
objectively present, as a road under construction breaks off, or it may

---

3. The difference between whole and sum, ὅλον and πᾶν, *totum* and *compositum* is familiar to us ever since Plato and Aristotle. Of course, the systematics of the categorical
transformation already contained in this division is not yet *recognized* and conceptualized. For the beginning of a detailed analysis, cf. E. Husserl, *Logische Untersuchungen*,
vol. 2, Third Investigation: "On the Doctrine of Wholes and Parts."

rather constitute the "finishedness" of something objectively present—
the painting is finished with the last stroke of the brush.

But ending as getting finished does not include fulfillment. On
the other hand, whatever has got to be fulfilled must reach its possible finishedness. Fulfillment is the mode of "finishedness," and is
founded upon it. Finishedness is itself possible only as a determination
of something objectively present or at hand.

Even ending in the sense of disappearing can still be modified
according to the kind of being of the being. The rain is at an end,
that is, it has disappeared. The bread is at an end, that is, used up,
no longer available as something at hand.

*None of these modes of ending are able to characterize death appro-
priately as the end of Dasein.* If dying were understood as being-at-an-
end [Zu-Ende-sein] in the sense of an ending of the kind discussed,
Dasein would be posited as something objectively present or at hand.
In death, Dasein is neither fulfilled nor does it simply disappear; it has
not become finished or completely available as something at hand.

Rather, just as Dasein constantly already *is* its not-yet as long as
it is, it also always already *is* its end. The ending that we have in view
when we speak of death, does not signify a being-at-an-end of Dasein,
but* rather a *being toward the end* [Sein zum Ende] of this being. Death
is a way to be that Dasein takes over as soon as it is. "As soon as a
human being comes into life, he is old enough to die."[4]

Ending, as being toward the end, must be clarified ontologically
in terms of the kind of being of Dasein. And supposedly the possibility of an existing being of the not-yet that lies "before" the "end" will
become intelligible only if the character of ending has been determined
existentially. The existential clarification of being toward the end first
provides the adequate basis for defining the possible meaning of our
discussion of a totality of Dasein, if indeed this totality is to be constituted by death as an "end."

The attempt to reach an understanding of the wholeness appropriate to Dasein by starting with a clarification of the not-yet and proceeding to a characterization of ending has not attained its goal. It
showed only *negatively* that the not-yet which Dasein always *is* resists
an interpretation as something outstanding. The end *toward* which
Dasein *is*, as existing, remains inappropriately defined by being-at-an-
end. At the same time, however, our reflections should make it clear

246

---

* death as dying.

4. *Der Ackermann aus Böhmen,* ed. A. Bernt and K. Burdach, in *Vom Mittelalter zur Ref-
ormation: Forschungen zur Geschichte der deutschen Bildung,* ed. K. Burdach, vol. 3, part 2
(1917), chap. 20, p. 46.

that their course must be reversed. A positive characterization of the phenomena in question (not-yet-being, ending, wholeness) can be successful only when it is unequivocally oriented toward the constitution of being of Dasein. This unequivocal character, however, is protected in a negative way from being sidetracked when we have an insight into the regional belonging together of the structures of end and wholeness which belong to Dasein ontologically.

The positive, existential, and ontological interpretation of death and its character of end are to be developed following the guideline of the fundamental constitution of Dasein that has been attained thus far: the phenomenon of care.

§ 49. How the Existential Analysis of Death Differs from Other Possible Interpretations of this Phenomenon

The unequivocal character of the ontological interpretation* of death should be made more secure by explicitly bringing to mind what this interpretation can *not* ask about, and where it would be useless to expect information and instruction.

In the broadest sense, death is a phenomenon of life. Life† must be understood as a kind of being to which belongs a being-in-the-world. It can only be ontologically defined in a privative orientation to Dasein. Dasein, too, can be considered as mere life. For the biological and physiological line of questioning, it then moves into the sphere of being [Sein] which we know as the world of animals and plants. In this field, dates and statistics about the life-span of plants, animals, and human beings can be ontically ascertained. Connections between the life-span, reproduction, and growth can be known. The "kinds" of death, the causes, "arrangements," and ways of its occurrence can be investigated.[5]

An ontological problematic underlies this biological and ontic investigation of death. We must still ask how the essence of death is defined in terms of the essence of life. In a certain sense the ontic inquiry into death has always already decided about this. More or less clarified preconceptions of life and death are operative in it. These preliminary concepts need to be sketched out in the ontology of Dasein. Within the ontology of Dasein, which has *priority* over an ontology of life, the existential analytic of death is *subordinate* to a characterization of the fundamental constitution of Dasein. We called the ending of

247

---

* That is, fundamental ontology.
† If we are talking about human life, otherwise not—'world.'

5. Cf. E. Korschelt's comprehensive portrayal, *Lebensdauer, Altern und Tod,* 3rd edition, 1924, especially the rich bibliography, pp. 414ff.

what is alive *perishing* [*Verenden*]. Dasein, too, "has" its physiological death of the kind appropriate to anything that lives; it has it not
ontically in isolation, but as also determined by its primordial kind of
being. Dasein, too, can end without authentically dying, though on the
other hand, qua Dasein, it does not simply perish. We call this inter-
mediate phenomenon its *demise* [*Ableben*]. Let the term *dying* [*Sterben*]
stand for the *way of being* in which Dasein *is toward* its death. Thus we
can say that Dasein never perishes. Dasein can suffer demise only as
long as it dies. The medical and biological inquiry into demising can
attain results which can also become significant ontologically if the
fundamental orientation is ensured for an existential interpretation of
death. Or must sickness and death in general—even from a medical
point of view—be conceived primarily as existential phenomena?

The existential interpretation of death is prior to any biology
and ontology of life. But it also is the foundation for any biographico-
historical or ethnologico-psychological inquiry into death. A "typology" of "dying" characterizing the states and ways in which a demise
is "experienced" already presupposes the concept of death. Moreover,
a psychology of "dying" gives information about the "life" of the
"dying person" rather than about dying itself. That is only a reflection
of the fact that when Dasein dies—and even when it dies authentically—it is not a matter of an experience of, or in, its factical demise.
Similarly, the interpretations of death in primitive peoples, of their
behavior toward death in magic and cult, throw light primarily on the
understanding of *Dasein*; but the interpretation of this understanding
already requires an existential analytic and a corresponding concept
of death.

The ontological analysis of being-toward-the-end, on the other
hand, does not anticipate any existentiell stance toward death. If death
is defined as the "end" of Dasein, that is, of being-in-the-world, no ontic
decision has been made as to whether "after death" another being [Sein]
is still possible, either higher or lower, whether Dasein "lives on" or
even, "outliving itself," is "immortal." Nor is anything decided ontically
about the "otherworldly" and its possibility, any more than about the
"this-worldly," as if norms and rules for behavior toward death should
be proposed for "edification." But our analysis of death remains purely
"this-worldly" in that it interprets the phenomenon solely with respect
to the question of how it *enters into* each and every Dasein as its possibility-of-being. We cannot even *ask* with any methodological assurance
about what *is after death* until death is understood in its full ontological
essence. Whether such a question presents a possible *theoretical* question
at all is not to be decided here. The this-worldly, ontological interpretation of death comes before any ontic, other-worldly speculation.

Finally, an existential analysis of death lies outside the scope of
what might be discussed under the rubric of a "metaphysics of death."

248

The questions of how and when death "came into the world," what "meaning" it can and should have as an evil and suffering in the whole of beings—these are questions that necessarily presuppose an understanding not only of the character of being of death, but the ontology of the universe of beings as a whole, as well as the ontological clarification of evil and negativity in particular.

The existential analysis is methodologically prior to the questions of a biology, psychology, theodicy, and theology of death. Taken ontically, the results of the analysis show the peculiar *formality* and emptiness of any ontological characterization. However, that must not make us blind to the rich and complex structure of the phenomenon. Since Dasein never becomes accessible at all as something objectively present, because being possible belongs in its own way to its kind of being, even less may we expect to simply read off the ontological structure of death, if indeed death is an eminent possibility of Dasein.

On the other hand, our analysis cannot be supported by an idea of death that has been devised arbitrarily and at random. We can restrain this arbitrariness only by giving beforehand an ontological characterization of the kind of being in which the "end" enters into the average everydayness of Dasein. For this we need to envisage fully the structures of everydayness worked out earlier. The fact that existentiell possibilities of being toward death have their resonance in an existential analysis of death is implied by the essence of any ontological inquiry. All the more explicitly, then, must an existentiell neutrality go together with the existential conceptual definition, especially with regard to death, where the character of possibility of Dasein can be revealed most clearly of all. The existential problematic aims solely at developing the ontological structure of the being-*toward*-the-end of Dasein.[6]

249

6. The anthropology developed in Christian theology—from Paul to Calvin's *meditatio futurae vitae*—has always already viewed death together with its interpretation of "life." Dilthey, whose true philosophical tendencies aimed at an ontology of "life," could not fail to recognize its connection with death. "And finally, the relation which most deeply and universally defines the feeling of our Dasein—that of life toward death, for the limitation of our existence by death is always decisive for our understanding and our estimation of life." *Das Erlebnis und die Dichtung*, 5th edition, p. 230. Recently G. Simmel has also explicitly related the phenomenon of death to the definition of "life," however without a clear separation of the biological and ontic from the ontological and existential problematic. Cf. *Lebensanschauung: Vier metaphysische Kapitel*, 1918, pp. 99–153. For the present inquiry, compare *especially* K. Jaspers, *Psychologie der Weltanschauungen*, 3rd edition, 1925, p. 299ff. and especially 259–70. Jaspers understands death by following the guidelines of the phenomenon of the "limit situation" developed by him, whose fundamental significance lies beyond any typology of "attitudes" and "worldviews."

R. Unger took up Dilthey's suggestions in his work *Herder, Novalis und Kleist: Studien über die Entwicklung des Todesproblems im Denken und Dichten von Sturm und Drang zur Romantik*, 1922. Unger offers a major reflection on Dilthey's questions in the lecture: *Literaturgeschichte als Problemgeschichte: Zur Frage geisteshistorischer Synthese, mit besonderer Beziehung auf W. Dilthey* (*Schriften der Königsberger Gelehrten Gesellschaft*, Geisteswiss. Klasse I.1, 1924). Unger (pp. 17ff.) sees clearly the significance of phenomenological investigation for a more radical foundation of the "problems of life."

§ 50. *A Preliminary Sketch of the Existential and Ontological Structure of Death*

From our considerations of something outstanding, end, and wholeness, the necessity of interpreting the phenomenon of death as being-toward-the-end in terms of the fundamental constitution of Dasein has emerged. Only in this way can it become clear how a wholeness constituted by being-toward-the-end is possible in Dasein itself in accordance with its structure of being. We have seen that care is the fundamental constitution of Dasein. The ontological significance of this expression was expressed in the "definition": being-ahead-of-itself-already-being-in (the world) as being-together-with beings encountered (within the world).[7] Thus the fundamental characteristics of the being of Dasein are expressed: in being-ahead-of-itself, existence, in already-being-in . . . , facticity, in being-together-with . . . , falling prey. Provided that death belongs to the being of Dasein in an eminent sense, it (or being-toward-the-end) must be able to be defined in terms of these chacacteristics.

250

First, we must make clear in a preliminary way how the existence, facticity, and falling prey of Dasein reveal themselves in the phenomenon of death.

The interpretation of the not-yet, and thus also of the most extreme not-yet, of the end of Dasein in the sense of something outstanding, was rejected as inappropriate. For it included the ontological distortion of Dasein as something objectively present. Being-at-an-end means existentially being-toward-the-end. The most extreme not-yet has the character of something *to which Dasein relates itself*. The end is imminent for Dasein. Death is not something not yet objectively present, nor the last outstanding element reduced to a minimum, but rather an *imminence* [*Bevorstand*].

However, many things can be imminent for Dasein as being-in-the-world. The character of imminence is not in itself distinctive for death. On the contrary, this interpretation could even make us suspect that death would have to be understood in the sense of an imminent event to be encountered in the surrounding world. For example, a thunderstorm can be imminent, remodeling a house, the arrival of a friend, accordingly, beings which are objectively present, at hand or Dasein-with. Imminent death does not have this kind of being.

But a journey, for example, can also be imminent for Dasein, or a discussion with others, or renouncing something which Dasein itself can be: its own possiblities-of-being which are founded in being-with others.

---

7. Cf. § 41.

Death is a possibility of being that Dasein always has to take upon itself. With death, Dasein stands before itself in its *ownmost* potentiality-of-being. In this possibility, Dasein is concerned about its being-in-the-world absolutely [schlechthin]. Its death is the possibility of no-longer-being-able-to-be-there. When Dasein is imminent to itself as this possibility, it is *completely* thrown back upon its ownmost potentiality-of-being. Thus imminent to itself, all relations to other Dasein are dissolved in it. This nonrelational ownmost possibility is at the same time the most extreme one. As a potentiality of being, Dasein is unable to bypass the possibility of death. Death is the possibility of the absolute impossibility of Dasein. Thus *death* reveals itself as one's *ownmost, nonrelational, and insuperable [unüberholbar] possibility*. As such, it is *an eminent* imminence. Its existential possibility is grounded in the fact that Dasein is essentially disclosed to itself, and it is disclosed as being-ahead-of-itself. This structural factor of care has its most primordial concretion in being-toward-death. Being-toward-the-end becomes phenomenally clearer as being toward the eminent possibility of Dasein which we have characterized.

251

The ownmost, nonrelational, and insuperable possibility is not created by Dasein subsequently and occasionally in the course of its being. Rather, when Dasein exists, it is already *thrown* into this possibility. Initially and for the most part, Dasein does not have any explicit or even theoretical knowledge of the fact that it is delivered over to its death, and that death thus belongs to being-in-the-world. Thrownness into death reveals itself to it more primordially and penetratingly in the attunement of anxiety.[8] Anxiety in the face of death is anxiety "in the face of" the ownmost, nonrelational, and insuperable potentiality-of-being. What anxiety is about is being-in-the-world itself. What anxiety is about is simply the potentiality-of-being of Dasein. Anxiety about death must not be confused with a fear of one's demise. It is not an arbitrary and chance "weak" mood of an individual, but, as a fundamental attunement [Grundbefindlichkeit] of Dasein, it is the disclosedness of the fact that Dasein exists as thrown being-*toward*-its-end. Thus the existential concept of dying is clarified as thrown being toward the ownmost, nonrelational, and insuperable potentiality-of-being. Precision is gained by distinguishing this from mere disappearance, and also from merely perishing, and finally from the "experience" ["Erleben"] of a demise.

Being-toward-the-end does not first arise through some attitude which occasionally turns up; rather, it belongs essentially to the thrownness of Dasein which reveals itself in attunement (mood) in various ways. The factical "knowledge" or "lack of knowledge" preva-

---

8. Cf. § 40.

lent in Dasein as to its ownmost being-toward-the-end is only the expression of the existentiell possibility of maintaining itself in this being in different ways. That factically many people initially and for the most part do not know about death must not be used to prove that being-toward-death does not "universally" belong to Dasein; rather, it only proves that Dasein, fleeing *from* it, initially and for the most part covers over its ownmost being-toward-death. Dasein dies factically as long as it exists, but initially and for the most part in the mode of *falling prey*. For factically existing is not only generally and without further differentiation a thrown potentiality-for-being-in-the-world, but it is always already absorbed in the "world" taken care of. In this entangled being together with . . . , the flight from uncanniness [Unheimlichkeit] makes itself known, that is, the flight *from* its ownmost being-toward-death. Existence, facticity, falling prey characterize being-toward-the-end, and are accordingly constitutive for the existential concept of death. *With regard to its ontological possibility, dying is grounded in care.**

But if being toward death belongs primordially and essentially to the being of Dasein, it must also be demonstrated in everydayness, although initially in an inauthentic way. And if being-toward-the-end is even supposed to offer the existential possibility for an existentiell wholeness of Dasein, this would give the phenomenal confirmation for the thesis that care is the ontological term for the wholeness of the structural totality of Dasein. However, for the complete phenomenal justification of this statement, a *preliminary sketch* of the connection between being-toward-death and care is not sufficient. Above all, we must be able to see this connection in the *concretion* nearest to Dasein, its everydayness.

§ 51. *Being-toward-Death and the Everydayness of Dasein*

The exposition of everyday, average being-toward-death was oriented toward the structures of everydayness developed earlier. In being-toward-death, Dasein is related *to itself* as an eminent potentiality-of-being. But the self of everydayness is the they[9] that is constituted in the public interpretedness which expresses itself in idle talk. Thus, idle talk must make manifest in what way everyday Dasein interprets its being-toward-death. Understanding, which is also always attuned, that is, mooded, always forms the basis of this interpretation. Thus we must ask how the attuned understanding lying in the idle talk of the they has disclosed being-toward-death. How is the they related in

---

* But care presences [west] out of the truth of beyng [Seyns].

9. Cf. § 27.

an understanding way to the ownmost, nonrelational, and insuperable possibility of Dasein? What attunement discloses to the they that it has been delivered over to death, and in what way?

The publicness of everyday being-with-one-another "knows" death as a constantly occurring event, as a "case of death." Someone or another "dies," be it a neighbor or a stranger. People unknown to us "die" daily and hourly. "Death" is encountered as a familiar event occurring within the world. As such, it remains in the inconspicuousness[10] characteristic of everyday encounters. The they has also already secured an interpretation for this event. The "fleeting" talk about this, which is either expressed or else mostly kept back, says: one also dies at the end, but for now one is not affected [unbetroffen].

The analysis of "one dies" reveals unambiguously the kind of being of everyday being toward death. In such talk, death is understood as an indeterminate something which first has to show up from somewhere, but right now is *not yet present* for oneself, and is thus no threat. "One dies" spreads the opinion that death, so to speak, strikes the they. The public interpretation of Dasein says that "one dies," because in this way everyone can convince him/herself that in no case is it I myself, for this one is *no one*. "Dying" is leveled down to an event which does concern Dasein, but which belongs to no one in particular. If idle talk is always ambiguous, so is this way of talking about death. Dying, which is essentially and irreplaceably mine, is distorted into a publicly occurring event which the they encounters. Characteristic talk speaks about death as a constantly occurring "case." It treats it as something always already "real," and veils its character of possibility and concomitantly the two factors belonging to it, that is, its non-relationality and its insuperability. With such ambiguity, Dasein puts itself in the position of losing itself in the they with regard to an eminent potentiality-of-being that belongs to its own self. The they justifies and increases the *temptation* of covering over[11] for itself its ownmost being-toward-death.

The evasion of death which covers over dominates everydayness so stubbornly that, in being-with-one-another, those "closest by" often try to convince the one who is "dying" that he will escape death and soon return again to the tranquillized everydayness of his world taken care of. This "concern" has the intention of thus "comforting" the "dying person." It wants to bring him back to Dasein by helping him to veil completely his ownmost nonrelational possibility. Thus, the they provides a *constant tranquillization about death*. But, basically, this tranquillization is not only for the "dying person," but just as much for

253

254

10. Cf. § 16.
11. Cf. § 38.

"those comforting him." And even in the case of demise, the carefreeness that the public has provided for itself is still not to be disturbed and made uneasy by the event. Indeed, the dying of others is seen often as a social inconvenience, if not a downright tactlessness, from which the public should be spared.[12]

But along with this tranquillization, which keeps Dasein away from its death, the they at the same time justifies itself and makes itself respectable by silently ordering the way in which *one* is supposed to behave toward death in general. Even "thinking about death" is regarded publicly as cowardly fear, a sign of insecurity on the part of Dasein and a gloomy flight from the world. *The they does not permit the courage to have anxiety about death.* The dominance of the public interpretedness of the they has already decided what attunement is to determine our stance toward death. In anxiety about death, Dasein is brought before itself as delivered over to its insuperable possibility. The they is careful to distort this anxiety into the fear of a future event. Anxiety, made ambiguous as fear, is moreover taken as a weakness which no self-assured Dasein is permitted to know. What is "proper" according to the silent decree of the they is the indifferent calm as to the "fact" that one dies. The cultivation of such a "superior" indifference *estranges* [*entfremdet*] Dasein from its ownmost nonrelational potentiality-of-being.

Temptation, tranquillization, and estrangement, however, characterize the kind of being of *falling prey*. Entangled, everyday being-toward-death is a constant *flight from death*. Being *toward* the end has the mode of *evading that end*—reinterpreting it, understanding it inauthentically, and veiling it. Factically one's own Dasein is always already dying, that is, it is in a being-toward-its-end. And it conceals this fact from itself by reinterpreting death as a case of death occurring every day with others, a case which always assures us still more clearly that "one" is "oneself" still "alive." But in the entangled flight *from* death, the everydayness of Dasein bears witness to the fact that the they itself is always already determined *as being toward death*, even when it is not explicitly engaged in "thinking about death." *Even in average everydayness, Dasein is constantly concerned with its ownmost, nonrelational, and insuperable potentiality-of-being, even if only in the mode of taking care of things in a mode of untroubled indifference [Gleichgültigkeit] that opposes the most extreme possibility of its existence.*

The exposition of everyday being-toward-death, however, gives us at the same time a directive to attempt to secure a complete existential concept of being-toward-the-end, by a more penetrating interpretation

255

12. L. N. Tolstoi in his story "The Death of Ivan Ilyich" has portrayed the phenomenon of the disruption and collapse of this "one dies."

in which entangled being-toward-death is taken as an evasion *of death*. *That before which* one flees has been made visible in a phenomenally adequate way. We should now be able to project phenomenologically how evasive Dasein itself understands its death.[13]

§ 52. *Everyday Being-toward-Death and the Complete Existential Concept of Death*

Being-toward-the-end was determined in a preliminary existential sketch as being toward one's ownmost, nonrelational, and insuperable potentiality-of-being. Existing being [Sein] toward this possibility brings itself before the absolute impossibility of existence. Beyond this seemingly empty characteristic of being-toward-death, the concretion of this being revealed itself in the mode of everydayness. In accordance with the tendency toward falling prey essential to everydayness, being-toward-death proved to be an evasion of it, an evasion that covers over. Whereas previously our inquiry made the transition from the formal preliminary sketch of the ontological structure of death to the concrete analysis of everyday being-toward-the-end, we now wish to reverse the direction and attain the complete existential concept of death by filling out the interpretation of everyday being-toward-the-end.

The explication of everyday being-toward-death stayed with the idle talk of the they: one also dies sometime, but for the time being not yet. Up to now we solely interpreted the "one dies" as such. In the "also sometime, but for the time being not yet," everydayness acknowledges something like a *certainty* of death. Nobody doubts that one dies. But this "not doubting" need not already imply *that* kind of being-certain that corresponds to the way death—in the sense of the eminent possibility characterized above—enters into Dasein. Everydayness stops with this ambiguous acknowledgment of the "certainty" of death—in order to weaken the certainty by covering dying over still more and alleviating its own thrownness into death.   256

By its very meaning, this evasive covering over of death is *not* capable of being authentically "certain" of death, and yet it *is*. How does it stand with this "certainty of death"?

To be certain of a being means to *hold* it for true [für wahr *halten*] as something true: but truth means discoveredness of beings. All discoveredness, however, is ontologically based in the most primordial truth, in the disclosedness of Dasein.[14] As a being that is disclosed and disclosing, and one that discovers, Dasein is essentially "in the

---

13. Cf., with regard to this methodological possibility, what was said about the analysis of anxiety, § 40.
14. Cf. § 44.

truth." *But certainty is based in truth or belongs to it equiprimordially.* The expression "certainty," like the expression "truth," has a double meaning. Primordially, truth means the same as being-disclosive as a mode of behavior of Dasein. From this comes the derivative meaning: discoveredness of beings. Accordingly, certainty is primordially tantamount to being-certain as a kind of being of Dasein. However, in a derivative significance, any being of which Dasein can be certain is also called "certain."

One mode of certainty is *conviction.* In conviction, Dasein lets the testimony of the thing itself that has been discovered (the true thing itself) be the sole determinant for its being toward that thing understandingly. Holding-something-for-true [Für-wahr-halten] is adequate as a way of keeping-oneself-in-the-truth, if it is based on the discovered beings themselves, and as being [Sein] toward the beings thus discovered, has become transparent to itself with regard to its appropriateness to them. Something like this is lacking in any arbitrary fiction or in the mere "opinion" about a being.

The adequacy of holding-for-true is measured by the truth claim to which it belongs. This claim gets its justification from the kind of being of the beings to be disclosed, and from the direction of the disclosure. The kind of truth and, along with it, the certainty, changes with the various kinds of beings, and accords with the leading tendency and scope of the disclosure. Our present considerations are limited to an analysis of being-certain with regard to death; and this being-certain will, in the end, present us with an eminent *certainty of Dasein.*

For the most part, everyday Dasein covers over its ownmost, nonrelational, and insuperable possibility of being. This factical tendency to cover over confirms our thesis that Dasein, as factical, is in 257   "untruth."[15] Thus the certainty which belongs to such a covering over of being-toward-death must be an inappropriate way of holding-for-true, and not an uncertainty in the sense of doubting. Inappropriate certainty keeps that of which it is certain covered over. If "one" understands death as an event encountered in the surrounding world, the certainty related to this does not get at being-toward-the-end.

They say that it is certain that "death" comes. *They* say it and overlook the fact that, in order to be able to be certain of death, Dasein itself must always be certain of its ownmost, nonrelational, and insuperable potentiality-of-being. They say that death is certain, and thus implant in Dasein the illusion that it is *itself* certain of its own death. And what is the ground of everyday being-certain? Evidently it is not just mutual persuasion. Yet one experiences daily the "dying" of others. Death is an undeniable "fact of experience."

---

15. Cf. § 44, b.

The way in which everyday being-toward-death understands the certainty thus grounded betrays itself when it tries to "think" about death, even when it does so with critical foresight—that is to say, in an appropriate way. So far as one knows, all human beings "die." Death is probable to the highest degree for every human being, yet it is not "unconditionally" certain. Strictly speaking, "only" an *empirical* certainty may be attributed to death. Such certainty falls short of the highest certainty, the apodictical one, which we attain in certain areas of theoretical knowledge.

In this "critical" determination of the certainty of death and its imminence, what is manifested in the first instance is, once again, the failure to recognize the kind of being of Dasein and the being-toward-death belonging to it, a failure characteristic of everydayness. *The fact that demise, as an event that occurs, is "only" empirically certain, in no way decides about the certainty of death.* Cases of death may be the factical occasion for the fact that Dasein initially notices death at all. But, remaining within the empirical certainty which we character-ized, Dasein cannot become certain at all of death as it "is." Although in the publicness of the they Dasein seemingly "talks" only of this "empirical" certainty of death, *basically* it does *not* keep exclusively and primarily to those cases of death that merely occur. *Evading its death,* everyday being-toward-the-end is indeed certain of death in another way than it would itself like to realize in purely theoretical consider-ations. Everydayness mostly hides this "in another way" from itself. It does not dare become transparent to itself in this way. We have already characterized the everyday attunement that consists in an air of superiority with regard to the certain "fact" of death—a superiority that is "anxiously" concerned, while seemingly free of anxiety. In this attunement, everydayness acknowledges a "higher" certainty than the merely empirical one. One *knows* about the certainty of death, and yet "is" not really certain about it. The entangled everydayness of Das-ein knows about the certainty of death, and yet avoids *being*-certain. But in the light of what it evades, this evasion itself bears witness phenomenally to the fact that death must be grasped as the ownmost nonrelational, insuperable, *certain* possibility.

One says that death certainly comes, but not right away. With this "but . . . ," the they denies that death is certain. "Not right away" is not a merely negative statement, but a self-interpretation of the they with which it refers itself to what is initially accessible to Dasein to take care of. Everydayness penetrates to the urgency of taking care of things, and divests itself of the fetters of a weary, "futile think-ing about death." Death is postponed to "sometime later," by relying on the so-called "general opinion." Thus the they covers over what is peculiar to the certainty of death, *that is possible in every moment.*

258

Together with the certainty of death goes the *indefiniteness* of its when. Everyday being-toward-death evades this indefiniteness by making it something definite. But this procedure cannot mean calculating when the demise is due to arrive. Dasein rather flees from such definiteness. Everyday taking care of things makes definite for itself the indefiniteness of certain death by interposing before it those manageable urgencies and possibilities of the everyday matters nearest to us.

But covering over this indefiniteness also covers over certainty. Thus the ownmost character of the possibility of death gets covered over: a possibility that is certain, and yet indefinite, that is, possible at any moment.

Now that we have completed our interpretation of the everyday talk of the they about death and the way death enters Dasein, we have been led to the characteristics of certainty and indefiniteness. The full existential and ontological concept of death can now be defined as follows: *as the end of Dasein, death is the ownmost, nonrelational, certain, and, as such, indefinite and insuperable possibility of Dasein.* As the end of Dasein, *death is* in the being [Sein] of this being [Seienden] *toward its end.*

The delineation of the existential structure of being-toward-the-end helps unfold a kind of being of Dasein in which it can, *as Dasein*, be whole. The fact that even everyday Dasein *is* always already *toward its end*, that is, is constantly coming to grips with its own death, even though "fleetingly," shows that this end, which concludes and defines being-whole, is not something which Dasein ultimately arrives at only in its demise. In Dasein, existing [seiende] toward its death, its most extreme not-yet, which everything else precedes, is always already included. So if one has given an ontologically inappropriate interpretation of the not-yet of Dasein as something outstanding, any formal inference from this to the lack of wholeness of Dasein, will be incorrect. *The phenomenon of the not-yet has been taken from the ahead-of-itself; no more than the structure of care in general can it serve as a higher court that would rule against a possible, existent [existentes] wholeness; indeed, this ahead-of-itself first makes possible such a being-toward-the-end.* The problem of the possible being whole [Ganzseins] of the being [Seienden], which we ourselves in each instance are, is legitimate if care, as the fundamental constitution of Dasein, "is connected" with death as the most extreme possibility of this being.

Yet it remains questionable whether this problem has been adequately worked out. Being-toward-death is grounded in care. As thrown being-in-the-world, Dasein is always already delivered over to its death. Being toward its death, it dies factically and constantly as long as it has not reached its demise. That Dasein dies factically means at the same time that it has always already decided in this or that way in its being-toward-death. Everyday, entangled evasion *of* death is an *inauthentic* being *toward* it. Inauthenticity has possible authenticity as its basis.[16] Inauthenticity characterizes the kind of being in which Dasein diverts itself and for the most part has always diverted itself, but it does not have to do this necessarily and constantly. Because Dasein exists, it determines itself as the kind of being it is, and it does so always in terms of a possibility which it itself *is* and understands.

Can Dasein *authentically understand* its ownmost, nonrelational, insuperable, certain possibility that is, as such, indefinite? That is, can it maintain itself in an authentic being-toward-its-end? As long as this   *260* authentic being-toward-death has not been set forth and ontologically determined, there is something essentially lacking in our existential interpretation of being-toward-the-end.

Authentic being-toward-death signifies an existentiell possibility of Dasein. This ontic potentiality-of-being must in its turn be ontologically possible. What are the existential conditions of this possibility? How are they themselves to become accessible?

§ 53. *Existential Project of an Authentic Being-toward-Death*

Factically, Dasein maintains itself initially and for the most part in an inauthentic being-toward-death. How is the ontological possibility of an *authentic* being-toward-death to be characterized "objectively," if, in the end, Dasein is never authentically related to its end, or if this authentic being must remain concealed from others in accordance with its meaning? Is not the project of the existential possibility of such a questionable existentiell potentiality-of-being a chimerical undertaking? What is needed for such a project to get beyond a merely poetizing, arbitrary construction? Does Dasein itself provide directives for this project? Can the grounds for its phenomenal justification be taken from Dasein itself? Can our analysis of Dasein up to now give us any prescriptions for the ontological task we have now formulated, so that what we have before us can be kept on a secure path?

The existential concept of death has been established, and thus we have also established that to which an authentic being-toward-the-end should be able to relate itself. Furthermore, we have also characterized inauthentic being-toward-death and thus we have prescribed how authentic being-toward-death cannot be in a negative way. The existential structure of an authentic being-toward-death must let itself be projected with these positive and prohibitive instructions.

Dasein is constituted by disclosedness, that is, by attuned understanding. *Authentic* being-toward-death can*not evade* its ownmost, nonrelational possibility or *cover* it *over* in this flight and *reinterpret* it for

---

16. Regarding the inauthenticity of Dasein, see § 9, § 27, and in particular § 38.

the common sense of the they. The existential project of an authentic being-toward-death must thus set forth the factors of such a being [Sein] which are constitutive for it as an understanding of death—in the sense of being toward this possibility without fleeing it or covering it over.

261   First of all, we must characterize being-toward-death as a *being toward a possibility*, toward an eminent possibility of Dasein itself. Being toward a possibility, that is, toward something possible, can mean to be out for something possible, as in taking care of its actualization. In the field of things at hand and objectively present, we constantly encounter such possibilities: what is attainable, manageable, viable, and so forth. Being out for something possible and taking care of it has the tendency of *annihilating* the *possibility* of the possible by making it available. The actualization of useful things at hand in taking care of them (producing them, getting them ready, readjusting them, etc.), is, however, always only relative, insofar as even that which has been actualized still has the character of being relevant. Even when actualized, as something actual it remains possible for . . . , it is characterized by an in-order-to. Our present analysis should simply make clear how being out for something and taking care of it is related to the possible. It does so not in a thematic and theoretical reflection on the possible as possible, or even with regard to its possibility as such, but rather in such a way that it *circum*spectly looks *away* from the possible to what it is possible for.

Obviously, being-toward-death, which is now in question, cannot have the character of being out for something and taking care of it with a view toward its actualization. For one thing, death as something possible is not a possible thing at hand or objectively present, but a possibility-of-being of *Dasein*. Then, however, taking care of the actualization of what is thus possible would have to mean bringing about one's own demise. But by doing this Dasein would deprive itself of the very basis for an existing being-toward-death.

Thus if being-toward-death is not meant as an "actualization" of death, neither can it mean to dwell near the end in its possibility. This kind of behavior would amount to "thinking about death," thinking about this possibility, how and when it might be actualized. Brooding over death does not completely take away from it its character of possibility. It is always brooded over as something coming, but we weaken it by calculating how to have death under our control [Verfügenwollen]. As something possible, death is supposed to show as little as possible of its possibility. On the contrary, if being-toward-death has to disclose understandingly the possibility which we have characterized as *such*, then in such being-toward-death this possibility must not be weakened, it must be understood *as possibility*, cultivated *as possibility*, and *endured as possibility* in our relation to it.

However, Dasein relates to something possible in its possibility, by *expecting* [*Erwarten*] it. Anyone who is intent on something possible, may encounter it unimpeded and undiminished in its "whether it comes or not, or whether it comes after all." But with this phenomenon of expecting has our analysis not reached the same kind of being toward the possible which we already characterized as being out for something and taking care of it? To expect something possible is always to understand and "have" it with regard to whether and when and how it will really be objectively present. Expecting is not only an occasional looking away from the possible to its possible actualization, but essentially a *waiting for that actualization*. Even in expecting, one leaps away from the possible and gets a footing in the real. It is for its reality that what is expected is expected. By the very nature of expecting, the possible is drawn into the real, arising from it and returning to it.

262

But being toward this possibility, as being-toward-death, should relate itself to *death* so that it reveals itself, in this being [Sein] and for it, *as possibility*. Terminologically, we shall formulate this being toward possibility as *anticipation* [*Vorlaufen*] *of this possibility*. But does not this mode of behavior contain an approach [Näherung] to the possible, and does not its actualization emerge with the nearness [Nähe] of the possible? In this kind of coming near, however, one does not tend toward making something real available and taking care of it, but as one comes nearer understandingly, the possibility of the possible only becomes "greater." *The nearest nearness of being-toward-death as possibility is as far removed as possible from anything real*. The more clearly this possibility is understood, the more purely does understanding penetrate to it *as the possibility of the impossibility of existence* [*Existenz*] *in general*. As possibility, death gives Dasein nothing to "be actualized" and nothing which it itself could *be* as something real. It is the possibility of the impossibility of every mode of behavior toward . . . , of every way of existing. In anticipating this possibility, it becomes "greater and greater"; that is, it reveals itself as something which knows no measure at all, no more or less, but means the possibility of the measureless impossibility of existence. Essentially, this possibility offers no support for becoming intent on something, for "picturing" for oneself the actuality that is possible and so forgetting its possibility. As anticipation of possibility, being-toward-death first *makes* this possibility *possible* and sets it free as possibility.

Being-toward-death is the anticipation of a potentiality-of-being of *that* being whose kind of being is anticipation itself. In the anticipatory revealing of this potentiality-of-being, Dasein discloses itself to itself with regard to its most extreme possibility. But to project oneself upon one's ownmost potentiality of being means to be able to understand

263   oneself in the being [Sein] of the being [Seienden] thus revealed: to exist. Anticipation shows itself as the possibility of understanding one's *ownmost* and extreme potentiality-of-being, that is, as the possibility of *authentic existence*. Its ontological constitution must be made visible by setting forth the concrete structure of anticipation of death. How is the phenomenal definition of this structure to be accomplished? Evidently by defining the characteristics of anticipatory disclosure which must belong to it so that it can become the pure understanding of the ownmost, nonrelational, insuperable, certain, and, as such, indefinite possibility. We must remember that understanding does not primarily mean staring at a meaning, but understanding oneself in the potentiality-of-being that reveals itself in the project.[17]

Death is the *ownmost* possibility of Dasein. Being toward it discloses to Dasein its *ownmost* potentiality-of-being in which it is concerned about the being of Dasein absolutely. Here it can become evident to Dasein that in the eminent possibility of itself it remains torn away from the they, that is, anticipation can always already have torn itself away from the they. The understanding of this "ability," however, first reveals its factical lostness in the everydayness of the they-self.

The ownmost possibility is *nonrelational*. Anticipation lets Dasein understand that it has to take over solely from itself the potentiality-of-being in which it is concerned absolutely about its ownmost being. Death does not just "belong" in an undifferentiated way to one's own Dasein, but it *lays claim* on it as something *individual*. The nonrelational character of death, understood in anticipation, individualizes Dasein down to itself. This individualizing is a way in which the "there" is disclosed for existence. It reveals the fact that any being-together-with what is taken care of and any being-with others fails when one's ownmost potentiality-of-being is at stake. Dasein can *authentically* be *itself* only when it makes this possible of its own accord. But if taking care and being concerned fail us, this does not, however, mean at all that these modes of Dasein have been cut off from its authentic being a self. As essential structures of the constitution of Dasein they also belong to the condition of the possibility of existence in general. Dasein is authentically itself only insofar as it projects itself, *as* being-together with things taken care of and concernful being-with . . . , primarily upon its ownmost potentiality-of-being, rather than upon the

264   possibility of the they-self. Anticipation of its nonrelational possibility forces the being that anticipates into the possibility of taking over its ownmost being from itself of its own accord.

The ownmost, nonrelational possibility is *insuperable*. Being toward this possibility lets Dasein understand that the most extreme possibility

17. Cf. § 31.

of existence, that of giving itself up, is imminent. But anticipation does not evade the impossibility of bypassing death, as does inauthentic being-toward-death, but *frees* itself *for* it. Becoming free *for* one's own death in anticipation liberates one from one's lostness in chance possibilities urging themselves upon us, so that the factical possibilities lying before the insuperable possibility can first be authentically understood and chosen. Anticipation discloses to existence that its extreme possibility lies in giving itself up, and thus it shatters all one's clinging to whatever existence one has reached. In anticipation, Dasein guards itself against falling back behind itself, or behind the potentiality-for-being that it has understood. It guards against "becoming too old for its victories" (Nietzsche). Free for its ownmost possibilities, which are determined by the *end*, and so understood as *finite*, Dasein averts the danger that it may, by its own finite understanding of existence, fail to recognize that it is getting overtaken by the existence-possibilities of others, or that it may misinterpret these possibilities by forcing them upon its own, thus divesting itself of its ownmost factical existence. As the nonrelational possibility, death individualizes, but only, as the insuperable possibility, in order to make Dasein as being-with understand the potentialities-of-being of others. Because anticipation of the insuperable possibility also disclosed all the possibilities lying before it, this anticipation includes the possibility of taking the *whole* of Dasein in advance in an existentiell way, that is, the possibility of existing as a *whole potentiality-of-being*.

The ownmost, nonrelational, and insuperable possibility is *certain*. The mode *of being* certain of it is determined by the truth (disclosedness) corresponding to it. But Dasein discloses the certain possibility of death as possibility only by making this possibility as its ownmost potentiality-of-being *possible* in anticipating it. The disclosedness of this possibility is grounded in a making possible that anticipates. Holding oneself in this truth, that is, being certain of what has been disclosed, demands anticipation above all. The certainty of death cannot be calculated in terms of ascertaining cases of death encountered. This certainty by no means holds itself in the truth of something objectively present. When something objectively present has been discovered, it is encountered most purely by just looking at it and letting it be encountered in itself. Dasein must first have lost itself in the factual circumstances (this can be one of care's own tasks and possibilities) if it is to gain the pure objectivity [Sachlichkeit], that is, the indifference, of apodictic evidence. If being-certain in relation to death does not have this character, that does not mean it is of a lower grade, but rather that *it does not belong at all to the order of degrees of evidence about things objectively present*.

Holding death for true—death *is* in each instance only one's own—exhibits a different kind of certainty, and is more primordial

265

than any certainty related to beings encountered in the world or to formal objects, for it is certain of being-in-the-world. As such, holding death for true requires not just *one* definite kind of behavior of Dasein, but requires Dasein in the complete authenticity of its existence.[18] In anticipation, Dasein can first make certain of its ownmost being in its insuperable totality. Thus, the evidence of the immediate given-ness of experiences, of the ego, or of consciousness, necessarily has to lag behind the certainty contained in anticipation. And yet this is not because the kind of apprehension belonging to it is not rigorous enough, but because it fundamentally cannot hold *for true* (disclosed) something that it basically insists upon "having there" as true: name-ly, the Dasein which I myself *am* and can be as potentiality-of-being authentically only in anticipation.

The ownmost, nonrelational, insuperable, and certain possibility is *indefinite* with regard to its certainty. How does anticipation dis-close this character of the eminent possibility of Dasein? How does understanding, anticipating, project itself upon a definite potentiality-of-being which is constantly possible in such a way that the when in which the absolute impossibility of existence becomes possible remains constantly indefinite? In anticipating the indefinite certainty of death, Dasein opens itself to a constant *threat* arising from its own there. Being-toward-the-end must hold itself in this very threat, and can so little phase it out that it rather has to cultivate the indefinite-ness of the certainty. How is it existentially possible for this constant threat to be genuinely disclosed? All understanding is attuned. Mood brings Dasein before the thrownness of its "that-it-is-there."[19] *But the attunement which is able to hold open the constant and absolute threat to itself arising from the ownmost individualized being of Dasein is anxiety.*[20] In anxiety, Dasein finds itself *faced* with the nothingness of the possible impossibility of its existence. Anxiety is anxious *about* the potentiality-of-being of the being thus determined, and thus discloses the most extreme possibility. Because the anticipation of Dasein absolutely indi-vidualizes and lets it, in this individualizing of itself, become certain of the wholeness of its potentiality-of-being, the fundamental attunement of anxiety belongs to this self-understanding of Dasein in terms of its ground. Being-toward-death is essentially anxiety.* This is attested unmistakably, although "only" indirectly, by being-toward-death as we characterized it, when it distorts anxiety into cowardly fear and, in overcoming that fear, only makes known its own cowardliness in the face of anxiety.

---

* I.e., but not only anxiety and certainly not anxiety as mere emotion [Emotion].

18. Cf. § 62.
19. Cf. § 29.
20. Cf. § 40.

What is characteristic about authentic, existentially projected being-toward-death can thus be summarized as follows: *anticipation reveals to Dasein its lostness in the they-self, and brings it face to face with the possibility to be itself, primarily unsupported by concern that takes care, but to be itself in passionate, anxious **freedom toward death**, which is free of the illusions of the they, factical, and certain of itself.*

All relations belonging to being-toward-death, up to the complete content of the most extreme possibility of Dasein, gather themselves together to reveal, unfold, and hold fast the anticipation that they constitute and that makes this possibility possible. The existential project in which anticipation has been delimited, has made visible the *ontological* possibility of an existentiell, authentic being-toward-death. But with this, the possibility then appears of an authentic potentiality-for-being whole—*but only as an ontological possibility.* Of course, our existential project of anticipation stayed with those structures of Dasein gained earlier and let Dasein itself, so to speak, project itself upon this possibility, without proffering to Dasein the "content" of an ideal of existence forced upon it "from the outside." And yet this existentially "possible" being-toward-death remains, after all, existentielly a fantastical demand. The ontological possibility of an authentic potentiality-for-being-a-whole of Dasein means nothing as long as the corresponding ontic potentiality-of-being has not been shown in terms of Dasein itself. Does Dasein ever project itself factically into such a being-toward-death? Does it *demand*, even only on the basis of its ownmost being, an authentic potentiality of being which is determined by anticipation?

Before answering these questions, we must investigate to what extent *at all* and in what way Dasein *bears witness* [*Zeugnis gibt*] to a possible *authenticity* of its existence from its ownmost potentiality-of-being, in such a way that it not only makes this known as *existentielly* possible, but *demands* it of itself.

The question hovering over us of an authentic wholeness of Dasein and its existential constitution can be placed on a viable, phenomenal basis only if that question can hold fast to a possible authenticity of its being attested [*bezeugte*] by Dasein itself. If we succeed in discovering phenomenologically such an attestation [*Bezeugung*] and what is attested to in it, the problem arises again of *whether the anticipation of death projected up to now only in its ontological possibility has an essential connection with that authentic potentiality-of-being attested to.*

267

CHAPTER TWO

The Attestation of Dasein of an Authentic Potentiality-of-Being and Resoluteness

§ 54. *The Problem of the Attestation of an Authentic Existentiell Possibility*

We are looking for an authentic potentiality-of-being of Dasein that is attested by Dasein itself in its existentiell possibility. First of all, this attestation must let itself be discovered.* If it is to "give Dasein to understand" itself in its possible authentic existence, it will have its roots in the being of Dasein. The phenomenal demonstration of such an attestation thus contains evidence of its origin in the constitution of being of Dasein.

The attestation is to give us to understand an authentic *potentiality-of-being-one's-self*. With the expression "self," we answered the question of the *who* of Dasein.[1] The selfhood of Dasein was defined formally as a *way of existing*, that is, not as a being objectively present. *I myself* am not for the most part the who of Dasein, rather the they-self is. Authentic being-a-self shows itself to be an existentiell modification of the they, which is to be defined existentially.[2] What does this modification imply, and what are the ontological conditions of its possibility?

With the lostness in the they, the nearest, factical potentiality-of-being of Dasein—tasks, rules, standards, the urgency and scope of being-in-the-world as concerned and taking care—has already been decided upon. The they has always already kept Dasein from taking hold of these possibilities-of-being. The they even conceals the way it has silently disburdened Dasein of the explicit *choice* of these pos-

268

---

* (1) What attests as such (2) What is attested by it

1. Cf. § 25.
2. Cf. § 27, p. 126ff, esp. p. 130.

sibilities. It remains indefinite who is "really" ["eigentlich"] choosing. So Dasein is taken along by the no one, without choice, and thus gets caught up in inauthenticity. This process can be reversed only in such a way that Dasein explicitly brings itself back to itself from its lostness in the they. But this bringing-back must have *the* kind of being *by the neglect of which Dasein* has lost itself in inauthenticity. When Dasein thus brings itself back from the they, the they-self is modified in an existentiell manner so that it becomes *authentic* being-one's-self. This must be accomplished by *making up for not choosing*. But making up for not choosing signifies *choosing to make this choice*—deciding for a potentiality-of-being, and making this decision from one's own self. In choosing to make this choice, Dasein *makes possible*, for the first time, its authentic potentiality-of-being.*

But because Dasein is lost in the "they," it must first *find* itself. In order to find *itself* at all, it must be "shown" to itself in its possible authenticity. In terms of its *possibility*, Dasein *is* already a potentiality-for-being-its-self, but it needs to have this potentiality attested.

In the following interpretation, we shall claim that this potentiality is attested by that which, in the everyday self-interpretation of Dasein, is familiar to us as the *"voice of conscience."*[3] That the very "fact" of conscience has been disputed, that its function as a higher court for Dasein's existence has been variously assessed, and that "what conscience says" has been interpreted in manifold ways—all this might only mislead us into dismissing this phenomenon if the very "doubtfulness" of this fact—or of the way in which it has been interpreted—did not precisely *prove* that here a *primordial* phenomenon of Dasein lies before us. In the following analysis, conscience will be posited within the thematic plan of an existential investigation† which has fundamental ontology as its aim.

We shall first trace conscience back to its existential foundations and structures, and make it visible *as* a phenomenon of Dasein, holding fast to what we have hitherto arrived at as that being's constitution of being. The ontological analysis of conscience started in this way is prior to any psychological description and classification of experiences of conscience, just as it lies outside any biological "explanation," that is, dissolution, of this phenomenon. But it is no less distant from a

---

* happening of being—philosophy, freedom
† More radically now in terms of the essence of philosophizing.

3. These observations, and those which follow, were communicated as theses on the occasion of a public lecture on the concept of time, which was given at Marburg in July 1924.

theological exegesis of conscience or any employment of this phenomenon for proofs of God's existence or an "immediate" consciousness
of God.

Still, in our restricted inquiry into conscience, we must neither
exaggerate its importance nor make distorted claims about it and less-
en its worth. As a phenomenon of Dasein, conscience is not a fact that
occurs and is occasionally present. It "*is*" only in the kind of being
of Dasein and makes itself known as a fact only in factical existence.
The demand for an "inductive, empirical proof" for the "factuality" of
conscience and for the legitimacy of its "voice" is based on an ontological distortion of the phenomenon. But this distortion is also shared by
every superior critique of conscience as something that occurs only at
times rather than as a "universally established and ascertainable fact."
The fact of conscience cannot be coupled with such proofs and coun-
ter-proofs at all. That is not a lack, but only the mark [Kennzeichen]
that identifies it as ontologically different in kind from things present
in the surrounding world.

Conscience gives us "something" to understand, it *discloses*. From
this formal characteristic arises the directive to take this phenomenon
back into the *disclosedness* of Dasein. This fundamental constitution of
the being that we ourselves actually are is constituted by attunement,
understanding, falling prey, and discourse. A more penetrating analysis
of conscience reveals it as a *call* [Ruf]. Calling is a mode of *discourse*.
The call of conscience has the character of *summoning* [Anruf] Dasein
to its ownmost potentiality-of-being-a-self by *summoning* [Aufruf] it to
its ownmost being-guilty.*

But this existential interpretation is necessarily a far cry from
the everyday, ontic understanding, although it does expose the ontological foundations of what the vulgar interpretation of conscience
has always understood in certain limits and has conceptualized as a
"theory" of conscience. Thus our existential interpretation needs to
be confirmed by a critique of the vulgar interpretation of conscience.
Once this phenomenon has been exhibited, we can bring out to what
extent it bears witness to an authentic potentiality-of-being of Dasein.
To the call of conscience there corresponds a possible hearing. Under-   *270*
standing the summons reveals itself as *wanting to have a conscience*.
But in this phenomenon lies that existentiell choosing of the choice of

---

* *Schuldigsein.* The basic meaning of *Schuld* here is ontological, lacking something ontologically. Since "being a lack" is often linguistically cumbersome, we retain the terms
"guilt" and "guilty," bearing in mind that they are ontological, not "ethical" and certainly
not "theological." [TR]

being-a-self which we are looking for and which we call *resoluteness*[4] in accordance with its existential structure. Thus we have the divisions of the analyses of this chapter: the existential and ontological foundations of conscience (§ 55); the character of conscience as a call (§ 56); conscience as the call of care (§ 57); understanding the summons and being a lack (§ 58); the existential interpretation of conscience and the vulgar interpretation of conscience (§ 59); the existential structure of the authentic potentiality-of-being attested to in conscience (§ 60).

§ 55. *The Existential and Ontological Foundations* * *of Conscience*

The analysis[†] of conscience will start out with a neutral fact about this phenomenon: that it somehow gives one something to understand. Conscience discloses, and thus belongs to the scope of those existential phenomena which constitute the *being [Sein] of the there* as disclosedness.[5] We have analyzed the most general structures of attunement, understanding, discourse, and falling prey. If we put conscience in this phenomenal context, this is not a matter of a schematic application of the structures gained there to a particular "case" of the disclosure of Dasein. Rather, our interpretation of conscience will not only continue the earlier analysis of the disclosedness of the there, but will grasp it more primordially with regard to the authentic being of Dasein.

Through disclosedness, the being that we call Dasein is in the possibility of *being* its there. It is there for itself, together with its world, initially and for the most part in such a way that it has disclosed its potentiality-of-being in terms of the "world" taken care of. The potentiality-of-being as which Dasein exists has always already given itself over to definite possibilities. And this is the case because it is a thrown being, and its thrownness is disclosed more or less clearly and penetratingly by being attuned. Understanding belongs equiprimordially to attunement (mood). In this way Dasein "knows"[‡] where it stands, since it has projected itself upon possibilities of itself, or, absorbed in the they, has let itself be given such possibilities as are prescribed by

---

* Horizon.
† Here many things are necessarily conflated: (1) The call of what we call conscience (2) Being called (3) The experience of this being [Sein] (4) The usual, traditional interpretation (5) The way of coming to terms with it.
‡ Or thinks it knows.

4. *Entschlossenheit*. Literally in Heidegger's primary meaning "unlockedness," the emphasis being on freed and open *for* something. [TR]
5. Cf. § 28 et seq.

its public interpretedness. But this prescription is existentially possible
through the fact that Dasein as understanding being-with can *listen*   *271*
[*hören*]* to others. Losing itself in the publicness of the they and its idle
talk, it *fails to hear* [*überhört*] its own self in listening to the they-self.
If Dasein is to be brought back from this lostness of failing to hear
itself, and if this is to be done through itself, it must first be able to
find itself, to find itself as something that has failed to hear itself and
continues to do so in *listening to* [*Hinhören*] the they. This listening
must be stopped, that is, the possibility of another kind of hearing that
interrupts that listening must be given by Dasein itself. The possibility of such a breach lies in being summoned without any mediation.
Dasein fails to hear itself, and listens to the they, and this listening
gets broken by the call if that call, in accordance with its character as
call, arouses another kind of hearing which, in relation to the hearing
that is lost, has a character in every way opposite. If this lost hearing
is numbed [benommen] by the "noise" of the manifold ambiguity of
everyday "new" idle talk, the call must call silently, unambiguously,
with no foothold for curiosity. *That which, by calling in this way, gives
us to understand, is conscience* [*Gewissen*].

We take calling as a mode of discourse. Discourse articulates
intelligibility. The characterization of conscience as a call is by no
means only an "image," like the Kantian representation of conscience
as a court of justice. We must only not overlook the fact that vocal
utterance is not essential to discourse, and thus not for the call either.
Every speaking and "calling out" already presupposes discourse.[6] If
the everyday interpretation knows about a "voice" of conscience, it is
thinking not so much about an utterance, which can factically never
be found,† but "voice" is understood as giving-to-understand. In the
tendency toward disclosure of the call lies the factor of a jolt [Stoßes],
of an abrupt arousal.‡ The call calls from afar to afar. It reaches one
who wants to be brought back.§

But with this characterization of conscience, only the phenomenal
horizon for the analysis of its existential structure has been outlined.
We are not comparing this phenomenon with a call, but we are understanding it as discourse, in terms of the disclosedness constitutive for
Dasein. Our reflection avoids from the very beginning the path which
initially offers itself for an interpretation of conscience: one traces

---

* Where does this listening and being able to listen come from? Sensuous listening with
the ears is a thrown mode of being affected.
† We do not "hear" it with the senses.
‡ But it stops us, too.
§ Who has distanced himself from his own self.

6. Cf. § 34.

conscience back to a faculty of the soul, understanding, will, or feeling, or explains it as the product of a mixture of these. Confronted by a phenomenon such as conscience,* what is ontologically and anthropologically inadequate about a free-floating framework of classified faculties of the soul or personal acts becomes painfully obvious.[7]

272

§ 56. *The Character of Conscience as a Call*

To discourse belongs what is talked about in it. Discourse gives information about something in a certain respect. It draws from what is thus talked about what it actually says as this discourse, what is said as such. In discourse as communication, this becomes accessible to the Dasein-with of others, mostly by way of utterance in language.

What is talked about in the call of conscience, what is summoned? Obviously Dasein itself. This answer is just as incontestable as it is indefinite. If the call had such a vague goal, it would at best be an occasion for Dasein to be attentive to itself. But to Dasein essentially belongs the fact that it is disclosed to itself with the disclosedness of its world, so that it always already *understands itself*. The call reaches Dasein in this always-already-understanding-itself in everyday, average taking care of things. The call reaches the they-self of heedful being-with with others.

273

And to what is one summoned? To one's *own self*. Not to what Dasein is, can do, and takes care of in everyday being-with-one-another, not even to what has moved it, what it has pledged itself to, what it has let itself be involved with. Dasein that is understood in a worldly

---

* Namely, in view of its origin in being-a-self, but is this not just an assertion so far?

7. Besides Kant's, Hegel's, Schopenhauer's, and Nietzsche's interpretations of conscience, we should note M. Kähler, *Das Gewissen*, erster geschichtlicher Teil, 1878, and the article by the same author in the *Realenzyklopädie f. prot. Theologie und Kirche*. Furthermore, A. Ritschl. *Über das Gewissen*, 1876, reprinted in *Gesammelte Aufsätze*, Neue Folge, 1896, pp. 177ff. Finally, cf. the monograph just published of H. G. Stoker, *Das Gewissen*, in *Schriften zur Philosophie und Soziologie*, ed. Max Scheler, vol. 2, 1925. This broadly conceived inquiry clarifies a manifold richness of phenomena of conscience, characterizes critically the various possible ways of treating the phenomenon, and notes further literature which is not complete with regard to the history of the concept of conscience. Stoker's monograph differs from our existential interpretation already in his initial position, and thus also in its conclusions, despite many points of agreement. Stoker underestimates from the beginning the hermeneutical conditions for a "description" of the "objectively real conscience," p. 3. Hand in hand with this goes the muddling of the borderlines between phenomenology and theology—to the detriment of both. With regard to the anthropological foundations of the inquiry which Scheler's personalism takes over, cf. the present inquiry, § 10. Still, Stoker's monograph signifies considerable progress as compared with the traditional interpretations of conscience, but more by the comprehensive treatment of the phenomenon of conscience and its ramifications, than by pointing out the ontological roots of the phenomenon.

way for others and for itself is *passed over in this call*. The call to the self does not take the slightest notice of all this. Because only the *self* of the they-self is summoned and made to hear, the *they* collapses. The fact that the call *passes over* both the they and the public interpreted-ness of Dasein by no means signifies that it has not been *reached too*. Precisely in *passing over* the they, the call pushed it (adamant as it is about public recognition) into insignificance. But, robbed of its refuge and this subterfuge by the summons, the self is brought to itself by the call.

The they-self is summoned to the self. However, this is not the self that can become an "object" for itself on which to pass judgment, not the self that unrestrainedly dissects its "inner life" with excited curiosity, and not the self that stares "analytically" at states of the soul and their backgrounds. The summons of the self in the they-self does not force it inwards upon itself so that it can close itself off from the "external world." The call passes over all this and disperses it, so as to summon solely the self which is in no other way than being-in-the-world.

But how are we to define *what is talked about* in this discourse? *What* does conscience call to the one summoned? Strictly speaking— nothing. The call does not say anything, does not give any informa-tion about events of the world, has nothing to tell. Least of all does it strive to open a "conversation with itself" in the self which has been summoned. "Nothing" is called *to* the self which is summoned, but it is *summoned* to itself, that is, to its ownmost potentiality-of-being. In accordance with its tendency as call, the call does not mandate a "trial" for the self which has been summoned, but as a summons to the ownmost *potentiality*-of-being-a-self, it calls Dasein forth (ahead-of-itself) to its most unique possibilities.

The call is lacking any kind of utterance. It does not even come to words, and yet it is not at all obscure and indefinite. *Conscience speaks solely and constantly in the mode of silence*. Thus it not only loses none of its perceptibility [Vernehmlichkeit], rather it forces the Dasein thus summoned and called upon into the reticence of itself. The fact that what is called in the call is lacking a formulation in words does not push this phenomenon into the indefiniteness of a mysterious voice, but only indicates that the understanding of "what is called" may not cling to the expectation of anything like a communication.

What the call discloses is nevertheless unequivocal, even if it gets interpreted in different ways in individual Dasein in accordance with its possibilities of being understood. Whereas the content of the call is seemingly indefinite, the direction it takes is a sure one and is not to be overlooked. The call does not need to search gropingly for someone to be summoned, nor does it need a sign showing whether it is he who is meant or not. "Deceptions" occur in conscience not by an oversight of the call (a mis-calling), but only because the call is *heard* in such a way that, instead of being understood authentically, it is drawn by the they-self into a conversation with one's self in which one "makes deals" and in which the disclosing character of the call is distorted.

We must remember that when we designate conscience as a call, this call is a summons to the they-self in its self. As this summons [Anruf], it is the summons [Aufruf] of the self to its potentiality-of-being-a-self and thus calls it Dasein forth to its possibilities.

But we shall not obtain an ontologically adequate interpretation of conscience until we can clarify not only *who* is called by the call but also *who calls*, how the one who is summoned is related to the caller, how this "relation" is to be grasped ontologically as a connection of being.

§ 57. *Conscience as the Call of Care*

Conscience calls the self of Dasein forth from its lostness in the they. The self summoned remains indifferent and empty in its what. The call passes over *what* Dasein, initially and for the most part, understands itself *as* in its interpretation in terms of taking care. And yet the self is unequivocally and unmistakably reached. Not only is the call meant for one who is summoned "without regard to his person," the caller, too, remains in a striking indefiniteness. It not only fails to answer questions about name, status, origin, and repute, but also leaves not the slightest possibility of making the call familiar for an understanding of Dasein with a "worldly" orientation. On the other hand, it by no means disguises itself in the call. The caller of the call—and this belongs to its phenomenal character—absolutely distances itself from 275 any kind of becoming familiar. It goes against its kind of being to be drawn into any consideration and talk. The peculiar indefiniteness and indefinability of the caller is not nothing, but rather a *positive* distinction. It lets us know that the caller uniquely coincides with summoning to . . . , that it wants to be *heard only as such*, and not be chattered about any further. But is it then not suitable to the phenomenon to leave unasked the question of who the caller is? Yes, for the existentiell way of listening to the factical call of conscience, but not for the existential analysis of the facticity of calling and the existentiality of hearing.

But is it even necessary to keep explicitly raising the question of *who* is calling? Is this not answered for Dasein just as unequivocally as the question about the one who is summoned in the call? *Dasein calls itself in conscience*. This understanding of the caller may be more or less awakened in factically hearing the call. However, ontologically it is not enough to answer that Dasein is the caller and the one summoned *at the same time*. When Dasein is summoned, *is* it not "there" in another way from that in which it does the calling? Is it perhaps the ownmost potentiality-of-being that functions as the caller?

The call is precisely something that *we ourselves* have neither planned, nor prepared for, nor willfully brought about. "It" calls, against our expectations and even against our will. On the other hand, the call without doubt does not come from someone else who is with me in the world. The calls comes *from* me, and yet *over* me.

These phenomenal findings are not to be explained away. They were also taken as the point of departure for interpreting the voice of conscience as an alien power invading [hereinragenden] Dasein. Continuing in this direction of interpretation, one supplies an owner for the power thus localized, or else one takes that power as a person (God) making himself known. Conversely, one tries to reject this interpretation of the caller as the expression of an alien power and at the same time to explain conscience away "biologically." Both interpretations hastily pass over the phenomenal findings. Such procedures are made easier by the unspoken, but ontologically guiding, dogmatic thesis that what *is* (that is, anything so factual as the call) must be *objectively present*; what cannot be demonstrated as *objectively present* just *is not* at all.

As opposed to this methodological hastiness, we want not only to hold on to the phenomenal findings in general—the fact that the call, coming from me and over me, reaches me—but also to the implication that this phenomenon is here delineated ontologically as a phenomenon *of Dasein*. The existential constitution of *this* being can offer the sole guideline for the interpretation of the kind of being of the "it" that calls.

276

Does our previous analysis of the constitution of the being of Dasein show a way of making ontologically intelligible the kind of being of the caller, and thus also that of calling? The fact that the call is not explicitly brought about *by me*, but rather, "it" calls, does not justify looking for the caller in a being unlike Dasein. Dasein, after all, always exists factically. It is not an unattached self-projection, but its character is determined by thrownness as a fact of the being that it is, and thus determined, it has always already been delivered over to existence, and remains so constantly. The facticity of Dasein is, however, essentially distinguished from the factuality of something objectively present. Existing Dasein does not encounter itself as something objectively present within the world. But neither is thrownness attached to Dasein as an inaccessible quality that is of no importance to its existence. As thrown, Dasein has been thrown *into existence*. It exists as a being that has to be as it is and can be.

That it factically is might be concealed with regard to its *why*, but the "*that-it-is*" has *itself* been disclosed to Dasein. The thrownness of this being belongs to the disclosedness of the "there," and reveals itself constantly in each and every attunement. Attunement brings Dasein, more or less explicitly and authentically, before its "that it is, and as the being that it is, has to be as a potentiality-of-being." But for the most part, mood *closes off* thrownness. Dasein flees from thrownness to the relief that comes with the supposed freedom of the they-self. We characterized this flight as a fleeing in the face of the uncanniness that fundamentally determines individualized being-in-the-world. Uncanniness reveals itself authentically in the fundamental attunement of anxiety, and, as the most elemental disclosedness of thrown Dasein, it confronts being-in-the-world with the nothingness of the world about which it is anxious in the anxiety about its ownmost potentiality-of-being. *What if Dasein, finding itself in the ground of its uncanniness, were the caller of the call of conscience?*

Nothing speaks against this; but all the phenomena that were set forth up to now in characterizing the caller and its calling speak for it.

In its who, the caller is definable by *nothing* [*nichts*] "worldly." It is Dasein in its uncanniness, primordially thrown being-in-the-world, as not-at-home, the naked "that" in the nothingness [Nichts] of the world. The caller is unfamiliar to the everyday they-self; it is something like an *alien* voice. What could be more alien to the they, lost in the manifold "world" of its heedfulness, than the self individualized to itself in uncanniness thrown into nothingness? "It" calls, and yet gives the heedfully curious ears nothing to hear that could be passed along and publicly spoken about. But what should Dasein even report from the uncanniness of its thrown being? *What* else remains for it than its own potentiality-of-being revealed in anxiety? How else should it call than by summoning to this potentiality-of-being about which it is solely concerned?

The call does not report any facts; it calls without uttering anything. The call speaks in the uncanny mode of *silence*. And it does this only because in calling the one summoned, it does not call him into the public idle chatter of the they, but *calls* him *back* from that *to the reticence of his existent potentiality-of-being*. When the caller reaches one who is summoned, it does so with a cold assurance that is uncanny and by no means obvious. Wherein lies the basis for this assurance, if not in the fact that Dasein, individualized to itself in its uncanniness, is absolutely unmistakable to itself? What is it that takes away from Dasein so radically the possibility of misunderstanding itself from some other direction and failing to recognize itself, if not the abandonment [Verlassenheit] in being delivered over [Überlassenheit] to itself?

Uncanniness is the fundamental kind of being-in-the-world, although it is covered over in everydayness. Dasein itself calls as conscience from the ground of this being. The "it calls me" is an eminent kind of discourse of Dasein. The call attuned by anxiety first makes possible for Dasein its project upon its ownmost potentiality-of-being. The call of conscience, existentially understood, first makes known what was simply asserted[8] before: uncanniness pursues Dasein and threatens its self-forgetful lostness.

The statement that Dasein is at the same time the caller and the one summoned has now lost its empty formal character and its obviousness. *Conscience reveals itself as the call of care*: the caller is Dasein, anxious in thrownness (in its already-being-in . . .) about its potentiality-of-being. The one summoned is also Dasein, called forth to its ownmost potentiality-of-being (in its already-ahead-of-itself . . .). And what is called by the summons—out of falling prey to the they (already-being-together-with-the-world-taken-care-of)—is Dasein. The call of conscience, that is, conscience itself, has its ontological possibility in the fact that Dasein, in the ground of its being, is care.

278

Thus we need not resort to powers unlike Dasein, especially since recourse to these is so far from explaining the uncanniness of the call that it rather annihilates it. In the end, does not the reason for the farfetched "explanations" of conscience lie in the fact that we have not looked *long* enough to establish the phenomenal findings regarding the call, and have tacitly presupposed Dasein, coincidentally, to be in some kind of ontological determination or indetermination. Why should we look to alien powers for information before we have made sure that in starting our analysis we have not given *too low* an assessment of the being of Dasein, that is, as an innocuous subject occurring somehow, endowed with personal consciousness [Bewußtsein]?

And yet if the caller—who is "no one" viewed from the perspective of the world—is interpreted as a power, this seems to be an unprejudiced recognition of something "objectively ascertainable." But rightly considered, this interpretation is only a flight from conscience, a way out for Dasein along which it slips away from the thin wall that separates the they, so to speak, from the uncanniness of its being. This interpretation of conscience pretends to recognize the call in the sense of a "universally" binding voice that "does not speak just subjectively." Better yet, this "universal" conscience gets exalted to a "world conscience," which still has the phenomenal character of an "it" and "no one," yet which speaks—there in the individual "subject"—as this indefinite something.

But what else is this "public conscience" than the voice of the they? Dasein can only come up with the dubious idea of a "world conscience" *because* at bottom conscience is, in its essence, *in each instance mine*. It is this not only in the sense that one's ownmost potentiality-

---

8. Cf. § 40.

of-being is always summoned, but because the call comes from the being that I myself always am.

With this interpretation of the caller, which absolutely follows the phenomenal character of calling, the "power" of conscience is not diminished and made "merely subjective." On the contrary, the inexorability and unequivocal quality of the call thus becomes free for the first time. The "objectivity" of the summons thus is first justified when the interpretation leaves it its "subjectivity" which, of course, denies the they-self its dominance.

279   Nevertheless, this interpretation of conscience as the call of care will be countered by the question of whether any interpretation of conscience can stand up if it distances itself so far from "natural experience." How is conscience as that which *summons* us to our ownmost potentiality-of-being supposed to function when it, after all, initially and for the most part, *reprimands* and *warns*? Does conscience speak in so indefinite and empty a way about our ownmost potentiality-of-being? Does it not rather speak definitely and concretely in relation to failures and omissions which have already occurred and which we intended? Does the alleged summons come from a *"bad"* conscience or a *"good"* one? Does conscience give us anything positive at all? Does it not function rather only critically?

Such second thoughts are incontestably justified. We can demand in any interpretation of conscience that "one" should recognize in it the phenomenon in question, as it is experienced daily. But to do justice to this demand does not mean that the vulgar, ontic understanding of conscience must be recognized as the first court of appeal for an ontological interpretation. On the other hand, the second thoughts we are having are premature as long as the analysis of conscience to which they pertain has not reached its goal. Up to now we have tried solely to trace conscience *as a phenomenon of Dasein* back to the ontological constitution of this being. This served to prepare the task of making conscience intelligible as *an attestation* in Dasein of *its ownmost potentiality-of-being*.

But what conscience attests to becomes completely definite only when we have stated with sufficient clarity what quality the *listening* must have that genuinely corresponds to calling. The *authentic* understanding "following" the call is not an addition annexed to the phenomenon of conscience, a process that can either occur or else be lacking. The *complete* experience of conscience can only be grasped *from* understanding the summons together *with* it. If the caller and the one who is summoned are *themselves at the same time* one's own Dasein, then *a definite kind of being* of Dasein lies in every failure to hear the call or in a mishearing *of oneself*. Viewed existentially, an unattached call from which "nothing ensues" is an impossible fiction. "That *nothing* ensues" means something *positive* with regard to Dasein.

Thus only an analysis of understanding the summons can lead to an explicit discussion of *what the call gives to understand*. But only with our foregoing, general ontological characterization of conscience is the possibility given to comprehend existentially conscience's call of "guilty." All interpretations and experiences of conscience agree that the "voice" of conscience somehow speaks of "guilt" ["Schuld"].

280

§ 58. *Understanding the Summons and Guilt*

In order to grasp phenomenally what is heard in understanding the summons, we shall take up this summons anew. Summoning the they-self means calling forth the authentic self to its potentiality-of-being, as Dasein, that is, as being-in-the-world taking care of things and being-with others. The existential interpretation of that to which the call calls forth thus cannot seek to define any concrete individual possibility of existence if it understands itself correctly in its methodological possibilities and tasks. What can be established, and what seeks to get established, is not what is called in and to each particular Dasein from an existentiell standpoint, but is rather what *belongs* to the *existential condition of the possibility* of each and every factical and existentiell potentiality-of-being.

When the call is understood with an existentiell kind of hearing, such understanding is the more authentic the more Dasein hears and understands *its* own being summoned in a nonrelational way, and the less the meaning of the call gets distorted by what one says is considered proper and fitting. In what then does the authenticity of understanding the summons essentially reside? What is essentially *given* to understand in any particular call, even if it is not always factically understood?

We have already answered this question in our thesis that the call "says" *nothing* which could be talked about, it does not give any information about events. The call directs Dasein *forward toward* its potentiality-of-being, as a call *out of* uncanniness. The caller is indeed indefinite, but where it calls from is not indifferent for the calling. Where it comes from—the uncanniness of thrown individuation—is also called in the calling, that is, is also disclosed. Where the call comes from in calling forth to . . . is that to which it is called back. The call does not give us to understand an ideal, universal potentiality-of-being; it discloses it as what is actually individualized in that particular Dasein. The disclosive character of the call has not been completely determined until we understand it as a calling back that calls forth. Only if we are oriented toward the call thus understood may we ask *what* it gives to understand.

But is the question of what the call says not answered more easily and certainly by the "simple" reference to what we generally hear or fail

281   to hear in any experience of conscience: namely, that the call addresses Dasein as "guilty" or, as in the warning conscience, refers to a possible "guilt" or, as a "good" conscience, confirms that one is "conscious of no guilt"? If only this "guilty" about which our experiences are in such "agreement" were not defined in such completely different ways in the experience and interpretation of conscience! And even if the meaning of this "guilty" could be conceived in a way all could agree with, the *existential concept* of this being-guilty would still be obscure. However, when Dasein addresses itself as guilty, where should its idea of guilt be drawn from if not from the interpretation of its own being? But the question arises again: *who says how we are guilty and what guilt means?* The idea of guilt cannot be arbitrarily thought up and forced upon Dasein. But if an understanding of the essence of guilt is possible at all, this possibility must have been sketched out in Dasein beforehand. How are we to find that trace that can lead to revealing this phenomenon? All ontological inquiries into phenomena such as guilt, conscience, and death must start from what everyday Dasein "says" about them. Because its kind of being is entangled, the way Dasein gets interpreted is for the most part *inauthentically* "oriented" and does not get at the "essence," since the primordially appropriate ontological kind of questioning remains alien to it. Whenever we see something wrongly, a directive as to the primordial "idea" of the phenomenon is also revealed. But where do we get our criterion for the primordial, existential meaning of "guilty"? From the fact that this "guilty" turns up as a predicate of the "I am." Does what is understood as "guilt" in inauthentic interpretation possibly lie in the being of Dasein as such, in such a way that it *is* also already guilty in that it actually, factically exists?

Thus by invoking the "guilty," which everyone agrees that one hears, one has not yet answered the question of the existential meaning of what is called in the call. This must first be defined if we are to make intelligible what the call of "guilty" means, and why and how it gets distorted in its significance by the everyday interpretation.

Everyday common sense initially takes "being guilty" in the sense of "owing something," "having something on account." One is supposed to return something to the other which is due to him. This "being guilty" as *"having debts"* [*"Schulden haben"*] is a way of being-with with others in the field of taking care of things, as in providing something or bringing it along. Further modes of taking care of things are depriving, borrowing, withholding, taking, robbing; that is, in some way not doing
282   justice to the claims that others have made as to their possessions. This kind of being guilty is related to *things that can be taken care of.*

Being guilty has the further significance of *"being responsible for"* [*"schuld sein an"*], that is, being the cause or author of something or "being the occasion" for something. In the sense of this "being responsible" for something, one can "be guilty" without "owing" anything to someone else or coming to "owe" him. Conversely, one can owe something to another without being responsible for it oneself. Another person can "incur debts" to others "for me."

These vulgar significations of being guilty as "having debts with . . ." and "being responsible for . . ." can go together and determine a kind of behavior which we call *"making oneself responsible"* [*"sich schuldig machen"*]; that is, by having the responsibility for having a debt, one may break a law and make oneself punishable. However, the requirement that one fails to satisfy need not necessarily be related to possessions, it can regulate public being-with-one-another in general. This "making oneself responsible" by breaking a law, as we have defined it, can, at the same time, have the character of *"becoming responsible to others"* [*"Schuldigwerden an Anderen"*]. That does not occur by breaking a law as such, but through my having the responsibility [Schuld habe] for the other's becoming jeopardized in his existence, led astray, or even destroyed. This becoming responsible to others is possible without breaking the "public" law. The formal concept of being responsible in the sense of having become responsible to others can be defined as *being-the-ground* for a lack in the Dasein of another, in such a way that this being-the-ground itself is defined as "lacking" in terms of that for which it is the ground. This kind of lacking is a failure to satisfy some demand placed on one's existing being-with with others.

It remains a question how such demands arise, and in what way the character of their demands and laws is to be conceived on the basis of this origin. In any case, *being guilty* in this latter sense of breaking a "moral requirement" is a *kind of being of Dasein*. Of course, that is also true of being guilty as "making oneself punishable," as "having debts," and of any "having responsibility for . . .". These, too, are modes of behavior of Dasein. Very little is said by taking "burdened with moral guilt" as a "quality" of Dasein. On the contrary, it only thus becomes evident that this characterization is not sufficient for distinguishing ontologically between this kind of "determination of being" of Dasein and the other ways of behaving just listed. The concept of moral guilt is thus so little clarified ontologically that interpretations *283* of this phenomenon which include in its concept, or even employ in its definition, the ideas of deserving punishment or of having debts to someone, could become and remain prevalent. But in this way the "guilty" is again forced aside into the realm of taking care of things in the sense of calculating and counterbalancing claims.

The clarification of the phenomenon of guilt which is not necessarily related to "having debts" and breaking the law, can be successful only if we ask beforehand in principle about the *being*-guilty of Dasein; that is, only if the idea of "guilty" is conceived in terms of the kind of being of Dasein.

For this purpose, the idea of "guilty" must be *formalized* to the extent that the vulgar phenomena of guilt, which one related to being-with others in taking care of things, drop out. The idea of guilt must not only be removed from the area of calculating and taking care of things, but must also be separated from the relationship to an ought and a law such that by failing to comply with it one burdens oneself with guilt. For here, too, guilt is still necessarily defined as a *lack*, when something which ought to be and can be is missing. But to be missing means not being present. A lack, as the not being present of what ought to be, is a determination of being of objective presence. In this sense nothing can be essentially lacking in existence, not because it is complete, but because its character of being is distinguished from any kind of objective presence.

Still, the character of the *not* [*Nicht*] is present in the idea of "guilty." If the "guilty" is to be able to define existence, the ontological problem arises here of clarifying existentially the *not-character* of this not. Furthermore, there belongs to the idea of "guilty" that which is expressed without differentiation in the concept of guilt as "being responsible for": being-the-ground for. . . . Thus we define the formal existential idea of "guilty" as being-the-ground for a being [Sein] which is determined by a not—that is, *being-the-ground of a nullity* [*Nichtigkeit*]. If the idea of the *not* present in the existentially understood concept of "guilt" excludes relatedness to anything objectively present which is possible or which ought to be, if thus Dasein is altogether incommensurate with something objectively present or valid which it itself is not, or which is not in the way Dasein is, that is, *exists*, then any possibility that, with regard to being-the-ground for a lack, the being that is itself such a ground might be reckoned as "deficient," is a possibility that is excluded. If a lack, such as a failure to fulfill some requirement, has been "caused" in a way characteristic of Dasein, we cannot simply calculate back to a deficiency of the "cause." Being-the-ground-for . . . need not have the same character of not as the *privativum*, which is grounded in it and arises from it. The ground need not acquire its nullity from that which grounds it. But this means that *being-guilty does not result from an indebtedness, but the other way around: indebtedness is possible only "on the basis" of a primordial being guilty.* Can something like this be pointed to in the being of Dasein, and how is it existentially possible at all?

284

The being of Dasein is care. It includes in itself facticity (thrownness), existence (project) and falling prey. Dasein exists as thrown, brought into its there *not* of its own accord. It exists as a potentiality-of-being which belongs to itself, and yet has *not* given itself to itself. Existing, it never gets back behind its thrownness so that it could ever release this "that it is and has to be" from *its being* a self and lead it into the there. But thrownness does not lie behind it as an event which actually occurred, something that happened to it and was again separated from Dasein. Rather, as long as it is, *Dasein is* constantly its "that" as care. *As this being*, delivered over to which it can exist uniquely as the being which it is, it is, *existing*, the ground of its potentiality-of-being. Even though it has *not* laid the ground *itself*, it rests in the weight of it, which mood reveals to it as a burden.

And how *is* Dasein this thrown ground? Only by projecting itself upon the possibilities into which it is thrown. The self, which as such has to lay the ground of itself, can *never* gain power over that ground, and yet it has to take over being the ground in existing. To be its own thrown ground is the potentiality-of-being about which care is concerned.

Being the ground [Grund-seiend], that is, existing as thrown, Dasein constantly lags behind its possibilities. It is never existent *before* its ground, but only *from it* and *as it*. Thus being the ground means *never* to gain power over one's ownmost being from the ground up. This *not* [*Nicht*] belongs to the existential meaning of thrownness. Being the ground [Grund-seiend], it itself *is* a nullity of itself. Nullity by no means signifies not being present or not subsisting, but means a not that constitutes this *being* of Dasein, its thrownness. The quality of this not as a not is determined existentially. Being [seiend] a *self*, Dasein, *as* self, is the thrown being. *Not through* itself, but *released to* itself from the ground in order to be *as the ground*. Dasein is not itself the ground of its being, because the ground first arises from its own project, but as a self, it is the *being* [*Sein*] of its ground. The ground is always ground only for a being whose being has to take over being-the-ground.   285

Dasein is its ground by existing, that is, in such a way that it understands itself in terms of possibilities and, thus understanding itself, is thrown being. But this means that, as a potentiality-of-being, it always stands in one possibility or another; it is constantly *not* other possibilities and has relinquished them in its existentiell project. As thrown, the project is not only determined by the nullity [Nichtigkeit] of being-the-ground, but is itself *as project* essentially *null* [*nichtig*]. Again, this definition by no means signifies the ontic property of being "unsuccessful" or "of no value" but an existential constituent of the structure of being of projecting. This nullity belongs to the being-free of Dasein for its existentiell possibilities. But freedom *is* only in the choice of the one, that is, in bearing the fact of not having chosen and not being able also to choose the others.

In the structure of thrownness, as well as in that of the project, essentially lies a nullity. And it is the ground for the possibility of the nullity of *in*authentic Dasein in its falling prey which it always already in each instance factically is. *Care itself is in its essence thoroughly permeated with nullity.* Care, the being of Dasein, thus means, as thrown project: being the (null) ground of a nullity. And that means that *Dasein is as such guilty*, if our formal existential definition of guilt as being-the-ground of a nullity is valid.

Existential nullity by no means has the character of a privation, of a lack as compared with an ideal which is set up but is not attained in Dasein; rather, the being of this being is already null *as project before* everything that it can project and usually attains. Thus this nullity does not occur occasionally in Dasein, attached to it as a dark quality that it could get rid of if it made sufficient progress.

Still, the *ontological meaning of the notness* [*Nichtheit*] of this existential nullity remains obscure. But that is true also of the *ontological essence of the not in general*. Of course, ontology and logic have expected much of the not, and thus at times made its possibilities visible without revealing it itself ontologically. Ontology found the not and made use of it. But is it then so self-evident that every not means a *negativum* in the sense of a lack? Does its positivity get depleted insofar as it constitutes a "transition"? Why does every dialectic take refuge in negation, without grounding negation *itself* dialectically, without even being able to locate it *as a problem*? Has anyone ever made the *ontological origin* of notness a problem at all, or, *before that*, even looked for the *conditions* on the basis of which the problem of the not and its notness, and the possibility of this notness, could be raised? And where else should they be found *than in a thematic clarification of the meaning of being in general*?

286

The concepts of privation and lack which, moreover, are hardly transparent, are insufficient for the ontological interpretation of the phenomenon of guilt, though if we take them formally enough, we can put them to considerable use. Least of all, can we get nearer to the existential phenomenon of guilt by taking our orientation toward the idea of evil, the *malum* as *privatio boni*. The *bonum* and the *privatio* have the same ontological provenance in the ontology of *objective presence* which also characterizes the idea of "value" derived from that.

Beings whose being is care can not only burden themselves with factical guilt, but they *are* guilty in the ground of their being. This being guilty first gives the ontological condition for the fact that Dasein can become guilty while factically existing. This essential being guilty is, equiprimordially, the existential condition of the possibility of the "morally" good and evil, that is, for morality in general and its possible factical forms. Primordial being guilty cannot be defined by morality because morality already presupposes it for itself.

But what experience speaks for this primordial being-guilty of Dasein? However, one should not forget the counter-question: "is" guilt "there" only if a consciousness of guilt is awakened, or does not the most primordial being-guilty make itself known in the very fact that guilt "is sleeping"? The fact that this primordial being-guilty initially and for the most part remains undisclosed and is kept closed off by the entangled being of Dasein *reveals* only this aforesaid nullity. *Being*-guilty is more primordial than any *knowing* [*Wissen*] about it. Only because Dasein is guilty in the ground of its being, and closes itself off from itself as thrown and fallen prey, is conscience possible, if indeed the call basically gives us to understand *this being guilty*.

The call is the call of care. Being guilty constitutes the being that we call care. Dasein stands primordially together with itself in uncanniness. Uncanniness brings this being face to face with its undisguised nullity, which belongs to the possibility of its ownmost potentiality-of-being. Insofar as Dasein—as care—is concerned about its being, it calls itself as a they that has factically fallen prey, and calls itself from its uncanniness to its potentiality-of-being. The summons calls back by calling forth: *forth* to the possibility of taking over in existence the thrown being that it is, *back* to thrownness in order to understand it as the null ground that it has to take up into existence. The calling back in which conscience calls forth gives Dasein to understand that Dasein itself—as the null ground of its null project, standing in the possibility of its being—must bring itself back to itself from its lostness in the they, and this means that it is *guilty*.

What Dasein thus gives itself to understand would then, after all, be a knowledge [*Kenntnis*] about itself. And the hearing corresponding to that call would be a *taking notice* [*Kenntnisnahme*] of the fact of being "guilty." But if the call is indeed to have the character of a summons, does not this interpretation of conscience lead to a complete distortion of its function? Summoning to being-guilty, is that not summoning to evil?

Even the most violent interpretation would not wish to impose upon conscience such a meaning for the call. But then what is "summoning to being-guilty" supposed to mean?

The meaning of the call becomes clear if our understanding of it keeps to the existential meaning of being-guilty, instead of making basic the derivative concept of guilt in the sense of an indebtedness "arising" from some deed done or left undone. Such a demand is not arbitrary if the call of conscience, coming from Dasein itself, is directed solely to this being. But then summoning to being-guilty means a calling forth to the potentiality-of-being that I always am as Dasein. Dasein need not first burden itself with "guilt" through failures or omissions; it must only *be authentically* the "guilty" that it is.

Then the correct hearing of the summons is tantamount to understanding oneself in one's ownmost potentiality-of-being, that is, in projecting oneself upon one's *ownmost* authentic potentiality for becoming guilty. When Dasein understandingly lets itself be called forth to this possibility, this includes its *becoming free* for the call: the readiness for being able to be summoned. Understanding the call, *Dasein listens to its ownmost possibility of existence.* It has chosen itself.

288   With this choice, Dasein makes possible its ownmost being-guilty, which remains closed off to the they-self. The common sense of the they recognizes only what satisfies and what fails to satisfy with respect to manageable rules and public standards. It calculates infractions of them and tries to balance them off. The they-self has slunk away from its ownmost being-guilty, and so it talks about mistakes all the more vociferously. But in the summons, the they-self is summoned to the ownmost being-guilty of the self. Understanding the call is choosing, but it is not a choosing of conscience, which as such cannot be chosen. What is chosen is *having* a conscience as being free for one's ownmost being-guilty. *Understanding the summons* means: *wanting to have a conscience.*

This does not mean wanting to have a "good conscience," nor does it mean willfully cultivating the "call"; it means solely the readiness to be summoned. Wanting to have a conscience is just as far away from searching out one's factical indebtedness as it is from the tendency to *liberation* from guilt in the sense of the essential "guilty."

Wanting to have a conscience is rather the most primordial existentiell presupposition for the possibility of factically becoming guilty. Understanding the call, Dasein lets its ownmost self *take action in itself* in terms of its chosen potentiality-of-being. Only in this way can it *be* responsible [verantwortlich]. But factically every action is necessarily "without conscience," not only because it does not avoid factical moral indebtedness, but because, on the basis of the null ground of its null project, it has always already become guilty toward others in being-with with them. Thus wanting to have a conscience takes over the essential lack of conscience within which alone there is the existentiell possibility of *being* "good."

Although the call does not give us any information, it is not merely critical, but *positive.* It discloses the most primordial potentiality-of-being of Dasein as being-guilty. Thus, conscience reveals itself as an *attestation* belonging to the being of Dasein—an attestation in which conscience calls Dasein forth to its ownmost potentiality-of-being. Can the authentic potentiality-of-being thus attested to be defined existentially in a more concrete way? But now that we have shown a potentiality-of-being that is attested to in Dasein itself, a preliminary question arises: can we claim sufficient evidential weight for the way we have shown this, as long as the strange feeling remains that the interpretation of conscience has been traced back one-sidedly to the constitution of Dasein while hastily passing over all of the findings that are familiar to the vulgar interpreta-

289   tion of conscience? Is the phenomenon of conscience still recognizable at all, as it "really" is, in our interpretation? Have we not been all too

sure of ourselves in the ingenuousness with which we deduced an idea of conscience from the constitution of being of Dasein?

The last step in our interpretation of conscience is the existential delimitation of the authentic potentiality-of-being to which conscience attests. If we are to assure ourselves of a way of access that will make such a step possible even for the vulgar understanding of conscience, then we need explicit evidence for the connection between the results of the ontological analysis and the everyday experiences of conscience.

§ 59. *The Existential Interpretation of Conscience and the Vulgar Interpretation of Conscience*

Conscience is the call of care from the uncanniness of being-in-the-world that summons Dasein to its ownmost potentiality-for-being-guilty. We showed that wanting-to-have-a-conscience corresponded to understanding the summons. Both of these characterizations are not immediately harmonious with the vulgar interpretation of conscience. Indeed, they seem to be in direct conflict with it. We call this interpretation of conscience vulgar because in characterizing this phenomenon and describing its "function" it keeps to what *they* know as conscience, how they follow it or fail to follow it.

But *must* the ontological interpretation agree with the vulgar interpretation at all? Should not the latter be, in principle, ontologically suspect? If Dasein initially and for the most part understands itself in terms of what it takes care of, and if it interprets all its modes of behavior as taking care of things, then will there not be falling prey and covering over in its interpretation of precisely *the* way of its being that, as a call, seeks to bring it back from its lostness in the cares of the they? Everydayness takes Dasein as something at hand that is taken care of, that is, is regulated and calculated. "Life" is a "business," whether or not it covers its costs.

With regard to the vulgar kind of being of Dasein itself, there is thus no guarantee that the interpretation of conscience arising from it, or the theories of conscience oriented toward it, have attained the appropriate ontological horizon for its interpretation. Nevertheless, even the vulgar experience of conscience must somehow—pre-ontologically—get at the phenomenon. Two things follow from this. On the one hand, the everyday interpretation of conscience cannot be valid as the ultimate criterion for the "objectivity" of an ontological analysis. On the other hand, such an analysis is not justified in elevating itself over the everyday understanding of conscience and passing over the anthropological, psychological, and theological theories of conscience based on it. *If* the existential analysis has exposed the phenomenon of conscience in its

290

ontological roots, then the vulgar interpretations must be intelligible precisely in terms of that analysis; they must be intelligible finally in the ways they miss the phenomenon and why they cover it over. However, since in the context of problems in this inquiry the analysis of conscience is only subservient to the ontological, fundamental question, the characterization of the connection between the existential interpretation of conscience and the vulgar interpretation of conscience will have to be satisfied with an indication of the essential problems.

In this vulgar interpretation of conscience there are four objections to our interpretation of conscience as the summons of care to being-guilty:

1. Conscience has an essentially critical function.
2. Conscience always speaks relative to a definite deed that has been done or wished for.
3. According to experience, the "voice" is never related so radically to the being of Dasein.
4. Our interpretation pays no attention to the basic forms of the phenomenon, to "evil" and "good" conscience, to what "reprimands" and "warns."

Let us begin our discussion with the last reservation. In all interpretations of conscience, it is the "evil" or "bad" conscience that has priority. Conscience is primarily "bad"; such a conscience makes known to us that in every experience of conscience something like a "guilty" gets experienced first. But in the idea of bad conscience how is this making itself known of evil understood? The "experience of conscience" turns up *after* the deed has been done or left undone. The voice follows up the transgression and points back to the event through which Dasein has burdened itself with guilt. If conscience makes known a "being-guilty," this cannot occur as a summons to . . . , but as a pointing that reminds us of the guilt incurred.

But does the "fact" that the voice comes later prevent the call from being basically a calling forth? The fact that the voice is grasped as a stirring of conscience that *follows after* is not yet evidence for a primordial understanding of the phenomenon of conscience. What if the factical indebtedness were only the occasion for the factical calling of conscience? What if the interpretation we described of "bad" conscience got stuck halfway? That this is true can be seen from the ontological forehaving within whose scope the phenomenon has been brought by this interpretation. The voice is something that turns up, it has its place in the series of present experiences, and it follows after the experience of the deed. But neither the call, nor the past deed, nor the guilt assumed are events with the character of something pres-

291

ent that runs its course. The call has the kind of being of care. In the call, Dasein "is" ahead of itself in such a way that at the same time it directs itself back to its thrownness. Only by first positing that Dasein is a serial connection of successive experiences, is it possible to take the voice as something coming afterwards, something later that necessarily refers back. The voice does call back, but it calls back beyond the past deed into thrown being-guilty, which is "earlier" than any indebtedness [Verschuldung]. But the call back at the same time calls forth a *being*-guilty, as something to be seized upon in one's own existence, in such a way that authentic, existentiell *being*-guilty precisely "comes after" the call, and not the other way around. Basically, bad conscience is so far from reproving and pointing back that it rather points forward by calling back into thrownness. *The order of succession in which experiences run their course does not yield the phenomenal structure of existing.*

If the characterization of "bad" conscience does not get at the primordial phenomenon, still less can this be done by characterizing the good conscience, whether one takes it as an independent form of conscience or as one essentially founded upon "bad" conscience. As the "bad" conscience makes known a "being evil," the good conscience would have to make known the "being good" of Dasein. One can easily see that conscience that used to be the "emanation of divine power" now becomes the slave of Pharisaism. It is supposed to let human beings say of themselves: "I am good." Who else can say this, and who would be less willing to affirm it, than one who is good? But from this impossible consequence of the idea of good conscience, the fact only becomes apparent that being-guilty is what conscience calls.

To escape this consequence, one has interpreted "good" conscience as a privation of the "bad" one, and defined it as an "experienced lack of bad conscience."[9] According to this, the "good" conscience would be an experience of the call not turning up, that is, that I have nothing to reproach myself with. But how is this "lack" *"experienced"* [*"erlebt"*]? The supposed lived-experience is not the experience [Erfahren] of a call at all, but the self-verification that a deed attributed to Dasein was not committed by it and that Dasein is *therefore* innocent. Becoming certain of not having done something does *not* have the character of a phenomenon of conscience *at all*. On the contrary, it can rather mean a forgetting of conscience, that is, that one moves out of the possibility of being able to be summoned. This "certainty" contains the tranquilizing suppression of wanting to have a conscience, that is, of understanding one's ownmost and constant being-guilty. "Good" conscience

292

---

9. Cf. M. Scheler, *Der Formalismus in der Ethik und die materiale Wertethik*, part 2. This *Jahrbuch*, vol. 2 (1916), p. 334.

is neither an independent form of conscience, nor a founded form of conscience, that is, it is not a phenomenon of conscience at all.

Since talk about a "good" conscience arises from the experience of conscience of everyday Dasein, the latter only betrays the fact that basically it does not get at the phenomenon, even when it speaks of "bad" conscience. For factically the idea of "bad" conscience is oriented toward that of the "good" conscience. The everyday interpretation maintains itself in the dimension of calculating and taking care of "guilt" and "innocence," and balancing them out. It is in this horizon that the voice of conscience is "experienced."

In characterizing the primordiality of the ideas of a "bad" and a "good" conscience, we have also already decided as to the distinction between a conscience that points ahead and warns, and one that points back and reprimands. It is true that the idea of the warning conscience comes closest to the phenomenon of summoning to. . . . It shares with the latter the character of pointing ahead. But this agreement is only an illusion, after all. The experience of a warning conscience again sees the voice only as oriented toward the willed deed from which it wants to deter us. As the suppression of what is wanted, the warning is thus possible only because the "warning" call aims at the potentiality-of-being of Dasein, that is, at its understanding of itself in being-guilty which is that on "what is wished for" first gets shattered. The warning conscience has the function of sporadically governing our staying free from indebtedness. The experience of a "warning" conscience sees the tendency of its call only to the extent that it remains accessible to the common sense of the they.

The third reservation appeals to the fact that the everyday experience of conscience *is not familiar with* anything like a being summoned to be guilty. This we must admit. But does the everyday experience of conscience then guarantee that the complete possible content of the call of the voice of conscience is heard in it? Does it follow from this that the theories of conscience based on the vulgar experience of conscience have secured for themselves the appropriate ontological horizon for the analysis of the phenomenon? Does not rather an essential kind of being of Dasein, falling prey, show that this being initially and for the most part understands itself ontically in terms of the horizon of taking care of things, but ontologically defines being in the sense of objective presence? But from this comes a twofold covering over of the phenomenon: the theory sees a series of experiences or "psychic processes" that are for the most part quite indefinite in their kind of being. Experience encounters conscience as a judge and an admonisher with whom Dasein calculatingly makes deals.

The fact that Kant takes the "idea of a court of justice" as the key idea for the basis of his interpretation of conscience is not a matter

293

of chance, but was suggested by the idea of moral *law*, although his concept of morality was far removed from utilitarianism and eudaemonism. Even the theory of value, whether it be formally or materially conceived, has a "metaphysics of morals," that is, an ontology of Dasein and of existence as its unspoken ontological presupposition. Dasein is conceived as a being to be taken care of, and this taking care of has the meaning of "actualizing values" or satisfying norms.

Appealing to the full range of what everyday conscience is familiar with as the sole higher court for the interpretation of conscience cannot be justified unless it has first considered whether conscience can become authentically accessible in this way at all.

Thus the further objection that the existential interpretation overlooks the fact that the call of conscience is always related to a definite "actualized" or willed deed, also loses its force. It cannot be denied that the call is frequently experienced as having such a tendency. It remains questionable only whether this experience of the call lets it "proclaim" itself fully. The commonsense interpretation might believe that it keeps itself to "facts," and yet in the end has restricted the call's scope of disclosure by its very common sense. As little as the "good" conscience can be placed in the service of a "Pharisaism," just as little may the function of the "bad" conscience be reduced to pointing out indebtednesses that are objectively present or to repressing possible ones. As if Dasein were a "household" whose indebtedness only needed to be balanced out in an orderly way for the self to be able to stand "by" as an uninvolved spectator as these experiences run their course.

But if what is primary in the call is not a relatedness to factically "present" guilt, or culpable deeds that have been factically willed, and if thus the "reprimanding" and "warning" types of conscience express no primordial functions of the call, then the ground is also taken out from under the feet of the first reservation, that the existential interpretation fails to recognize the "essentially" *critical* accomplishment of conscience. This reservation, too, arises from a view of the phenomenon which is genuine within certain limits. For, indeed, in the content of the call, nothing can be shown that the voice "positively" recommends and commands. But how is this positivity that is missing from what conscience does to be understood? Does it follow from this that conscience has a "negative" character?

We miss a "positive" content in what is called *because of the expectation that we will be given information that is actually useful about assured possibilities of "action" that are avoidable and calculable.* This expectation is based on the horizon of interpretation of the commonsense way of taking care of things, which forces the existence of Dasein to be subsumed under the idea of a governable course of business. Such expectations (which also in part inexplicitly underlie the demands of a *material*

294

ethics of value as opposed to a "merely" formal one) are, however, disappointed by conscience. Such "practical" directions are not given by the call of conscience *for the sole reason* that it summons Dasein to existence, to its ownmost potentiality-of-being-a-self. With its unequivocally calculable maxims that one is led to expect, conscience would deny to existence nothing less than the *possibility of acting*. Because conscience evidently cannot be "positive" in this way, neither does it function in the same way "only negatively." The call discloses nothing that could be positive or negative as *something to be taken care of*, because it has to do with an ontologically completely different being, namely, *existence*. On the contrary, the correctly understood call gives the "most positive thing of all" in the existential sense—the ownmost possibility that Dasein can give itself as a calling back that calls it forth to its factical potentiality-of-being-a-self. To hear the call authentically means to bring oneself to factical action. But only by setting forth the existential structure implied in our understanding of the summons when we hear *it authentically*, shall we attain a completely adequate interpretation of what is called in the call.

We wanted to show how the phenomena that alone are familiar to the vulgar interpretation of conscience point back to the primordial meaning of the call of conscience when they are understood in an ontologically appropriate way; then, since the vulgar interpretation arises from the limitations of the entangled self-interpretation of Dasein, and since falling prey belongs to care itself, we must show that this interpretation, *even though it is self-evident*, is *by no means accidental*.

295   The ontological critique of the vulgar interpretation of conscience could be subject to the misunderstanding that, by showing the lack of *existential* primordiality of the everyday experience of conscience, one wanted to pass judgment upon the *existentiell* "moral quality" of Dasein. Just as existence is not necessarily and directly jeopardized by an ontologically insufficient understanding of conscience, the existentiell understanding of the call is not guaranteed by an existentially adequate interpretation of conscience either. Seriousness is no less possible in the vulgar experience of conscience than is a lack of seriousness in a more primordial understanding of conscience. Still, the existentially more primordial interpretation also discloses *possibilities* of a more primordial existentiell understanding, as long as our ontological concepts do not get cut off from ontic experience.

§ 60. *The Existential Structure of the Authentic Potentiality-of-Being Attested to in Conscience*

The existential interpretation of conscience should expose an *existent* [*seiende*] attestation in Dasein itself of its ownmost potentiality-of-being.

Conscience attests not by making something known in an undifferentiated way, but by a summons that calls forth to being-guilty. What is thus attested to is "grasped" in the listening which understands without distortion the call in the sense it has itself intended. Understanding the summons, as a mode of *being* of Dasein, first gives the phenomenal content of what is attested to in the call of conscience. We characterized authentically understanding the call as wanting to have a conscience. Letting one's ownmost self act in itself of its own accord in its being-guilty represents phenomenally the authentic potentiality-of-being attested to in Dasein itself. Its existential structure must now be exposed. Only in this way can we penetrate to the fundamental constitution, disclosed in Dasein itself, of the *authenticity* of its existence.

As self-understanding in one's ownmost potentiality of being, wanting-to-have-a-conscience is a mode of *disclosedness* of Dasein. Disclosedness is constituted by attunement and discourse as well as by understanding. Existentiell understanding means to project oneself upon one's ownmost factical possibility of having the potentiality-for-being-in-the-world. But the *potentiality*-of-being is understood only by existing in this possibility.

What mood corresponds to such understanding? Understanding the call discloses one's own Dasein in the uncanniness of its individuation. The uncanniness revealed in understanding is genuinely disclosed by the attunement of anxiety belonging to it. The fact of the *anxiety of conscience* is a phenomenal confirmation of the fact that in understanding the call Dasein is brought face to face with its own uncanniness. Wanting to have a conscience becomes a readiness for anxiety.

The third essential element of disclosedness is *discourse*. The call itself is a primordial discourse of Dasein, but there is no corresponding counter-discourse in which, for example, one talks about what conscience has said and tries to deal with it. In hearing the call understandingly, one denies oneself any counter-discourse, not because one has been overcome by an "obscure power," which suppresses one's hearing, but because this hearing appropriates the content of the call in an unconcealed way. The call introduces the fact of constantly being-guilty and thus brings the self back from the loud idle chatter of the they's common sense. Thus the mode of articulative discourse belonging to wanting to have a conscience is *reticence [Verschwiegenheit]*. We characterized silence [Schweigen] as an essential possibility of discourse.[10] Whoever wants to give something to understand in silence must "have something to say." In the summons, Dasein gives itself to understand its ownmost potentiality-of-being. Thus this calling is a keeping silent.

296

---

10. Cf. § 34.

The discourse of conscience never comes to utterance. Conscience only calls silently; that is, the call comes from the soundlessness of uncanniness and calls Dasein thus summoned back to become still in the stillness of itself. Wanting to have a conscience thus understands this silent discourse appropriately only in reticence. It takes the words away from the commonsense idle chatter of the they.

The commonsense interpretation of conscience, which "strictly adheres to facts," takes the silent discourse of conscience as the occasion to pass it off as something not ascertainable or present at all. The fact that *they*, hearing and understanding only loud idle chatter, cannot "confirm" any call, is attributed to conscience with the excuse that it is "mute" and obviously not objectively present. With this interpretation, the they only covers over its own failure to hear the call and the fact that its "hearing" does not reach very far.

The disclosedness of Dasein in wanting-to-have-a-conscience is thus constituted by the attunement of anxiety, by understanding as projecting oneself upon one's ownmost being-guilty, and by discourse as reticence. The eminent, authentic disclosedness attested in Dasein itself by its conscience—*the reticent projecting oneself upon one's ownmost being-guilty which is ready for* anxiety—we call *resoluteness* [*Entschlossenheit*].

Resoluteness is an eminent mode of the disclosedness of Dasein. But, in an earlier passage[11] disclosedness was interpreted existentially as *primordial truth*. This is not primarily a quality of "judgment" or of any particular mode of behavior at all, but an essential constituent of being-in-the-world as such. Truth must be understood as a fundamental existential. Our ontological clarification of the statement that "Dasein is in the truth" has pointed to the primordial disclosedness of this being as the *truth of existence*; and for its delineation we have referred to the analysis of the authenticity of Dasein.[12]

Now, in resoluteness the most primordial truth of Dasein has been reached, because it is *authentic*. The disclosedness of the there discloses equiprimordially the whole of being-in-the-world—the world, being-in, and the self that is this being [Seiende] as "I am." With the disclosedness of world, innerworldly beings have always already been discovered. The discoveredness of things at hand and objectively present is grounded in the disclosedness of the world;[13] for if the actual totality of relevance of things at hand is to be freed, this requires a pre-understanding of significance. In understanding significance, Dasein taking care of things is circumspectly referred to the things encountered at hand. The understanding of significance as the disclosedness

297

---

11. Cf. § 44.
12. Cf. § 44.
13. Cf. § 18.

of the actual world is again grounded in the understanding of the for-the-sake-of-which, to which discovering of the totality of relevance goes back. In seeking shelter, sustenance, and livelihood, we do so for-the-sake-of the constant possibilities of Dasein that are near to it; upon these, this being which is concerned about its being has always already projected itself. Thrown into its "there," Dasein is always factically dependent on a definite "world"—its "world." At the same time those closest factical projects are guided by the *lostness* in the they taking care of things. This lostness can be summoned by one's own Dasein; the summons can be understood in the mode of resoluteness. But *authentic* disclosedness then modifies equiprimordially the discoveredness of "world" grounded in it and the disclosedness of being-with with others. The "world" at hand does not become different as far as "content," the circle of the others is not exchanged for a new one, and yet the being toward things at hand which understands and takes care of things, and the concerned being-with with others is now defined in terms of their ownmost potentiality-of-being-a-self.

298

As *authentic being a self*, resoluteness does not detach Dasein from its world, nor does it isolate it as free floating ego. How could it, if resoluteness as authentic disclosedness is, after all, nothing other than *authentically being-in-the-world*? Resoluteness brings the self right into its being together with things at hand, actually taking care of them, and pushes it toward concerned being-with with the others.

In the light of the for-the-sake-of-which of the potentiality-of-being which it has chosen, resolute Dasein frees itself for its world. The resoluteness toward itself first brings Dasein to the possibility of letting others who are with it "be" in their ownmost potentiality-of-being, and also discloses that potentiality in concern which leaps ahead and frees. Resolute Dasein can become the "conscience" of others. It is from the authentic being a self of resoluteness that authentic being-with-one-another first arises, not from ambiguous and jealous stipulations and talkative fraternizing in the they and in what one wants to undertake.

In accordance with its ontological essence, resoluteness always belongs to a particular factical Dasein. The essence of this being is its existence. Resoluteness "exists" only as a resolution that projects itself understandingly. But upon what does Dasein resolve itself in resoluteness? To what should it to resolve itself? *Only* the resolution itself can answer this. It would be a complete misunderstanding of the phenomenon of resoluteness if one were to believe that it is simply a matter of incorporating and seizing possibilities that have been presented and suggested. *Resolution is precisely the disclosive projection and determination of the actual factical possibility*. The *indefiniteness* that characterizes every factically projected potentiality-of-being of Dasein *belongs* necessarily to resoluteness. Resoluteness is certain of itself only as resolution. But the *existentiell indefiniteness* of resoluteness never makes itself definite except in a resolution; it nevertheless has its *existential definiteness*.

What one resolves upon [Wozu] in resoluteness is prefigured ontologically in the existentiality of Dasein in general as a potentiality-of-being in the mode of heedful concern. But, as care, Dasein is determined by facticity and falling prey. Disclosed in its "there," it stays equiprimordially in truth and in untruth.[14] This "really" is true, in particular, for resoluteness as authentic truth. Thus resoluteness appropriates untruth authentically. Dasein is in each instance already in irresoluteness, and perhaps will be soon again. The term irresoluteness merely expresses the phenomenon that was interpreted as being at the mercy of the dominant interpretedness of the they. As the they-self, Dasein is "lived" by the commonsense ambiguity of publicness in which no one resolves, but which has always already made its decision. Resoluteness means letting oneself be summoned out of one's lostness in the they. The irresoluteness of the they nevertheless remains dominant, but it cannot challenge resolute existence. As the counter-concept to existentially understood resoluteness, irresoluteness does not refer to an ontic, psychical quality in the sense of being burdened with inhibitions. Even resolutions are dependent upon the they and its world. Understanding this is one of the things that resolution discloses, insofar as resoluteness first gives to Dasein its authentic transparency. In resoluteness, Dasein is concerned with its ownmost potentiality-of-being that, as thrown, can project itself only upon definite, factical possibilities. Resolution does not escape from "reality," but first discovers what is factically possible in such a way that it grasps it as it is possible as one's ownmost potentiality-of-being in the they. The existential definiteness of possible resolute Dasein includes the constitutive moments of the existential phenomenon, which we have hitherto passed over, that we call *situation* [*Situation*].

In the term situation (position [Lage]—"to be in the position of"), there is an overtone of a spatial significance. We shall not attempt to eliminate it from the existential concept. For such an overtone is also implied in the "there" of Dasein. Being-in-the-world has a spatiality of its own that is characterized by the phenomena of de-distancing and directionality. Dasein "makes room" in factically existing.[15] But the spatiality of Dasein, on the basis of which existence actually determines its "place," is grounded in the constitution of being-in-the-world, for which disclosedness is primarily constitutive. Just as the spatiality of

299

---

14. Cf. § 44.
15. Cf. §§ 23 and 24.

the there is grounded in disclosedness, so too does situation have its basis in resoluteness. Situation is the there disclosed in resoluteness— as which the existing being is there. It is not an objectively present framework in which Dasein occurs or into which it could even bring itself. Far removed from any objectively present mixture of the circum- stances and accidents encountered, situation *is* only through and in resoluteness. The actual factical relevant character of the circumstances is disclosed to the self only when that relevant character is such that one is resolute for the there which that self, in existing, has to be. What we call accidents [Zufälle] in the with-world and the surround- ing world can only *be-fall* [*zu-fallen*] resoluteness.

For the they, however, situation [*Situation*] *is essentially closed off.* The they knows only the *"general situation"* [*"allgemeine Lage"*], loses itself in the closest *"opportunities,"* and settles its Dasein by calculating the *"accidents"* which it misjudges as its own achievement and passes off as such.

Resoluteness brings the being of the there to the existence of its situation. But resoluteness delineates the existential structure of the authentic potentiality-of-being attested to in conscience, that is, of wanting to have a conscience. In this potentiality we recognized the appropriate understanding of the summons. This makes it quite clear that the call of conscience does not dangle an empty ideal of existence before us when it summons us to our potentiality-of-being, but *calls forth to the situation.* This existential positivity of the correctly under- stood call of conscience at the same time makes us see how it is that, in limiting the inclination of the call to actual and planned incidents of indebtednesses, we fail to recognize the disclosive character of con- science. It also makes us see how the concrete understanding of the voice of conscience is only seemingly transmitted to us if this restric- tion is made. The existential interpretation of understanding the sum- mons as resoluteness reveals conscience as the kind of being contained in the ground of Dasein, in which it makes its factical existence possible for itself, attesting to its ownmost potentiality-of-being.

The phenomenon set forth with the term *resoluteness* can hardly be confused with an empty "habitus" and an indefinite "velleity." Res- oluteness does not first represent and acknowledge a situation to itself, but has already placed itself in it. Resolute, Dasein is already *acting.* We are purposely avoiding the term "action." For in the first place, it would have to be so broadly conceived that activity also encom- passes the passivity of resistance. In the second place, that term sug- gests a misinterpretation of the ontology of Dasein as if resoluteness were a special mode of behavior of the practical faculty as opposed to the theoretical one. But, as concernful taking care, care includes the being of Dasein so primordially and completely that it must be already

300

presupposed as a whole when we distinguish between theoretical and practical behavior; it cannot first be put together from these faculties with the help of a dialectic that is necessarily groundless because it is *301* existentially unfounded. *But resoluteness is only the authenticity of care itself, cared for in care and possible as care.*

To portray the factical existentiell possibilities in their general features and connections, and to interpret them according to their existential structure, belongs to the scope of tasks of a thematic existential anthropology.[16] For the purpose of our inquiry as a study of fundamental ontology, it will be sufficient to outline existentially the authentic potentiality-of-being attested to in conscience for Dasein itself from out of Dasein itself.

Now that resoluteness has been worked out as a self-projection upon one's ownmost being-guilty in which one is reticent and ready for anxiety, we are prepared to define the ontological meaning of the *authentic* potentiality-of-being-whole of Dasein which we have been looking for. The authenticity of Dasein is neither an empty term nor a fabricated idea. But even so, as an authentic potentiality-of-being-whole, the authentic being-toward-death which we have deduced existentially remains a purely existential project for which the attestation appropriate to Dasein is lacking. Only when we have found this attestation, will our inquiry suffice to set forth (as its problematic requires) an authentic potentiality-of-being-a-whole of Dasein, existentially confirmed and clarified. For only when this being [Seiende] has become phenomenally accessible in its authenticity and its wholeness will the question of the meaning of the being of *this* being, to whose existence belongs an understanding of being in general, be based upon something that will stand a test.

---

16. K. Jaspers explicitly conceived and carried out the task of a doctrine of worldviews for the first time in the direction of this problematic. Cf. his *Psychologie der Weltanschauungen*, 3 ed., 1925. "What man is" is here questioned and determined in terms of what one can essentially be (cf. the Preface to the first edition). From this the existential ontological significance of "limit situations" becomes clear. The philosophical tendency of this work is completely missed if one uses it solely as an encyclopedia of "types of worldviews."

CHAPTER THREE
The Authentic Potentiality-for-Being-a-Whole of Dasein, and Temporality as the Ontological Meaning of Care

§ 61. *Preliminary Sketch of the Methodological Step from Outlining the Authentic Being-a-Whole of Dasein to the Phenomenal Exposition of Temporality*

We projected existentially an authentic potentiality-for-being-whole of Dasein. Analyzing this phenomenon revealed authentic being-toward-death as *anticipation*.[1] In its existentiell attestation, the authentic potentiality-of-being of Dasein was shown to be *resoluteness*, and at the same time was interpreted existentially. How are we to bring these phenomena of anticipation and resoluteness together? Did our ontological projection of the authentic potentiality-for-being-whole not lead us to a dimension of Dasein that is far removed from the phenomenon of resoluteness? What is death supposed to have in common with the "concrete situation" of acting? Does not the attempt to bring resoluteness and anticipation forcibly together lead us astray into an intolerable, completely unphenomenological construction which may no longer even claim to have the character of an ontological project that is phenomenally grounded?

Externally binding both phenomena together is intrinsically out of the question. There is still one way to bring these together, and this is the only possible method; namely, to start from the phenomenon of resoluteness, attested to in its existentiell possibility, and to ask: *does resoluteness, in its ownmost existentiell tendency of being [Seinstendenz], itself point ahead to anticipatory resoluteness as its ownmost authentic possibility?* What if resoluteness, following its own meaning, were brought into its authenticity only when it no longer projects itself upon

302

---

1. Cf. § 53.

arbitrary possibilities merely lying near by, but rather upon the most extreme possibility that lies ahead of every factical potentiality of being of Dasein, and, as such, more or less enters without distortion every potentiality-of-being of Dasein factically seized upon? What if resoluteness, as the *authentic* truth of Dasein, reached the *certainty authentically belonging to it* only in the anticipation of death? What if all the factical *"anticipatoriness"* of resolve were authentically understood, that is, existentielly *caught up with* only in the *anticipation* of death?

As long as our existential interpretation does not forget that the being given to it as its theme has the kind of being of Dasein, and that it cannot be joined together out of objectively present pieces into something objectively present, its steps must be guided by the idea of *existence*. For the question of the possible connection between anticipation and resoluteness, this means nothing less than the demand that we should project these existential phenomena upon the existentiell possibilities prefigured in them and think these possibilities "through to the end" in an existential way. Thus the development of anticipatory resoluteness as an existentielly possible authentic potentiality-for-being-whole loses the character of an arbitrary construction. It becomes the interpretation that frees Dasein *for* its most extreme possibility of existence.

With this step, the existential interpretation at the same time makes known its ownmost methodological character. Apart from occasional, necessary remarks, we have until now deferred explicit discussions of method. We wanted first of all to "proceed" to the phenomena. *Before* exposing the meaning of being of the being revealed in its fundamental phenomenal content, the course of our inquiry needs to pause, not in order to "rest," but in order to gain new momentum.

Any genuine method is grounded in the appropriate preview of the fundamental constitution of the "object" or domain of objects to be disclosed. Any genuine reflection on method, which is to be distinguished from empty discussions of mere technique, thus at the same time tells us something about the kind of being of the being in question.[*] The clarification of methodological possibilities, requirements, and limits of the existential analytic in general can alone secure the transparency that is necessary if we are to take the basic step of revealing the meaning of being of care. *But the interpretation of the ontological meaning of care must be made on the basis of a complete and constant phenomenological reconsideration of the existential constitution of Dasein set forth up to now.*

Ontologically, Dasein is in principle different from everything objectively present and real. Its "content" is not founded in the substantiality of a substance, but in the *"self-constancy"* [*"Selbständigkeit"*] of the existing self whose being was conceived as care. The phenom-

---

[*] Distinguish between scientific method and the advance of thinking.

enon of the *self* included in care needs a primordial and authentic existential definition, in contrast to our preparatory demonstration of the inauthentic they-self. Along with this, we must establish what possible ontological questions are to be directed toward the "self," if it is neither substance nor subject.

The phenomenon of care, thus sufficiently clarified, can then be interrogated as to its ontological meaning. Determining this meaning will lead to the exposition of temporality. In exhibiting this, we are not led into remote, distinct regions of Dasein, rather, we merely get a conception of the total phenomenal content of the existential fundamental constitution of Dasein in the ultimate foundations of its own ontological intelligibility. *Temporality is experienced as a primordial phenomenon in the authentic being-whole of Dasein, in the phenomenon of anticipatory resoluteness.* If temporality makes itself known primordially here, the temporality of anticipatory resoluteness is presumably a distinctive mode of that temporality. Temporality can *temporalize* [*zeitigen*] itself in various possibilities and various ways. The fundamental possibilities of existence, the authenticity and inauthenticity of Dasein, are ontologically grounded in possible temporalizations of temporality.

If the ontological character of its own being is remote from Dasein because of the dominance of its entangled understanding of being (being as objective presence), then this entangled understanding keeps it even more remote from the primordial foundations of this being [Sein]. Thus one must not be surprised if at first glance temporality does not correspond to what is accessible to the vulgar understanding as "time." Thus neither the concept of time belonging to the vulgar experience of time, nor the problematic arising from it can function uncritically as a criterion for the appropriateness of an interpretation of time. Rather, our inquiry must become familiar with the primordial phenomenon of temporality *beforehand*, so that *in terms of this* we may cast light on the necessity, the source, and the reason for the dominance of the vulgar understanding of time.

The primordial phenomenon of temporality will be made secure by demonstrating that all the fundamental structures of Dasein exposed up to now are to be basically conceived "temporally" with regard to their possible wholeness, unity, and development, and as modes of the temporalizing of temporality. Thus, when temporality has been exposed, the task arises for the existential analytic of repeating the analysis of Dasein in the sense of interpreting the essential structures with a view to their temporality. Temporality itself sketches out the fundamental directions of the analyses thus required. Thus the chapter has the following divisions: Anticipatory resoluteness as the existentielly authentic potentiality-for-being-a-whole of Dasein (§ 62); the hermeneutical situation at which we have arrived for interpret-

304

ing the meaning of being of care and the methodological character
of the existential analytic in general (§ 63); care and selfhood (§ 64);
temporality as the ontological meaning of care (§ 65); the temporality
of Dasein and the tasks arising from it of a primordial repetition of
the existential analytic (§ 66).

§ 62. *The Existentielly Authentic Potentiality-for-Being-Whole of Dasein as Anticipatory Resoluteness*

To what extent does resoluteness, "thought out to its end" in accordance
with its ownmost tendency of being, lead us to authentic being-toward-
death? How is the connection between wanting to have a conscience
and the existentially projected, authentic potentiality-of-being-whole
of Dasein to be conceived? Does welding the two together result in
a new phenomenon? Or are we left with the resoluteness attested
to in its existentiell possibility in such a way that it can undergo an
existentiell modalization through being-toward-death? But what does it
mean "to think through to the end" the phenomenon of resoluteness
existentially?

    Resoluteness was characterized as the reticent self-projecting
upon one's ownmost being-guilty, and as demanding anxiety of oneself. Being-guilty belongs to Dasein and means: null *being* the ground
of a nullity. The "guilty" that belongs to the being of Dasein admits
neither of increase nor decrease. It lies *before* all quantification, if the
latter has any meaning at all. Being essentially guilty, Dasein is not
just guilty *occasionally* and *other times not*. Wanting-to-have-a-conscience
resolves itself for this being-guilty. The intrinsic sense of resoluteness
is to project upon itself this being-guilty that Dasein is *as long as it is*.
Taking over this "guilt" existentielly in resoluteness occurs authentically only if, in its disclosing of Dasein, resoluteness has become *so*
transparent that it understands being-guilty *as something constant*. But
this understanding is made possible only in such a way that Dasein
discloses to itself its potentiality-of-being "up to its end." The *being*-at-
an-end of Dasein, however, means existentially being-*toward*-the-end.
Resoluteness becomes authentically what it can be as *being-toward-the-
end-that-understands*, that is, as anticipation of death. Resoluteness does
not simply "have" a connection with anticipation as something other
than itself. *It harbours in itself authentic being-toward-death as the possible
existentiell modality of its own authenticity.* We want now to clarify this
"connection" phenomenally.

    Resoluteness means: letting oneself be called forth to one's own-
most *being*-guilty. Being-*guilty* belongs to the being of Dasein itself,
which we defined primarily as potentiality-of-being. The statement
that Dasein "is" constantly guilty can only mean that it always main-

tains itself in this being either as authentic or inauthentic existence.   *306*
Being-guilty is not just a lasting quality of something constantly
objectively present, but the *existentiell possibility* of *being* authentically
or inauthentically guilty. "Guilty" *is* always only in the actual factical potentiality-of-being. Thus, being-guilty must be conceived as a
potentiality-for-being-guilty, because it belongs to the *being* of Dasein.
Resoluteness projects itself upon this potentiality-of-being, that is,
understands itself in it. Thus, this understanding maintains itself in
a primordial possibility of Dasein. It stays *in it authentically* when
resoluteness is primordially what it tends to be. But we revealed the
primordial being of Dasein toward its potentiality-of-being as being-
toward-death, that is, toward that which we characterized as the eminent possibility of Dasein. Anticipation disclosed this possibility as
possibility. Thus, resoluteness becomes a primordial being toward the
ownmost potentiality-of-being of Dasein only *as anticipatory*. Resoluteness understands the "can" of its potentiality-for-being-guilty only
when it "qualifies" itself as being-toward-death.

Resolutely, Dasein takes over authentically in its existence the fact
that it *is* the null ground of its nullity. We conceived of death existentially as what we characterized as the possibility of the *im*possibility of
existence, that is, as the absolute nothingness of Dasein. Death is not
tacked on to Dasein as its "end," but, as care, Dasein is the thrown (that
is, null) ground of its death. The nothingness primordially dominant in
the being of Dasein is revealed to it in authentic being-toward-death.
Anticipation makes being-guilty evident only on the basis of the *whole*
being of Dasein. Care contains death and guilt equiprimordially. Only
anticipatory resoluteness understands the potentiality-for-being-guilty
authentically and wholly, that is, *primordially*.[2]

Understanding the call of conscience reveals the lostness in the   *307*
they. Resoluteness brings Dasein back to its ownmost potentiality-of-
being-a-self. One's own potentiality-of-being becomes authentic and
transparent in the understanding being-toward-death as one's *ownmost*
possibility.

The call of conscience passes over all "worldly" status and abili-
ties of Dasein in its summons. It uncompromisingly individualizes

---

2. The being guilty that belongs primordially to the constitution of the being of Dasein is
to be distinguished from the *status corruptionis* as it is understood by theology. Theology
can find an ontological condition of the factical possibility in being guilty as it is defined
existentially. The guilt contained in the idea of this *status* is a factical indebtedness of
a completely unique kind. It has its own attestation that remains fundamentally closed
off to every philosophical experience. The existential analysis of being-guilty does not
prove anything *for* or *against* the possibility of sin. Strictly speaking, one cannot even
say that the ontology of Dasein leaves this possibility open at all *of its own accord* since,
as philosophical questioning, it "knows" nothing about sin in principle.

294 BEING AND TIME   II.III

Dasein down to its potentiality-for-being-guilty which it expects it
to be authentically. The relentless severity with which Dasein is thus
essentially individualized down to its ownmost potentiality-of-being
discloses anticipation of death as the possibility which is *nonrelational*. Anticipatory resoluteness lets the potentiality-for-being-guilty,
as its ownmost nonrelational possibility, completely drive into its
conscience.

Wanting-to-have-a-conscience signifies the readiness for the summons to one's ownmost being-guilty that always already determined
factical Dasein *before* any factical indebtedness and *after* that indebtedness has been settled. This prior and constant being-guilty, which is
constantly with us, does not show itself without being covered over
in its character as prior until that priority is placed in the possibility
which is for Dasein absolutely *insuperable* [*unüberholbar*]. When resoluteness, anticipating, has caught up with the possibility of death in its
potentiality-of-being, the authentic existence of Dasein can no longer
be *overtaken* [*überholt*] by anything.

With the phenomenon of resoluteness we were led to the primordial *truth* of existence. Resolute, Dasein is revealed to itself in its
actual factical potentiality-of-being in such a way that it itself *is* this
revealing and being revealed. To any truth, there belongs a corresponding holding-for-true. The explicit appropriation of what is disclosed or
discovered is *being*-certain. The primordial truth of existence requires
an equiprimordial being-certain in which one holds oneself in what resoluteness discloses. It *gives* itself the actual factical situation and *brings*
itself into that situation. The situation cannot be calculated in advance
and pregiven like something objectively present waiting to be grasped.
It is disclosed only in a free act of resolve that has not been determined
beforehand, but is open to the possibility of such determination. *What,
then, does the certainty belonging to such resoluteness mean?* This certainty
must hold itself in what is disclosed in resolution. But this means that
it simply cannot become *rigid* about the situation, but must understand
that the resolution must be *kept* free and *open* for the actual factical
308   possibility in accordance with its own meaning as a disclosure. The
certainty of the resolution means *keeping oneself free for* the possibility
of *taking it back*, a possibility that is always factically necessary. This
holding-for-true in resoluteness (as the truth of existence), however,
by no means lets us fall back into irresoluteness. On the contrary, this
holding-for-true, as a resolute holding oneself free for taking back, is
the *authentic resoluteness to repeat itself*. But thus one's very lostness
in irresoluteness is existentielly undermined. The holding-for-true that
belongs to resoluteness tends, in accordance with its meaning, toward
constantly keeping oneself free, that is, to keep itself free for the *whole*

potentiality-of-being of Dasein. This constant certainty is guaranteed to resoluteness only in such a way that it relates to that possibility of which it can *be* absolutely certain. In its death, Dasein must absolutely "take itself back." Constantly certain of this, that is, *anticipating*, resoluteness gains its authentic and whole certainty.

But Dasein is equiprimordially in untruth. Anticipatory resoluteness at the same time gives Dasein the primordial certainty of its being closed off. In anticipatory resoluteness, Dasein *holds* itself open for its constant lostness in the irresoluteness of the they—a lostness which is possible from the very ground of its own being. As a constant possibility of Dasein, irresoluteness is *co-certain*. Resoluteness, transparent to itself, understands that the *indefiniteness* of its potentiality-of-being is always determined only in a resolution with regard to the actual situation. It knows about the indefiniteness that prevails in a being that exists. But this knowledge must itself arise from an authentic disclosure if it is to correspond to authentic resoluteness. Although it always becomes certain in resolution, the *indefiniteness* of one's own potentiality-of-being always reveals itself *completely* only in being-toward-death. Anticipation brings Dasein face to face with a possibility that is constantly certain and yet remains indefinite at every moment as to when possibility becomes impossibility. Anticipation makes manifest that this being has been thrown into the indefiniteness of its "limit situation"; resolved upon this, Dasein attains its authentic potentiality-of-being-whole. The indefiniteness of death discloses itself primordially in anxiety. But this primordial anxiety strives to expect resoluteness of itself. It clears away every covering over of the fact that Dasein is itself left to itself. The nothingness before which anxiety brings us reveals the nullity that determines Dasein in its *ground*, which itself is as thrownness into death.

Our analysis revealed in order the *moments of modalization* toward which resoluteness tends of itself and which stem from authentic being-toward-death as the ownmost, nonrelational, insuperable, certain and yet indefinite possibility. It is authentically and completely what it can be only as *anticipatory resoluteness*.

But, on the other hand, our interpretation of the "connection" between resoluteness and anticipation first attained the complete existential understanding of anticipation itself. Until now, it was valid only as an ontological projection. Now we see that anticipation is not a fictitious possibility that we have forced upon Dasein, but rather the *mode* of a potentiality-of-being existentielly attested to in Dasein, which it demands of itself, if indeed it authentically understands itself as resolute. Anticipation "is" not some kind of free-floating behavior, but must rather be conceived of as *the possibility of the authenticity of that*

309

*resoluteness existentielly attested to in such resoluteness—a possibility con-
cealed and thus also attested.* Authentic "thinking about death" is wanting to have a conscience, which has become existentielly transparent
to itself.

If resoluteness, as authentic, tends toward the mode delimited by
anticipation, and if anticipation constitutes the authentic potentiality-
of-being-whole of Dasein, then in resoluteness which is existentielly
attested to there is co-attested to an authentic potentiality-of-being
whole of Dasein. *The question of the potentiality-of-being-whole is a factical, existentiell one. It is answered by Dasein as resolute.* The question of
the potentiality-of-being-whole of Dasein has now completely cast off
the character which we initially[3] pointed out when we treated it as if it
were just a theoretical, methodological question of the analytic of Dasein, arising from the attempt to have the whole of Dasein completely
"given." The question of the wholeness of Dasein, initially discussed
only with regard to ontological method, has its justification, but only
because the ground for that justification goes back to an ontic possibility of Dasein.

Our clarification of the "connection" between anticipation and
resoluteness in the sense of a possible modalization of resoluteness by
anticipation, turned into the phenomenal demonstration of an authentic
potentiality-of-being-whole of Dasein. If with this phenomenon a mode
of being of Dasein has been grasped in which it brings itself to and
before itself, it must remain ontically and ontologically unintelligible
to the everyday, commonsense interpretation of Dasein by the they. It
310   would be a misunderstanding to put this existentiell possibility aside
as being "unproven" or to want to "prove" it theoretically. Nevertheless, the phenomenon must be shielded from the crudest distortions.

Anticipatory resoluteness is not a way out fabricated for the purpose of "overcoming" death, but it is rather the understanding that
follows the call of conscience and that frees for death the possibility
of *gaining power over* the *existence* of Dasein and of fundamentally dis-
persing every fugitive self-covering-over. Nor does wanting to have a
conscience, which we defined as being-toward-death, mean a detach-
ment in which one flees from the world; rather, it brings one without
illusions to the resoluteness of "acting." Nor does anticipatory resoluteness stem from "idealistic" expectations soaring above existence and its
possibilities, but arises from the sober understanding of the basic factical possibilities of Dasein. Together with the sober anxiety that brings
us before our individualized potentiality-of-being goes the unshakable
joy in this possibility. In it Dasein becomes free of the entertaining
"incidentals" that busy curiosity provides for itself, primarily in terms
of the events of the world. However, the analysis of these fundamental

3. Cf. § 45.

moods goes beyond the limits drawn for our present inquiry by aiming toward fundamental ontology.

But does not a definite ontic interpretation of authentic existence, a factical ideal of Dasein, underlie our ontological interpretation of the existence of Dasein? Indeed. But not only is this fact one that must not be denied and that we are forced to concede, it must be understood in its *positive necessity*, in terms of the thematic object of our inquiry. Philosophy will never seek to deny its "presuppositions," but neither may it merely admit them. It conceives them and develops with more and more penetration both the presuppositions themselves and that for which they are presuppositions. This is the function that the methodological considerations now demanded of us have.

§ 63. *The Hermeneutical Situation at Which We Have Arrived for Interpreting the Meaning of Being of Care, and the Methodological Character of the Existential Analytic in General*

In its anticipatory resoluteness, Dasein has been made phenomenally visible with regard to its possible authenticity and totality. The hemeneutical situation[4] which was previously insufficient for the interpretation of the meaning of being of care, now has the required primordiality. Dasein has been placed in our fore-having primordially, that is, with regard to its authentic potentiality-of-being-whole; the guiding fore-sight, the idea of existence, has attained its definiteness through the clarification of the ownmost potentiality-of-being; with the concretely developed structure of being of Dasein, its ontological peculiarity, as opposed to everything objectively present, has become so clear that our fore-conception of the existentiality of Dasein possesses sufficient articulation to guide securely the conceptual development of the existentials.

311

The path of the analytic of Dasein which we have traversed so far has led us to a concrete demonstration of the thesis[5] only suggested at the beginning: *The being [Seiende] that we ourselves in each instance are is ontologically farthest from us*. The reason for this lies in care itself. Entangled being together with those things of the "world" that are taken care of guides the everyday interpretation of Dasein and covers over ontically the authentic being of Dasein, thus denying the appropriate basis for an ontology oriented toward this being.* Thus the primordial phenomenal parameters of this being [Seienden] are not at all self-evident, even if ontology initially follows the course

---

* Wrong! As if ontology could be taken from genuinely ontic investigation. For what is a genuinely ontic account if it is not genuinely taken from a pre-ontological project—if all of this is to remain in this distinction.

4. Cf. § 45.
5. Cf. § 5.

of the everyday interpretation of Dasein. Rather, freeing the primordial being of Dasein must be *wrested* from Dasein by moving in the *opposite direction* from the entangled, ontic, and ontological tendency of interpretation.

Not only the demonstration of the most elemental structures of being-in-the-world—the definition of the concept of world, the clarification of the nearest and most average who of this being, of the they-self, the interpretation of the "there"—but above all the analyses of care, death, conscience, and guilt show *how*, in Dasein itself, the commonsense way of taking care has taken over the potentiality-of-being of Dasein and of its disclosure, which amounts to its closure.

Thus the *kind* of being of Dasein *requires* of an ontological interpretation that has set as its goal the primordiality of the phenomenal demonstration *that it overcome* [erobert] *the being of this being in spite of this being's own tendency to cover things over*. Thus the existential analytic constantly has the character of *doing violence*, whether for the claims of the everyday interpretation or for its complacency and its tranquillized obviousness. Of course, the ontology of Dasein is particularly distinguished by this characteristic, but it belongs as well to any interpretation, because the understanding that unfolds in interpretation has
312   the structure of a project. But is not anything of this sort *guided* and *regulated* in a way of its own? Where are ontological projects to get the evidence that their "findings" are phenomenally appropriate? Ontological interpretation projects the beings given to it upon the being [Sein] appropriate to them, so as to bring them to a concept with regard to their structure. Where are the guideposts to direct the projection so that being will be reached at all? And what if the being [Seiende] that is thematic for the existential analytic conceals the being which belongs to it and does so *in* its very way of being? To answer these questions we must initially restrict ourselves to clarifying the analytic of Dasein, as the questions themselves demand.

Self-interpretation belongs to the being of Dasein. In the circumspect discovery of the "world" that takes care, taking care is sighted too. Dasein always already understands itself factically in definite existentiell possibilities, even if its projects arise only from the common sense of the they. Whether explicitly or not, whether appropriately or not, existence is somehow understood too. Every ontic understanding "includes" certain things, even if only *pre*-ontologically, that is, even if they are not grasped theoretically and thematically. Every ontologically explicit question about the being of Dasein has already had the way prepared for it by the kind of being of Dasein.

Nevertheless, how are we to find out what constitutes the "authentic" existence of Dasein? Without an existentiell understanding, all analysis of existentiality remains baseless. Does not an ontic conception of existence underlie our interpretation of the authenticity and totality of Dasein, an ontic interpretation that might be possible, but need not be binding for every one? Existential interpretation will never seek to take over by *fiat* those things that, from an existentiell point of view, are possible or binding. But must it not justify itself with regard to *those* existentiell possibilities that it uses to give the ontic basis for the ontological interpretation? If the being of Dasein is essentially potentiality-of-being and being-free for its ownmost possibilities, and if it always exists only in freedom or unfreedom for them, can the ontological interpretation take as its basis anything other than *ontic possibilities* (modes of potentiality-of-being) and project *these* upon *their ontological possibility*? And if Dasein mostly interprets itself in terms of its lostness in taking care of the "world," is not the determination of the ontic and existentiell possibilities and the existential analysis based upon them (in opposition to that lostness) the mode of its disclosure appropriate to this being? *Does not then the violence of this project amount to freeing the undisguised phenomenal content [Bestand] of Dasein?*

313

The "violent" presentation of possibilities of existence may be required for our method, but can it escape being merely arbitrary? If our analytic takes anticipatory resoluteness as its basis, as an existentielly authentic potentiality-of-being, and if Dasein itself summons this possibility right out of the ground of its existence, is this possibility then an *arbitrary* one?* Is the mode of being in accordance with which the potentiality-of-being of Dasein relates to its eminent possibility, death, picked up by chance? *Does being-in-the-world have a higher instance of its potentiality-of-being than its own death?*

The ontic and ontological project of Dasein upon an authentic potentiality-of-being-a-whole may not be arbitrary, but is the existential interpretation of these phenomena then already justified? Where does this interpretation get its guidelines if not from a "presupposed" idea of existence in general? How are the steps of the analysis of inauthentic everydayness regulated, if not by the concept of existence that we have posited? And if we say that Dasein "falls prey," and that thus the authenticity of its potentiality-of-being is to be wrested from this tendency of being—from what perspective are we speaking here? Is not everything illuminated by the light of the "presupposed" idea of existence, even if rather dimly? Where does this idea get its justification? Has our initial project, in which we called attention to it, led us nowhere? By no means.

The formal indication of the idea of existence was guided by the understanding of being in Dasein itself. Without any ontological transparency, it was, after all, revealed that I myself am always the being

---

* Probably not; but "not arbitrary" does yet not mean "necessary and binding."

which we call Dasein, as the potentiality-of-being that is concerned to
be this being. Dasein understands itself as being-in-the-world, although
without sufficient ontological definiteness. Thus existing, it encounters
beings of the kind of being of things at hand and objectively present.
No matter how far removed from an ontological concept the distinction
between existence and reality may be, even if Dasein initially understands existence as reality, Dasein is not just objectively present, but
has always already *understood itself*, however mythical or magical its
interpretations may be. For otherwise, Dasein would not "live" in a
myth and would not take heed of its magic in rites and cults. The idea
of existence which we have posited gives us an outline of the formal
structure of the understanding of Dasein in general, and does so in a
way that is not binding from an existentiell point of view.

314   Under the guidance of this idea the preparatory analysis of the
everydayness nearest to us has been carried out as far as a first conceptual definition of care. This phenomenon enabled us to get a more
precise grasp of existence and of the relations to facticity and falling
prey belonging to it. The definition of the structure of care has given
us a basis on which to distinguish ontologically between existence and
reality for the first time.[6] This led to the thesis: the substance of human
being is existence.[7]

But even this formal idea of existence, which is not binding in an
existentiell way, already contains a definite though unprofiled ontological "content" that "presupposes" an idea of being in general—just like
the idea of reality contrasted with it. Only in the horizon of *that* idea
of being can the distinction between existence and reality be made.
After all, both mean *being* [*Sein*].

But is not the ontologically clarified idea of being in general first
to be attained by working out the understanding of being that belongs
to Dasein? However, this understanding can be grasped primordially
only on the basis of a primordial interpretation of Dasein guided by
the idea of existence. Does it not thus finally become evident that this
problem of fundamental ontology that we have set forth is moving in
a "circle"?

We already showed, in the structure of understanding in general, that what is criticized with the inappropriate expression "circle"
belongs to the essence and the distinctiveness of understanding itself.[8]
Still, our inquiry must now return explicitly to this "circular" argument if the problematic of fundamental ontology is to have its hermeneutical situation clarified. When it is objected that the existential
interpretation is "circular," it is said that the idea of existence and of

6. Cf. § 43.
7. Cf. §§ 44 and 26.
8. Cf. § 32.

being in general is "presupposed," and that Dasein gets interpreted "according to this" presupposition so that the idea of being may be obtained from it. But what does "presupposing" mean? In positing the idea of existence, do we also posit some proposition from which we can deduce further propositions about the being of Dasein according to the formal rules of consistency? Or does this pre-supposing have the character of a projection that understands in such a way that the interpretation from which this understanding is formed *lets* what is to be interpreted be *put in words for the very first time, so that it may decide of its own accord whether, as this being [Seiende], it will provide the constitution of being for which it has been disclosed in the projection with regard to its formal indication*? Is there any other way that beings can put themselves into words with regard to their being at all? A "circle" in the proof cannot be "avoided" in the existential analytic, because that analytic is *not* proving anything according to the rules of the logic of consequence *at all*. What common sense wishes to get rid of by avoiding the "circle," believing that it does justice to the loftiest rigor of scientific investigation, is nothing less than the basic structure of care. Primordially constituted by care, Dasein is always already ahead of itself. Existing [seiend], it has always already projected itself upon definite possibilities of its existence, and in these existentiell projects it has also projected pre-ontologically something like existence and being. But can one deny this projecting of *that* research essential to Dasein, which *like all research itself is a kind of being of disclosive Dasein*, that wants to develop and conceptualize the understanding of being belonging to Dasein?

    315

But the "charge of circularity" itself comes from a kind of being of Dasein. Something like projecting, especially ontological project- ing, necessarily remains foreign for the common sense of our heedful absorption in the they because common sense barricades itself against it "in principle." Whether "theoretically" or "practically," common sense only takes care of beings that are in view of its circumspection. What is distinctive about common sense is that it believes that it experiences only "factual" beings in order to be able to rid itself of its understand- ing of being. It fails to recognize that beings can be "factually" expe- rienced only when being has already been understood, even if this understanding is not conceptualized. Common sense misunderstands understanding. And *for this reason* it must also necessarily proclaim as "violent" anything lying beyond the scope of its understanding as well as any move in that direction.

Talk about the "circle" in understanding expresses the failure to recognize two things: (1) That understanding itself constitutes a basic kind of being of Dasein. (2) That this being [Sein] is constituted as care. To deny the circle, to make a secret of it, or even to wish to overcome it means to anchor this misunderstanding once and for all. Rather, our attempt must aim at leaping into this "circle" primordially and completely, so that even at the beginning of our analysis of Dasein we make sure that we have a complete view of the circular being of Dasein. Not too much, but *too little* is "presupposed" for the ontology of Dasein, if one "starts out with" a worldless I in order then to provide that I with an object and an ontologically groundless relation to that object. *Our view is too short-sighted* if we make "life" a problem *and then occasionally* take death into account too. The thematic objection is *artificially* and *dogmatically* cut out if one limits oneself "initially" to a "theoretical subject," in order to then complement it "on the practical side" with an additional "ethic."

This will suffice to clarify the existential meaning of the hermeneutical situation of a primordial analytic of Dasein. With the exposition of anticipatory resoluteness Dasein has been brought before us with regard to its authentic wholeness. The authenticity of the potentiality-of-being-a-self guarantees the fore-sight of primordial existentiality, and this assures us that we have coined the appropriate existential concepts.

At the same time, the analysis of anticipatory resoluteness led us to the phenomenon of primordial and authentic truth. Earlier we showed how the understanding of being that prevails initially and for the most part conceives being in the sense of objective presence and thus covers over[9] the primordial phenomenon of truth. But if "there is" ["es gibt"] being only when truth "is," and if the understanding of being always varies according to the kind of truth, then primordial and authentic truth must guarantee the understanding of the being of Dasein and of being in general. The ontological "truth" of the existential analysis is developed on the basis of primordial, existentiell truth. Yet the latter does not necessarily need the former. The most primordial and basic existential truth, for which the problematic of fundamental ontology strives in preparing the question of being in general is the *disclosure of the meaning of being of care*. In order to reveal this meaning, we need to hold in readiness, undiminished, the full structural content of care.

§ 64. *Care and Selfhood*

The unity of the constitutive moments of care, existentiality,[*] facticity, and falling prey made possible a first ontological definition of the wholeness of the structural whole of Dasein. The structure of care was given an existential formula: being-ahead-of-oneself-already-being-in

---

* Existence: (1) For the whole of the being [Sein] of Dasein; (2) only for "understanding."

9. Cf. § 44.

(a world) as being-together-with (innerworldly beings encountered). The totality of the structure of care does not first arise from a coupling together, yet it is *articulated*.[10] In assessing this ontological result, we have had to estimate how well it satisfies[11] the requirements of a *primordial* interpretation of Dasein. We found that neither the *whole* of Dasein nor its *authentic* potentiality-of-being had been made thematic. However, the attempt to grasp phenomenally the whole of Dasein seemed to get stranded precisely on the structure of care. The ahead-of-itself presented itself as a not-yet. But the ahead-of-itself, characterized in the sense of something outstanding, revealed itself to our genuine existential reflection as *being toward the end*, something that every Dasein in the depths of its being is. We also made it clear that care summons Dasein to its ownmost potentiality-of-being in the call of conscience. Understanding the summons revealed itself—primordially understood—as anticipatory resoluteness, which includes an authentic potentiality-of-being-whole of Dasein. The structure of care does not speak *against* the possibility of being-whole, but is the *condition of the possibility* of such an existentiell potentiality-of-being. In the course of these analyses it became clear that the existential phenomena of death, conscience, and guilt are anchored in the phenomenon of care. *The articulation of the wholeness of the structural whole has become still richer, and thus the existential question of the unity of this wholeness has become more urgent.*

How are we to grasp this unity? How can Dasein exist as a unity in the ways and possibilities of its being that we mentioned? Evidently only in such a way that it *itself is* this being in its essential possibilities, that *I* am always* this being [Seiende]. The "I" seems to "hold together" the wholeness of the structural whole. The "I" and the "self" have been conceived for a long time in the "ontology" of this being as the supporting ground (substance or subject). Even in its preparatory characterization of everydayness, our analytic also already encountered the question of the who of Dasein. We found that Dasein is initially and for the most part *not* itself, but is lost in the they-self.† The they-self is an existentiell modification of the authentic self. The question of the ontological constitution of selfhood remained unanswered. It is true that we already fundamentally established the guidelines for the problem:[12] if the self belongs to the essential qualities of Dasein, whose

318

---

* Dasein *itself is* this being [Seiende].
† The "I" as what is in a sense "closest" in the foreground, and thus seemingly the self.

10. Cf. § 41.
11. Cf. § 45.
12. Cf. § 25.

"essence" however lies in *existence*, then I-hood and selfhood must be conceived *existentially*. Negatively, we also saw that our ontological characterization of the they ruled out any application of the categories of objective presence (substance). In principle it became clear that care cannot be derived ontologically from reality or be constructed with the categories of reality.[13] Care already contains the phenomenon of self, if indeed the thesis is correct that the expression "care for self" would be *tautological*[14] if it were proposed in conformity with concern as care for others. But then the problem of the ontological definition of the selfhood of Dasein gets sharpened to the question of the existential "connection" between care and selfhood.

To clarify the existentiality of the self, we take as our "natural" point of departure the everyday self-interpretation of Dasein that expresses "itself" in *saying-I*. Utterance is not necessary. With the "I," this being means itself.* The content of this expression is taken to be absolutely simple. It always means only me, and nothing further. As this simple thing, the "I" is not a definition of other things: it is *itself* not a predicate, but the absolute "subject." What is expressed and addressed in saying-I is always met with as the same persisting thing. The characteristics of "simplicity," "substantiality," and "personality," which Kant, for example, takes as the foundation for his doctrine found in "On the Paralogisms of Pure Reason,"[15] arise from a genuine "pre-phenomenological" experience. The question remains whether what was experienced in such a way ontically may be interpreted ontologically with the aid of the "categories" mentioned.

In strict conformity with the phenomenal content given in saying-I, Kant did show that the ontic theses about the substance of the soul inferred from these characteristics are without justification. But in so doing, he merely rejects a wrong *ontic* explanation of the I.† He has by no means attained an *ontological* interpretation of selfhood, nor has he obtained some assurance of it and made positive preparations for it. Although Kant attempts, more strictly than his predecessors, to hold on to the phenomenal content of saying I, he still slips back into the *same* inappropriate ontology of the substantial, whose ontic foundations he theoretically rejected for the I. We must show this more precisely, in order to establish what it means ontologically to take saying I as the

319

---

* Clarify more precisely; *saying-I* and *being a self*.
† And being intent upon ontic-suprasensuous statements (*Metaphysica specialis*).

13. Cf. § 43, C.
14. Cf. § 41.
15. Cf. Kant, *Kritik der reinen Vernunft*, B399, and especially the treatment in the 1st edition, A348ff.

point of departure for the analysis of selfhood. The Kantian analysis of the "I think" should now be referred to as an illustration, but only to the extent that it is required for the clarification of the problematic in question.[16]

The "I" is a bare consciousness that accompanies all concepts. In the I, nothing more is represented than a transcendental subject of thoughts. "Consciousness in itself (is) not a representation . . . , but a form of representation in general."[17] The "I think" is "the form of apperception that adheres to every experience and precedes it."[18]

Kant grasps the phenomenal content of the "I" correctly in the expression "I think" or—if the relation of the "practical person" to "intelligence" is also considered—in the expression "I act." In Kant's sense saying I must be conceived as saying-I-think. Kant attempts to establish the phenomenal content of the I as *res cogitans*. If he then calls this I a "logical subject," that does not mean that the I in general is a concept gained merely by logical means. Rather, the I is the subject of logical behavior, of binding together [Verbinden]. The "I think" means: I bind together. All binding together is an "*I* bind together." In any taking together and relating, the I always already underlies—ὑποκείμενον. Thus, the subject is "consciousness in itself," not a representation,* but rather the "form" of representation. This means that the I think is not something represented, but the formal structure of representing as such, and this formal structure alone makes it possible for anything to be represented. The form of representation means neither a framework [Rahmen] nor a universal concept, but that which, as εἶδος, makes everything represented and every representing be what it is. If the I is understood as the form of representation, this amounts to saying that it is the "logical subject."

Kant's analysis has two positive aspects: on the one hand, he sees the impossibility of ontically reducing the I to a "substance." On the other hand, he holds fast to the I as "I think." Nevertheless, he conceives this I again as subject, thus in an ontologically inappropriate sense. For the ontological concept of the subject does *not* characterize *the selfhood of the I qua self, but the sameness and constancy of something always already objectively present.* To define the I ontologically as a *subject*

320

---

* Not something re-presented, but what represents as *what-places-something-before-itself* in representing—but *only* in this, and the I "is" only as this *before-itself*, only as this *itselfness.*

16. On the analysis of transcendental apperception, cf. M. Heidegger, *Kant und das Problem der Metaphysik*, 2nd ed., 1951, Division III.
17. *Kritik der reinen Vernunft*, B404.
18. Ibid., A354.

means to posit it as something always already objectively present. The being of the I is understood as the reality* of the *res cogitans*.[19]

321   What is the reason that while the "I think" gives Kant a genuine phenomenal point of departure, he cannot exploit it ontologically, but is forced to fall back upon the "subject," that is, something substantial? The I is not only an "I think," but an "I think something." However, does not Kant himself emphasize again and again that the I remains related to its representations, and would be nothing without them?

But for Kant these representations are the "empirical" which is "accompanied" by the I—the appearances to which the I is "connected." But nowhere does Kant show the kind of being of this "connection" and "accompanying." At bottom, however, their kind of being is understood as the constant objective presence of the I together with

---

* "Presence"; constant "accompanying."

19. The fact that Kant in principle conceived the ontological character of the self of the person within the horizon of an inappropriate ontology of things objectively present in the world as "substantial things" becomes clear in the material that H. Heimsoeth worked on in his essays *Persönlichkeitsbewußtsein und Ding an sich in der Kantischen Philosophie* (reprint from *Immanuel Kant: Festschrift zur zweiten Jahrhundertsfeier seines Geburtstages*, 1924). The tendency of the essay goes beyond a historiographical report and aims at the "categorial" problem of personality. Heimsoeth (pp. 31f.) says:

There is still too little consideration of the close working together of theoretical and practical reason as Kant practised and planned it. One does not sufficiently notice how even the categories (in contrast to their naturalistic fulfillment in the "principles") are here to receive explicit validity and find a new application free of natural rationalism with the primacy of practical reason (for example, substance in the "person" and the duration of personal immortality, causality as "causality through freedom," reciprocity in the "community of reasonable beings," etc.). As a means of establishing things in thought, they serve as a new access to the unconditioned without wishing to give rationalizing knowledge of objects.

But, after all, the real ontological problem has been *passed over* here. The question cannot be omitted whether these "categories" can maintain primordial validity and only need to be applied differently, or whether they do not in principle *distort* the ontological problematic of Dasein. Even if theoretical reason is included in practical reason, the existential and ontological problematic of the self remains not only unsolved, but *unasked*. On what ontological basis is the "working together" of theoretical and practical reason supposed to occur? Does theoretical behavior determine the kind of being of the person, or is it the practical reason or neither of the two—and which one then? Do not the paralogisms, in spite of their fundamental significance, reveal the lack of ontological foundation of the problematic of the self from Descartes' *res cogitans* to Hegel's concept of the Spirit? One does not even need to think "naturalistically" or "rationalistically" and can yet be in subservience to an ontology of the "substantial" that is only all the more fatal because it is seemingly self-evident. Cf. as an essential complement to the above-mentioned essay, Heimsoeth, "Metaphysische Motive in der Ausbildung des kritischen Idealismus," Kantstudien 29 (1924): 121ff. For the critique of Kant's concept of the I, cf. also Max Scheler, *Der Formalismus in der Ethik und die materiale Wertethik*, part II of this *Jahrbuch*, vol. 2 (1916), 246ff., On "Person und das 'Ich' der transcendentalen Apperception."

its representations. Kant did avoid cutting off the I from thinking, but without positing the "I think" itself in its full essential content as "I think something," and above all without seeing the ontological "presupposition" for the "I think something" as the fundamental determination of the self.* For even the point of departure of the "I think something" is not definite enough ontologically, because the "something" remains indefinite. If by this something we understand an *innerworldly* being, it tacitly implies that *world* has been presupposed; this very phenomenon of the world also determines the constitution of being of the I, if indeed it is to be possible for the I to be something like an "I think something." Saying-I means the being [Seiende] that I always am as "I-am-in-a-world." Kant did not see the phenomenon of world and was consistent enough to keep the "representations" at a distance from the *a priori* content of the "I think." But thus the I again was forced back to an *isolated* subject that accompanies representations in a way that is ontologically quite indefinite.[20]

In saying-I, Dasein, expresses itself as being-in-the-world [in-der-Welt-sein]. But does everyday saying-I take *itself as* being-in-the-world [in-der-Welt-seiend]? Here we must make a distinction. Surely in saying-I Dasein means the being [Seiende] that it itself always is. But the everyday interpretation of the self has the tendency to understand itself in terms of the "world" taken care of. When Dasein has itself in view ontically, it *fails to see* itself in relation to the kind of being of the being that it itself is. And this is particularly true of the fundamental constitution of Dasein, being-in-the-world.[21]

How is this "fleeting" ["flüchtige"] saying-I motivated? By the   322 entanglement of Dasein, for as falling prey it *flees* [*flieht*] from itself to the they. The "natural" talk about the I takes place in the they-self. What expresses itself in the "I" is that self which, initially and for the most part, I am *not* authentically. When one is absorbed in the everyday multiplicity and rapid succession of what is taken care of, the self of the self-forgetful "I take care of" shows itself as what is constantly and identically simple, but indefinite and empty. One *is*, after all, *what* one takes care of. The fact that the "natural" ontic way of saying-I overlooks the phenomenal content of Dasein that one has in view in the I does *not* give the ontological interpretation of the I the *right* to *go along with this oversight* and to force an inappropriate "categorial" horizon upon the problematic of the self.

---

* That is, temporality.

20. Cf. the phenomenological critique of Kant's "Refutation of Idealism" § 43a.
21. Cf. §§ 12 and 13.

Of course, by refusing to go along with the everyday way in which the I talks, our ontological interpretation of the "I" has by no means *solved* the problem; but it has indeed *prescribed the direction* for further questioning. The I is the being [Seiende] that one is in "being-in-the-world" ["in-der-Welt-seiend"]. Already-being-in-the-world [Schon-sein-in-einer-Welt] as being-together-with-innerworldly-things-at-hand means, however, equiprimordially being-ahead-of-oneself. "I" means the being [Seiende] that is concerned *about* the being of the being which it is. Care expresses itself with the "I" initially and for the most part in the "fleeting" talk about the I in taking care of things. The they-self keeps on saying I most loudly and frequently, because at bottom it *is not authentically* itself and evades its authentic potentiality-of-being. If the ontological constitution of the self can neither be reduced to a substantial I nor to a "subject," but if, on the contrary, the everyday, fleeting saying-I must be understood in terms of our *authentic* potentiality-of-being, the statement still does not follow that the self is the constantly objectively present ground of care. Existentially, selfhood is only to be found in the authentic potentiality-of-being-a-self, that is, in the authenticity of the being of Dasein *as care.* In terms of care the *constancy* [*Ständigkeit*] *of the self,* as the supposed persistence of the subject, gets its clarification. The phenomenon of this authentic potentiality-of-being, however, also opens our eyes to the *constancy of the self* in the sense of its having gained a steadiness. The *constancy of the self* in the double sense of constancy and steadfastness is an *authentic* counter-possibility to the unself-constancy [Unselbst-ständigkeit] of irresolute falling prey. Existentially, the *constancy of the self* [*Selbst-ständigkeit*] means nothing other than anticipatory resoluteness. Its ontological structure reveals the existentiality of the selfhood of the self.

323   Dasein *is authentically itself* in the mode of the primordial individuation of reticent resoluteness that expects anxiety* of itself. *In keeping silent,* authentic *being*-a-self does not keep on saying "I," but rather *"is"* in reticence the thrown being that it can authentically be. The self that is revealed by the reticence of resolute existence is the primordial phenomenal basis for the question of the being of the "I." Only if we are phenomenally oriented toward the meaning of being of the authentic-potentiality-of-being-a-self are we put in a position to discuss what ontological justification there is for treating substantiality, simplicity, and personality as characteristics of selfhood. The ontological question of the being of the self must be extricated from the fore-having, constantly suggested by the predominant way of saying-I, of a persistently objectively present self-thing.

Care does not need a foundation [*Fundierung*] *in a self. Rather, exis-tentiality as a constituent of care provides the ontological constitution of the*

---

* That is, the clearing of being as being.

self-constancy of Dasein to which there belongs, corresponding to the complete structural content of care, the factical falling prey to unself-constancy. The structure of care, conceived in full, includes the phenomenon of self-hood. This phenomenon is clarified by interpreting the meaning of care which we defined as the wholeness of being of Dasein.

§ 65. *Temporality as the Ontological Meaning of Care*

In characterizing the "connection" between care and selfhood, our aim was not only to clarify the special problem of I-hood, but also to help in the final preparation for phenomenally grasping the wholeness of the structural whole of Dasein. We need the *unwavering discipline* of the existential line of questioning if, for our ontological viewpoint, the kind of being of Dasein is not finally to be distorted into a mode of objective presence, even if it is a completely indifferent mode. Dasein becomes "essential" in authentic existence that is constituted as anticipatory resoluteness. This mode of the authenticity of care contains the primordial self-constancy and totality of Dasein. We must take an undistracted look at these and understand them existentially if we are to expose the ontological meaning of the being of Dasein.

What are we looking for ontologically with the meaning of care? What does *meaning* [*Sinn*] signify [bedeutet]? Our inquiry encountered this phenomenon in the context of the analysis of understanding and interpretation.[22] According to that analysis, meaning is that in which the intelligibility of something keeps itself, without coming into view explicitly and thematically itself. Meaning signifies that upon which the primary project is projected, that in terms of which something can be conceived in its possibility as what it is. Projecting discloses possibilities, that is, it discloses what makes something possible.

To expose that upon which a project is projected, means to disclose what makes what is projected possible. This exposure requires that we methodologically pursue the project (usually one that is unexpressed) underlying an interpretation in such a way that what is projected in the project is disclosed and conceivable with regard to its upon which. To set forth the meaning of care, then, means to pursue the project underlying and guiding the primordial existential interpretation of Dasein in such a way that its upon which becomes visible in what is projected. What is projected is the being of Dasein, disclosed in what constitutes it as an authentic potentiality-for-being-whole. The upon which of what is projected, of the disclosed being thus constituted, is what itself makes possible this constitution of being as care. With the question of the meaning of care, we are asking *what makes*

324

---

22. Cf. § 32.

possible the wholeness of the articulated structural whole of care in the unity
of its unfolded articulation?

Strictly speaking, meaning signifies the upon-which of the primary project of the understanding of being. Being-in-the-world that is disclosed to itself equiprimordially understands the being of innerworldly
beings together with the being of the being that it itself is, although
unthematically and not yet differentiated into its primary modes of
existence and reality. All ontic experience of beings, the circumspect
calculation of things at hand, as well as the positive scientific cognition
of things objectively present, is always grounded in the more or less
transparent projects of the being of the beings in question. But these
projects hide in themselves an upon-which from which, so to speak,
the understanding of being nourishes itself.

If we say that beings "have meaning," this signifies that they
have become accessible *in their being*, and this being, projected upon its
upon-which, is what "really" "has meaning" first of all. Beings "have"
meaning only because, as being that has been disclosed beforehand,
they become intelligible in the project of that being [Sein], that is, in
325 terms of the upon-which of this project. The primary project of the
understanding of being "gives" meaning. The question of the meaning of the being of a being takes as its theme the upon-which of the
understanding of being that underlies all *being* of beings.

Dasein is disclosed to itself authentically or inauthentically with
regard to its existence. Existing, it understands itself in such a way
that this understanding does not just grasp something, but constitutes
the existentiell being of its factical potentiality-of-being. The being that
is disclosed is that of a being that is concerned about its being. The
meaning of this being—that is, care—is what makes care possible in
its constitution, and it is what primordially makes up the being [Sein]
of this potentiality-of-being. The meaning of being of Dasein is not
something different from it, unattached and "outside" of it, rather, it
is self-understanding Dasein itself. What makes possible the being of
Dasein, and thus its factical existence?

What is projected in the primordial existential project of existence
revealed itself as anticipatory resoluteness.* What makes possible this
authentic being-whole of Dasein with regard to the unity of its articulated structural whole? Expressed formally and existentially, without constantly naming the complete structural content, anticipatory resoluteness
is the *being toward* one's ownmost, eminent potentiality-of-being. Something like this is possible only in that Dasein *can* come toward itself *at all*
in its ownmost possibility and hold itself in this possibility as possibility

---

* Ambiguous: existentiell project and the existential and projecting transfer of oneself
into that project go together.

in this letting-itself-come-toward-itself; in other words, in that it exists. This letting-*come-toward-itself* of the eminent possibility that it endures is the primordial phenomenon of the *future*. If authentic or inauthentic *being-toward-death* belongs to the being of Dasein, this is possible only as *futural* in the sense indicated now and to be more closely defined later. Here "future" does not mean a now that has *not yet* become "actual" and that sometime *will be* for the first time, but the coming in which Dasein comes toward itself in its ownmost potentiality-of-being. Anticipation makes Dasein *authentically* futural in such a way that anticipation itself is possible only in that Dasein, *as existing*, always already comes toward itself, that is, is futural in its being in general.

Anticipatory resoluteness understands Dasein in its essential being-guilty. This understanding means: taking over being-guilty in existing, to *be* the thrown ground of nullity. But to take over thrown-ness means to authentically *be* Dasein in the *way that it always already was*. Taking over throwness, however, is possible only in such a way that futural Dasein can *be* its ownmost "how it always already was," that is, its "having-been." Only because Dasein in general *is* as I *am*-having-been, can it come futurally toward itself in such a way that it comes-*back*. Authentically futural, Dasein is authentically *having-been*. Anticipation of the most extreme and ownmost possibility comes back understandingly to one's ownmost *having-been*. Dasein can *be* authentically having-been only because it is futural. In a certain sense, having-been arises from the future.   326

Anticipatory resoluteness discloses the actual situation of the there in such a way that existence circumspectly takes care of the factical things at hand in the surrounding world in action. Resolute being together with what is at hand in the situation, that is, letting *what presences* [*Anwesenden*] in the surrounding world be encountered in action, is possible only in a *making* that being *present* [*Gegenwärtigen*]. Only as the *present* [*Gegenwart*], in the sense of making present, can resoluteness be what it is; namely, letting what it takes hold of in action be encountered undistortedly.

Coming back to itself, from the future [zukünftig], resoluteness brings itself to the situation in making it present. Having-been arises from the future in such a way that the future that has-been (or better, is in the process of having-been) releases the present from itself. This unified phenomenon of the future that makes present in the process of having-been is what we call *temporality*. Only because Dasein is determined as temporality does it make possible for itself the authentic potentiality-of-being-a-whole of anticipatory resoluteness which we characterized. *Temporality reveals itself as the meaning of authentic care.*

The phenomenal content of this meaning, drawn from the constitution of being of anticipatory resoluteness, fulfills the significance of the term temporality. We must now keep the terminological use of this expression at a distance from all of the meanings of "future," "past," and "present" initially urging themselves upon us from the vulgar concept of time. That is also true of the concepts of a "subjective" and an "objective," or an "immanent" and "transcendent" "time." Since Dasein understands itself initially and for the most part inauthentically, we may suppose that the "time" of the vulgar understanding of time indeed presents a genuine phenomenon, but a derivative one. It arises from inauthentic temporality that has an origin of its own. The concepts of "future," "past," and "present" initially grew out of the inauthentic understanding of time. The terminological definition of the corresponding primordial and authentic phenomena battles with the same difficulty in which all ontological terminology is stuck. In this field of inquiry, forcing things is not an arbitrary matter, but a necessity rooted in facts. However, in order to demonstrate seamlessly the origin of inauthentic temporality from primordial and authentic temporality, we first need to work out the primordial phenomenon concretely, which we have thus far only sketched out roughly.

327

If resoluteness constitutes the mode of authentic care, and if it is itself possible only through temporality, the phenomenon at which we arrived by considering resoluteness must itself only present a modality of temporality, which makes care possible in general. The wholeness of the being of Dasein as care means: ahead-of-itself-already-being-in (a world) as being-together-with (beings encountered within the world). When we first established this articulated structure, we referred to the fact that with regard to this articulation the ontological question had to be taken back further to the exposition of the unity of the wholeness of the structural manifold.[23] *The primordial unity of the structure of care lies in temporality.*

Being-ahead-of-oneself is grounded in the future. Already-being-in . . . makes known having-been. Being-together-with . . . is made possible in making present. After what we have said, it is automatically ruled out to conceive the "ahead" in the "ahead-of-itself" and the "already" in terms of the vulgar understanding of time. The "ahead" does not mean the "before" in the sense of a "not-yet-now, but later." Nor does the "already" mean a "no-longer-now, but earlier." If the expressions "ahead of" and "already" had *this* temporal meaning, which they can also have, then we would be saying about the temporality of care that it is something that is "earlier" and "later," "not yet" and "no longer" at the same time. Then care would be conceived as a being that occurs and elapses "in time." The *being* of a being having the character of Dasein would then turn into *something objectively present.* If something like

---

23. Cf. § 41.

this is impossible, then the temporal significance of these expressions must be a different one. The "before" and the "ahead of" indicate the future that first makes possible in general the fact that Dasein can be in such a way that it is concerned *about* its potentiality-of-being. The self-project grounded in the "for the sake of itself" in the future is an essential quality of *existentiality. Its primary meaning is the future.*

Similarly, the "already" means the existential, temporal meaning of being [Seinssinn] of the being that, in that it *is*, is always already something thrown. Only because care is grounded in having-been, can Dasein exist as the thrown being that it is. "As long as" Dasein factically exists, it is never past, but is always already *having-been* in the sense of "I-*am*-as-having-been." And only as long as Dasein is, *can* it *be* as having-been. On the other hand, we call beings past that are no longer objectively present. Thus existing Dasein can never identify [feststellen] itself as an objectively present fact that comes into being and passes away "with time," and is already partially past. It always "finds itself" only as a thrown fact. In *attunement* Dasein is invaded by itself as the being that, still existing [seiend] it, already was, that is, that it constantly *is* as having been. The primary existential meaning of facticity lies in having-been. The formulation of the structure of care indicates the temporal meaning of existentiality of facticity with the expressions "before" and "already."

On the other hand, such an indication is lacking for the third constitutive factor of care: entangled being-together-with.... That is not supposed to mean that falling prey is not also grounded in temporality; it should instead intimate that *making present*, as the *primary* basis for the *falling prey* to things at hand and objectively present that we take care of, remains *included* in the future and in having-been in the mode of primordial temporality. Resolute, Dasein has brought itself back out of falling prey in order to be all the more authentically "there" for the disclosed situation in the "Moment" ["Augen*blick*"].*

Temporality makes possible the unity of existence, facticity, and falling prey and thus constitutes primordially the wholeness of the structure of care. The factors of care are not pieced together cumulatively,

---

* The word "Augenblick"—literally, "blink of the eye"—is rightly translated as "Moment." It is a commonly used word—"ein Augenblick, bitte" means "a moment, please"—but Heidegger emphasizes this word, as well as its component words ("Augen" and "Blick"), in a way that gives it a somewhat uncommon resonance and emphasis. The italicized "*blick*," which refers to a "look," links this word with a host of other words—Umsicht, Nachsicht, Rücksicht, Sicht—that also refer to a sight or look characterizing Dasein's way of being-in-the-world (see as well the marginal note on H. 61). While it is the temporal sense of the word, the momentariness that it names, that remains its dominant sense, it also needs to be stressed that it "cannot be clarified in terms of the *now*" (H. 338). Here, the English word "Moment" is capitalized in order to call attention to the importance placed upon the various senses of this word. [TR]

any more than temporality itself has first been put together out of future, past, and present "in the course of time." Temporality "is" not a *being* [*Seiendes*] at all. It is not, but rather *temporalizes* itself. Nevertheless, we still cannot avoid saying that "temporality 'is' the meaning of care," "temporality 'is' determined thus and so." The reason for this can be made intelligible only when we have clarified the idea of being and the "is" as such. Temporality temporalizes, and it temporalizes possible ways of itself. These make possible the multiplicity of the modes of being of Dasein, in particular the fundamental possibility of authentic and inauthentic existence.

329

Future, having-been, and present show the phenomenal characteristics of "toward itself," "back to," "letting something be encountered." The phenomena of toward . . . , to . . . , together with . . . reveal temporality as the ἐκστατικόν *par excellence. Temporality is the primordial, "outside of itself" in and for itself.* Thus we call the phenomena of future, having-been, and present the *ecstasies* of temporality. Temporality is not, prior to this, a being that first emerges from *itself*; rather, its essence is temporalizing in the unity of the *ecstasies.* What is characteristic of the "time" accessible to the vulgar understanding consists, among other things, precisely in the fact that it is a pure succession of nows, without beginning and without end, in which the ecstatic character of primordial temporality is leveled down. But this very leveling down, in accordance with its existential meaning, is grounded in the possibility of a definite kind of temporalizing, in conformity with which temporality temporalizes as inauthentic the kind of "time" we have just mentioned. Thus if we demonstrate that the "time" accessible to the common sense of Dasein is *not* primordial, but arises rather from authentic temporality, then, according to the principle *a potiori fit denominatio*, we are justified in calling the *temporality* now set forth *primordial time.*

In enumerating the ecstasies, we have always mentioned the future first. That should indicate that the future has priority in the ecstatic unity of primordial and authentic temporality, although temporality does not first originate through a cumulative sequence of the ecstasies, but always temporalizes itself in their equiprimordiality. But within this equiprimordiality, the modes of temporalizing are different. And the difference lies in the fact that temporalizing can be primarily determined out of the different ecstasies. Primordial and authentic temporality temporalizes itself out of the authentic future, and indeed in such a way that, futurally having-been, it first arouses the present. *The primary phenomenon of primordial and authentic temporality is the future.* The priority of the future will itself vary according to the modified temporalizing of inauthentic temporality, but it will still make its appearance in derivative "time."

Care is being-toward-death. We defined anticipatory resoluteness as authentic being toward the possibility that we characterized
as the absolute impossibility of Dasein. In this being-toward-the-end,
Dasein exists authentically and wholly as the being that it can be when
"thrown into death." It does not have an end where it just stops, but
it *exists finitely*. The authentic future, which is temporalized primarily
by *that* temporality which constitutes the meaning of anticipatory reso-   *330*
luteness, thus reveals itself *as finite*. But "does time not go on" despite
the no-longer-being-there of myself? And can there not be an unlimited
number of things that still lie "in the future" and arrive from it?

These questions are to be answered in the affirmative. Nevertheless, they do not contain any objection to the finitude of primordial
temporality because they no longer deal with that at all. The question is
not how many things can still occur "in a time that goes on," or about
what kind of a "letting-come-toward-oneself" we can encounter "out
of this time," but about how the coming-toward-oneself *itself* is to be
primordially determined *as such*. Its finitude does not primarily mean a
stopping, but is a characteristic of temporalizing itself. The primordial
and authentic future is the toward-oneself, toward *oneself*, existing as
the possibility of an insuperable nullity. The ecstatic quality of the primordial future lies precisely in the fact that it closes the potentiality-of-
being, that is, the future is itself closed and as such makes possible the
resolute existentiell understanding of nullity. Primordial and authentic
coming-toward-oneself is the meaning of existing in one's ownmost
nullity. With the thesis of the primordial finitude of temporality, we are
not contesting the fact that "time goes on." We are simply holding fast
to the phenomenal quality of primordial temporality that shows itself
in what is projected in the primordial existential project of Dasein.

The temptation to overlook the finitude of the primordial and
authentic future and thus the finitude of temporality, or to think that
it is *a priori* impossible, arises from the constant intrusion of the vulgar
understanding of time. If the latter is justifiably familiar with an endless time, and only with that, it has not yet been demonstrated that
it also understands this time and its "infinity." What does it mean to
say that time "goes on" and "keeps passing away?" What does "in
time" mean in general, and "in the future" and "out of the future"
in particular? In what sense is "time" endless? These things require
clarification if the vulgar objections to the finitude of primordial time
are not to remain without foundation. But we can clear them up only
if we have obtained the appropriate line of questioning with regard to
finitude and infinitude. However, this line of questioning arises only
if we view the primordial phenomenon of time understandingly. The
problem is not *how does "derivative,"* infinite time, "in which" objectively present things come into being and pass away, *become primordial,*   *331*

finite temporality, but rather, how does *in*authentic temporality, as *in*authentic, temporalize an *in*finite time out of finite time? Only because primordial time is *finite* can "derivative" time temporalize itself as *infinite*. In the order in which we grasp things through the understanding, the finitude of time does not become fully visible until we have set forth "endless time" so that these may be contrasted.

Let us summarize the analysis of primordial temporality in the following theses. Time is primordial as the temporalizing of temporality, and makes possible the constitution of the structure of care. Temporality is essentially ecstatic. Temporality temporalizes itself primordially out of the future. Primordial time is finite.

Yet the interpretation of care as temporality cannot remain restricted to the narrow basis we have so far attained, even if it did take the first steps toward the primordial, authentic being-whole of Dasein. The thesis that the meaning of Dasein is temporality must be confirmed by the concrete content of the fundamental constitution of this being which we have set forth.

§ 66. *The Temporality of Dasein and the Tasks of a*
More Primordial Repetition of the Existential Analysis Arising from It

Not only does the phenomenon of temporality which we have set forth require a more wide-ranging confirmation of its constitutive power, but only thus does this phenomenon itself come to view with regard to the fundamental possibilities of temporalization. We shall briefly call the demonstration of the possibility of the constitution of being of Dasein on the basis of temporality, the "temporal" interpretation, although this is only a provisional term.

Our next task is to go beyond the temporal analysis of the authentic potentiality-of-being-whole of Dasein and a general characterization of the temporality of care, so that the *inauthenticity* of Dasein may be made visible in its specific temporality. Temporality first showed itself in anticipatory resoluteness. This anticipatory resoluteness is the authentic mode of disclosedness that, for the most part, maintains itself within the inauthenticity of the entangled self-interpretation of the they. The nature of the temporality of disclosedness in general leads to the temporal understanding of that heedful being-in-the-world closest to us, and thus of the average indifference of Dasein from which the existential analytic first started.[24] We called the average kind of being of Dasein, in which it initially and for the most part stays, everydayness. By repeating our earlier analysis, *everydayness* must be revealed

332

---

24. Cf. § 9.

in its *temporal* meaning so that the problematic included in temporality may come to light and the seeming "obviousness" of our preparatory analyses may disappear completely. Indeed, confirmation is to be found for temporality in all the essential structures of the fundamental constitution of Dasein. Yet this does not lead us to run through our analyses again, superficially and schematically, in the same order of presentation. The course of our temporal analysis has a different direction. It is to make the connection of our earlier reflections clearer and to remove what is contingent and seemingly arbitrary. However, beyond these necessities of method, there are valid motives in the phenomenon itself that compel us to articulate our analysis in a different way when we repeat it.

The ontological structure of the being that I *myself* in each instance am is centered in the self-constancy of existence. Because the self cannot be conceived either as substance or as subject, but is rather grounded in existence, our analysis of the inauthentic self, the they, was left completely as part of the preparatory interpretation of Dasein.[25] Now that selfhood has been *explicitly* taken back into the structure of care, and thus of temporality, the temporal interpretation of self-constancy and the lack of self-constancy acquires an importance of its own. It requires a special thematic development. This not only provides us with the right safeguards against the paralogisms and the ontologically inappropriate question about the being of the I in general; at the same time it provides us, in accordance with its central function, with a more primordial insight into the *structure of the temporalizing of temporality*. Temporality reveals itself as the *historicity* of Dasein. The statement that Dasein is historical is confirmed as a fundamental existential and ontological proposition. It is far removed from merely ontically ascertaining the fact that Dasein occurs in a "world history." The historicity of Dasein, however, is the ground of a possible historiographical understanding that in its turn harbors the possibility of getting a special grasp of the development of historiography as a science.

The temporal interpretation of everydayness and historicity secures the view of primordial time sufficiently to uncover it as the condition of the possibility and necessity of the everyday experience of time. Dasein *expends itself* primarily *for itself* as a being that is concerned about its being, whether explicitly or not. Initially and for the most part, care is circumspect taking care. Expending itself for the sake of itself, Dasein "uses itself up." Using itself up, Dasein uses itself, that is, its time. Using its time, it reckons with it. Taking care, which is circumspect and reckoning, initially discovers time and develops a

333

---

25. Cf. §§ 25 et seq.

measurement of time. Measurement of time is constitutive for being-in-the-world. Measuring its time, the discovering of circumspection which takes care lets what it discovers at hand and objectively present be encountered in time. Innerworldly beings thus become accessible as "existing in time." We shall call the temporal quality of innerworldly beings "*within-time-ness*." The "time" initially found therein ontically becomes the basis for the development of the vulgar and traditional concept of time. But time as within-time-ness arises from an essential kind of temporalization of primordial temporality. This origin means that the time "in which" objectively present things come into being and pass away is a genuine phenomenon of time; it is not an external-ization of a "qualitative time" into space, as Bergson's interpretation of time—which is ontologically completely indeterminate and insuf-ficient—would have it.

The elaboration of the temporality of Dasein as everydayness, historicity, and within-time-ness first gives uncluttered insight into the *complexities* of a primordial ontology of Dasein. As being-in-the-world, Dasein exists factically together with beings encountered within the world. Thus the being of Dasein gets its comprehensive ontological transparency only in the horizon of the clarified being of beings unlike Dasein, that is, even of what is not at hand and not objectively present, but only "subsists." But if the variations of being are to be interpreted for everything of which we say that it *is*, we need beforehand a suf-ficiently clarified idea of being in general. As long as we have not reached this, the repetition of the temporal analysis of Dasein will remain incomplete and marred by lack of clarity—to say nothing of the factual difficulties. The existential-temporal analysis of Dasein requires in its turn a new repetition in the context of a fundamental discussion of the concept of being.

CHAPTER FOUR

Temporality and Everydayness

§ 67. *The Basic Content of the Existential Constitution of Dasein, and the Preliminary Sketch of Its Temporal Interpretation*

Our preparatory analysis[1] has made accessible a multiplicity of phenomena that must not be allowed to disappear from our phenomenological view, despite our concentration on the foundational structural wholeness of care. Far from excluding such a multiplicity, the *primordial* wholeness of the constitution of being of Dasein *as articulated* demands it. The primordiality of the constitution of Dasein does not coincide with the simplicity and uniqueness of an ultimate structural element. The ontological origin of the being of Dasein is not "inferior" to that which arises from it, but exceeds it in power from the beginning; any "arising" in the field of ontology is degeneration. The ontological penetration to the "origin" does not arrive at things which are ontically self-evident for the "common understanding;" rather it is precisely this that opens up the questionability of everything self-evident.

In order to bring the phenomena at which we arrived in our preparatory analysis back to a phenomenological view, a reference to the stages we have gone through will suffice. The delineation of care emerged from our analysis of the disclosedness that constitutes the being of the "there." The clarification of this phenomenon signified that a preliminary interpretation of the basic constitution of Dasein of being-in-the-world was necessary. Our inquiry began with a characterization of being-in-the-world, and then went on to secure from the very beginning an adequate phenomenal horizon as opposed to the inappropriate and typically unexpressed ontological predeterminations of Dasein. Being-in-the-world was initially characterized with regard to the phenomenon of the world. Indeed, our explication moved

---

1. Cf. Division I, Chapters One–Four.

from an ontic and ontological characterization of the things at hand and objectively present *"in"* the surrounding world to a delineation of innerworldliness, thus making the phenomenon of worldliness in general visible in innerworldliness. But the structure of worldliness, significance, turned out to be coupled with that upon which the understanding essentially belonging to disclosedness projects itself, that is, with the potentiality-of-being of Dasein *for the sake of which* it exists.

335   The temporal interpretation of everyday Dasein must begin with the structures in which disclosedness constitutes itself. These are: understanding, attunement, entanglement, and discourse. The modes of the temporalizing of temporality to be exposed with regard to these phenomena provide the basis for defining the temporality of being-in-the-world. This leads again to the phenomenon of the world and permits us to delineate the specifically temporal problematic of worldliness. It must be confirmed by characterizing the everyday being-in-the-world closest to us—by entangled, circumspect taking care of things. The temporality of taking care makes it possible for circumspection to be modified into a perceiving that looks at things and the theoretical knowledge based on such perceiving. The temporality of being-in-the-world that thus emerges at the same time turns out to be the foundation of the specific spatiality of Dasein. The temporal constitution of de-distancing and directionality must be shown. The whole of these analyses reveals a possibility of temporalizing of temporality in which the inauthenticity of Dasein is ontologically grounded, and leads to the question of how the temporal nature of everydayness—the temporal meaning of the "initially and for the most part," which we have continually been using—is to be understood. Fixing upon this problem makes clear that, and to what extent, the clarification of the phenomenon that we have so far attained is not sufficient.

The present chapter thus has the following structure: the temporality of disclosedness in general (§ 68); the temporality of being-in-the-world and the problem of transcendence (§ 69); the temporality of the spatiality commensurate with Dasein (§ 70); the temporal meaning of the everydayness commensurate with Dasein (§ 71).

§ 68. *The Temporality of Disclosedness in General*

Resoluteness, which we characterized with regard to its temporal meaning, represents an authentic disclosedness of Dasein. Disclosedness constitutes a being in such a way that, existing [existierend], it can itself be its "there." Care was characterized with respect to its temporal meaning only in its basic features. To demonstrate its concrete temporal constitution means to interpret temporally its individual structural moments, that is, understanding, attunement, entanglement, and discourse. Every understanding has its mood. Every attunement understands. Attuned understanding has the characteristic of entanglement. Entangled, attuned understanding articulates itself with regard to its intelligibility in discourse. The actual temporal constitution of these phenomena always leads back to that *one* temporality that holds within itself the possible structural unity of understanding, attunement, entanglement, and discourse.

(a) The Temporality of Understanding[2]

336

With the term understanding we mean a fundamental existential; neither a definite *kind of cognition*, as distinct from explaining and conceiving, nor a cognition in general in the sense of grasping something thematically. Understanding constitutes the being of the there in such a way that, on the basis of such understanding, a Dasein in existing can develop the various possibilities of sight, of looking around, and of just looking. All explanation, as discovery that understands, is rooted in the primary understanding of Dasein.

Formulated primordially and existentially, understanding means: *to be projecting toward a potentiality-of-being for the sake of which Dasein always exists*. Understanding discloses one's own potentiality-of-being in such a way that Dasein always somehow knows understandingly what is going on with itself. This "knowing" ["Wissen"], however, does not mean that it has discovered some fact, but that it holds itself in an existentiell possibility. The ignorance [Nichtwissen] corresponding to this does not consist in a failure to understand, but must be taken as a deficient mode of the projectedness of one's potentiality-of-being. Existence can be questionable. If it is to be possible for something "to be in question," a disclosedness is necessary. When one understands oneself projectively in an existentiell possibility, the future underlies this understanding, and it does so as a coming-toward-oneself from the actual possibility as which Dasein in each instance exists. The future makes ontologically possible a being that is in such a way that it exists understandingly in its potentiality-of-being. Projecting, which is fundamentally futural, does not primarily grasp the projected possibility thematically in an intention, but throws itself into it as possibility. Understandingly, Dasein in each instance *is* as it can be. Resoluteness turned out to be primordial and authentic existing. Of course, initially and for the most part Dasein remains irresolute, that is, it remains

---

2. Cf. § 31.

closed off from its ownmost potentiality-of-being to which it in each instance brings itself only in individuation. This means that temporality does not temporalize itself constantly out of the authentic future. However, this inconstancy does not mean that temporality at times lacks the future, but rather that the temporalizing of the future is variable [abwandelbar].

We shall retain the expression *anticipation* [*Vorlaufen*] for the terminological characterization of the authentic future. It indicates that Dasein, authentically existing, lets itself come toward itself as its ownmost potentiality-of-being—that the future must first win itself, not from a present, but from the inauthentic future. The formally neutral term for the future lies in the designation of the first structural factor of care, in *being-ahead-of-itself* [*Sich-vorweg*]. Factically, Dasein is constantly ahead-of-itself, but in its existentiell possibility inconstantly anticipatory.

How is the inauthentic future to be contrasted with this? Since the authentic future is revealed in resoluteness, the inauthentic future, as an ecstatic mode, can reveal itself only by going back ontologically from the everyday, inauthentic understanding taking care of things to its existential and temporal meaning. As care, Dasein is essentially ahead-of-itself. Initially and for the most part, the being-in-the-world that takes care understands itself in terms of *what* it takes care of. Inauthentic *understanding* projects itself upon what can be taken care of, what can be done, what is urgent or indispensable in the business of everyday activity. But what is taken care of is as it is for the sake of the potentiality-of-being that cares. This potentiality lets Dasein come toward itself in its heedful being together with what is to be taken care of. Dasein does not come toward itself primarily in its ownmost, nonrelational potentiality-of-being, but it *awaits this* heedfully *in terms of that which what is taken care of produces or denies*. Dasein comes toward itself in terms of what is taken care of. The inauthentic future has the character of *awaiting* [*Gewärtigens*]. The self-understanding of the they that takes care in terms of what one is doing, has the "ground" of its possibility in this ecstatic mode of the future. And *only because* factical Dasein is thus *awaiting* its potentiality-for-being in terms of what is taken care of, can it *expect* and wait for. . . . Awaiting must always already have disclosed the horizon and scope in terms of which something can be expected. *Expecting is a mode of the future founded in awaiting that temporalizes itself authentically as anticipation.* Thus a more primordial being-toward-death lies in anticipation than in the heedful expecting of it.

Existing in the potentiality-of-being, however it may be projected, understanding is *primarily* futural. But it would not temporalize itself if it were not temporal, that is, equiprimordially determined by having-been and the present. The way in which the ecstasy of the present co-constitutes inauthentic understanding was already made roughly clear. Everyday taking care of things understands itself in terms of the potentiality-of-being that confronts it as coming from its possible success or failure with regard to what is actually taken care of. Corresponding to the inauthentic future (awaiting), it has its own way of being *together with* what is taken care of. The ecstatic mode of this present reveals itself if we adduce for comparison this very same ecstasy, but in the mode of authentic temporality. To the anticipation of resoluteness there belongs a present in accord with which a resolution discloses the situation. In resoluteness, the present is not only brought back from the dispersion in what is taken care of closest at hand, but is held in the future and having-been. We call the *present* that is held in authentic temporality, and is thus *authentic*, the *Moment* [*Augenblick*]. This term must be understood in the active sense as an ecstasy. It means the resolute raptness of Dasein in what is encountered as possibilities and circumstances to be taken care of in the situation, but this rapture is *held* in resoluteness. The phenomenon of the Moment can *in principle not* be clarified in terms of the *now*. The now is a temporal phenomenon that belongs to time as within-time-ness: the now "in which" something comes into being, passes away, or is objectively present. "In the Moment" nothing can happen, but as an authentic present it lets us *encounter for the first time* what can be "in a time" as something at hand or objectively present.[3]

In contrast to the Moment as the authentic present, we shall call the inauthentic present *making present* [*Gegenwärtigen*]. Understood formally, every present makes present, but not every present is "in the moment." When we use the expression making present with no additional qualification, we always mean the inauthentic kind, which is irresolute and lacking the Moment. Making present will become clear only in terms of the temporal interpretation of falling prey to the "world" taken care of; this falling prey has its existential meaning in making present. But since inauthentic understanding projects its potentiality-of-being in terms of what can be taken care of, this means that it temporalizes itself in terms of making present. The Moment, on the other hand, temporalizes itself out of the authentic future.

Inauthentic understanding temporalizes itself as an awaiting that makes present—an awaiting to whose ecstatic unity a corresponding

338

339

---

3. S. Kierkegaard saw the *existentiell* phenomenon of the Moment in the most penetrating way, which does not mean that he was also as successful in the existential interpretation of it. He gets stuck in the vulgar concept of time and defines the Moment with the help of the now and eternity. When Kierkegaard speaks of "temporality," he means human being's being-in-time. Time as within-time-ness knows only the now, but never a moment. But if the moment is experienced existentially, a more primordial temporality is presupposed, although existentially inexplicit. In relation to the "Moment," cf. K. Jaspers, *Psychologie der Weltanschauungen*, 3rd edition 1925, p. 108ff and the "Referat Kierkegaards," pp. 419–432.

having-been must belong. The authentic coming-toward-itself of anticipatory resoluteness is at the same time a coming back to the ownmost self thrown into its individuation. This ecstasy makes it possible for Dasein to be able to take over resolutely the being [*Seiende*] that it already is. In anticipation, Dasein *brings* itself *forth again* [*holt sich . . . wieder . . . vor*] to its ownmost potentiality-of-being. We call authentic having-*been repetition* [*Wiederholung*]. But when one projects oneself inauthentically upon the possibilities drawn from what is taken care of in making it present, this is possible only because Dasein has *forgotten* itself in its ownmost *thrown* potentiality-of-being. This forgetting is not nothing, nor is it just a failure to remember; it is rather a "positive," ecstatic mode of having-been, a mode with a character of its own. The ecstasy (rapture [*Entrückung*]) of forgetting has the character of backing away *from* [*Ausrücken vor*] one's ownmost having-been in a way that is closed off from oneself. This backing away from . . . ecstatically closes off what it is backing away from, and thus closes itself off, too. As inauthentic having-been, forgetfulness is thus related to its own thrown *being* [*Sein*]. It is the temporal meaning of the kind of being that I initially and for the most part *am* as having-been. And only on the basis of this forgetting can the making present that takes care of and awaits *retain* beings unlike Dasein encountered in the surrounding world. To this retention corresponds a nonretention that presents us with a kind of "forgetting" in the derivative sense.

Just as expectation is possible only on the basis of awaiting, *remembering* is possible only on the basis of forgetting, *and not the other way around*. In the mode of forgottenness, having-been primarily "discloses" the horizon in which Dasein, lost in the "superficiality" of what is taken care of, can remember. *Awaiting that forgets and makes present* is an ecstatic unity in its own right, in accordance with which inauthentic understanding temporalizes itself with regard to its temporality. The unity of these ecstasies closes off one's authentic potentiality-of-being, and is thus the existential condition of the possibility of irresoluteness. Although inauthentic heedful understanding is determined in the light of making present what is taken care of, the temporalizing of understanding comes about primarily in the future.

(b) The Temporality of Attunement[4]

Understanding is never free floating, but always attuned. The there is equiprimordially disclosed by mood, or else closed off. Attunement brings Dasein *before* its thrownness in such a way that the latter is not known as such, but is disclosed far more primordially in "how one

340

---

4. Cf. § 29.

is." *Being*-thrown means existentially to find oneself in such and such a way. Thus attunement is grounded in thrownness. Mood represents the way in which I am always primarily the being [Seiende] that has been thrown. How can the temporal constitution of attunement become visible? How can we gain insight into the existential connection between attunement and understanding in terms of the ecstatic unity of actual temporality?

Mood discloses by turning away from and toward one's own Dasein. Whether authentically revealing or inauthentically concealing, *bringing* Dasein *before* the That of its own thrownness is existentially possible only if the being of Dasein, by its very meaning, *is* as constantly havingbeen. Having-been does not first bring one face to face with the thrown being that one is oneself, but the ecstasy of having-been first makes possible finding oneself in the mode of how-I-find-myself. Understanding is primarily grounded in the future; *attunement*, on the other hand, temporalizes itself *primarily* in having-been. Mood temporalizes itself, that is, its specific ecstasy belongs to a future and a present, but in such a way that having-been modifies the equiprimordial ecstasies.

We emphasized the fact that whereas moods are ontically familiar, they are not cognized in their primordial and existential function. They are taken as fleeting experiences that "color" one's whole "psychical condition." Whatever can be observed as having the nature of a fleeting appearance and disappearance belongs to the primordial constancy of existence. But, nevertheless, what should moods have in common with "time"? It is a trivial observation that these "experiences" come and go, that they run their course "in time"; it is certainly an ontic and psychological observation. But our task is to demonstrate the ontological structure of attunement in its existential and temporal constitution. And this is initially only a matter of first making the temporality of mood visible. The thesis that "attunement is primarily grounded in having-been" means that the existential fundamental nature of mood is a *bringing back to.* . . . This does not first produce having-been, but attunement always reveals a mode of having-been for the existential analysis. Thus the temporal interpretation of attunement cannot have the intention of deducing moods from temporality and dissolving them *341* in the pure phenomena of temporalizing. We simply want to show that moods are *not possible* in what they "signify" existentielly or how they "signify" it *except on the basis of temporality*. Our temporal interpretation will restrict itself to the phenomena of fear and anxiety, which were already analyzed in a preparatory way.

We shall begin our analysis by exhibiting the temporality of *fear*.[5] Fear was characterized as inauthentic attunement. Why does the

---

5. Cf. § 30.

existential meaning that makes such an attunement possible lie in what
has been? What mode of this ecstasis characterizes the specific temporality of fear? Fear is a fear *of* something threatening—of something
that is detrimental to the factical potentiality-of-being of Dasein, and
that approaches within the scope of the things at hand and objectively
present being taken care of in the manner described. Fearing discloses something threatening in the mode of everyday circumspection. A
subject that merely looks could never discover anything like this. But
is not this disclosing of fear of . . . a letting something-come-toward-
oneself? Has not fear been justifiably described as the expectation of
a coming evil (*malum futurum*)? Is not the primary temporal meaning
of fear the future and not at all having-been? Fearing is incontestably
"related" not only to "something futural" in the sense of what first
arrives "in time," but this relatedness itself is futural in the primordially temporal sense. *Awaiting* obviously *also* belongs to the existential
and temporal constitution of fear. But this initially only means that the
temporality of fear is *inauthentic*. Is the fear of . . . merely the expectation of something futurally threatening? The expectation of something
futurally threatening need not already be fear, and it is so far from
being fear that it lacks the specific mood character of fear. In fear, the
awaiting of fear lets what is threatening *come back* to one's potentiality-of-being factically taking care of things. Only if that to which this
comes back is already ecstatically open, can what is threatening be
awaited *back to* the being that I am, and only thus can Dasein be threatened. The character of the mood and *affect* of fear lies in the fact that
the awaiting that fears is afraid "for itself," that is, fear of is a fearing
about. The existential and temporal meaning of fear is constituted by
a self-forgetting: the confused backing away from one's own factical
potentiality-of-being, which is threatened being-in-the-world taking
care of what is at hand. Aristotle correctly defines fear as λύπη τις ἢ
ταραχή, as depression or confusion.[6] Depression forces Dasein back
to its thrownness, but in such a way that its thrownness is precisely
closed off. Confusion is based upon forgetting. When one forgets and
backs away from a factical, resolute potentiality-of-being, one keeps to
those possibilities of self-preservation and evasion that have already
been circumspectly discovered beforehand. Taking care which fears for
itself leaps from one thing to the other, because it forgets itself and thus
cannot take hold of any *definite* possibility. All "possible" possibilities
offer themselves, and that means impossible ones too. One who fears
for oneself stops at none of these—the "surrounding world" does not
disappear—but one encounters it in the mode of no longer knowing
one's way around in *it*. This *confused making present* of what seems to

342

---

6. Cf. *Rhetoric* B 5, 1382a21.

be the best thing close by belongs to forgetting oneself in fear. Thus, for example, it is known that the inhabitants of a burning house often "save" the most unimportant things nearby. When one has forgotten oneself and makes present a jumble of unattached possibilities, one thus makes possible the confusion that constitutes the nature of the mood of fear. The forgetfulness of confusion also modifies awaiting, and characterizes it as a depressed or confused awaiting that is different from pure expectation.

The specific, ecstatic unity that makes fearing for oneself existentially possible, temporalizes itself primarily out of the forgetting we described that, as a mode of having-been, modifies its present and its future in their temporalizing. The temporality of fear is a forgetting that awaits and makes present. In accordance with its orientation toward things encountered within the world, the commonsense interpretation of fear initially seeks to determine the "approaching evil" as what it is afraid of and to define its relation to that evil as expectation. What belongs to the phenomenon beyond that remains a "feeling of pleasure or pain."

How is the temporality of anxiety related to that of fear? We called the phenomenon of anxiety a fundamental attunement.[7] It brings Dasein before its ownmost thrownness and reveals the uncanniness of everyday, familiar being-in-the-world. Just like fear, anxiety is formally determined by something *in the face of* which one is anxious and something *about* which one is anxious. However, our analysis showed that these two phenomena coincide. That is not supposed to mean that their structural characteristics are fused, as if anxiety were anxious neither in the face of anything nor about anything. That the "in the face of which" and the "about which" coincide means that the being exhibiting these structures is the same, namely Dasein. In particular, that in the face of which one has anxiety is not encountered as something definite to be taken care of; the threat does not come from something at hand and objectively present, but rather from the fact that everything at hand and objectively present absolutely has nothing more to "say" to us. Beings in the surrounding world are no longer relevant. The world in which I exist has sunk into insignificance, and the world thus disclosed can set free only beings as having the character of irrelevance. The nothingness of the world in the face of which anxiety is anxious does not mean that an absence of innerworldly things objectively present is experienced in anxiety. They must be encountered in just such a way that they are of *no* relevance *whatsoever*, but can show themselves in a barren mercilessness. However, this means that our heedful awaiting finds nothing in terms of which it could understand itself; it grasps

343

---

7. Cf. § 40.

at the nothingness of the world. But, thrust toward the world, under-standing is brought by anxiety to being-in-the-world as such. Being-in-the-world is both what anxiety is anxious in the face of and what it is anxious about. Being anxious in the face of . . . neither has the nature of an expectation nor of an awaiting at all. That in the face of which one has anxiety is, after all, already "there"; it is Dasein itself. Does this not mean that anxiety is constituted by a future? Certainly, yet not by the inauthentic one of awaiting.

The insignificance of the world disclosed in anxiety reveals the nullity of what can be taken care of, that is, the impossibility of project-ing oneself upon a potentiality-of-being primarily based upon what is taken care of. But the revelation of this impossibility means to let the possibility of an authentic potentiality-of-being shine forth. What is the temporal meaning of this revealing? Anxiety is anxious about naked Dasein thrown into uncanniness. It brings one back to the sheer That of one's ownmost, individuated thrownness. This bringing back has neither the character of an evasive forgetting, nor that of a remember-ing. But neither does anxiety imply that one has already taken over one's existence in resolution and is retrieving it. On the contrary, anxiety brings one back to thrownness *as something to be possibly repeated*. And thus it *also* reveals the possibility of an authentic potentiality-of-being that must, as something futural in repetition, come back to the thrown There. *Bringing before the possibility of repetition is the specific ecstatic mode of the attunement of the having-been that constitutes anxiety.*

The forgetting constitutive for fear confuses Dasein and lets it stray back and forth between ungrasped "worldly" possibilities. In contrast to this frantic making present, the present of anxiety is *main-tained* in bringing oneself back to one's ownmost thrownness. In accor-dance with its existential meaning, anxiety cannot lose itself in what can be taken care of. If something like this happens in an attunement similar to it, this is fear, which everyday understanding mixes up with anxiety. Although the present of anxiety is *maintained*, it does not as yet have the character of the Moment that temporalizes itself in resolution. Anxiety only brings one into the mood for a *possible* resolution. The present of anxiety holds the Moment *in readiness* [*auf dem Sprung*], as which it, and only it, is possible.

In the peculiar temporality of anxiety, that it is primordially grounded in having-been and that out of it future and present tempo-ralize themselves, the possibility was shown of the powerfulness that distinguishes the mood of anxiety. In it, Dasein is taken back fully to its naked uncanniness and stunned [benommen] by it. But this feeling of being stunned not only *takes* Dasein back from its *"worldly"* pos-sibilities, but at the same time *gives* it the possibility of an *authentic* potentiality-of-being.

Yet neither of these moods, fear and anxiety, ever "occurs," just isolated in the "stream of experience," but always attunes an understanding or is attuned by it. Fear is occasioned by beings taken care of in the surrounding world. In contrast, anxiety arises from Dasein itself. Fear comes over us from innerworldly beings. Anxiety arises from being-in-the-world as thrown being-toward-death. Understood temporally, this "arising" of anxiety from Dasein means that the future and the present of anxiety temporalize themselves out of a primordial having-been in the sense of bringing us back to the possibility of repetition. But anxiety can arise authentically only in a resolute Dasein. One who is resolute knows no fear, but understands the possibility of anxiety as *the* mood that does not hinder and confuse him. Anxiety frees one *from* "nullifying" ["nichtigen"] possibilities and lets one become free *for* authentic possibilities.

Although both modes of attunement, fear and anxiety, are primarily grounded in *having-been*, their origin is different with regard to the temporalizing belonging to each in the totality of care. Anxiety arises from the *future* of resoluteness, while fear arises from the lost present of which fear is fearfully apprehensive, thus falling prey to it more than ever.

345

But is not the thesis of the temporality of moods perhaps valid only for the phenomena that we selected? How is a temporal meaning to be found in the pallid lack of mood that dominates the "grey everyday"? And how about the temporality of moods and affects such as hope, joy, enthusiasm, and gaiety? As soon as we mention phenomena such as weariness [Überdruß], sadness, melancholy, and despair, it becomes clear that not only fear and anxiety are founded existentially in having-been, but that other moods are as well. However, these must be interpreted on the broader basis of a developed existential analytic of Dasein. But even such a phenomenon as hope, which seems to be completely founded in the future, must be analyzed in a way similar to fear. In contrast to fear which is related to a *malum futurum*, hope has been characterized as the expectation of a *bonum futurum*. But what is decisive for the structure of hope as a phenomenon is not so much the "futural" character of that *to which* it is related as the existential meaning of *hoping itself*. Here, too, the mood character lies primarily in hoping as *hoping something for oneself*. One who hopes takes oneself, so to speak, *along* in the hope and brings oneself toward what is hoped for. But that presupposes having-achieved-oneself. The fact that hope *brings relief* from depressing apprehensiveness only means that even this attunement remains related to a burden in the mode of having-been. Elevated or elevating moods are ontologically possible only in an ecstatic-temporal relation of Dasein to the thrown ground of itself.

The pallid lack of mood of complete indifference, which clings to nothing and urges to nothing, and which goes along with what the day brings, yet in a way takes everything with it, demonstrates in the *most penetrating* fashion the power of *forgetting* in the everyday moods of taking care of what is nearby. Just barely living, which "lets everything alone" as it is, is grounded in giving oneself over to thrownness and forgetting. It has the ecstatic meaning of an inauthentic having-been. Indifference [Gleichgültigkeit], which can go along with falling over oneself with busyness, is to be sharply distinguished from equanimity [Gleichmut]. *This* mood arises from the resoluteness that, in the Moment [augen*blicklich*], has its view to the possible situations of the potentiality-of-being-whole disclosed in the anticipation of death.

346   Only beings that in accordance with the meaning of their being [Seinssinne] are attuned—that is, beings which, as existing, have in each instance already been and exist in a constant mode of having-been—can be affected. Ontologically, affection presupposes making present in such a way that in it Dasein can be brought back to itself as having-been. How the *stimulation* and *touching* of the senses in beings that are simply alive are to be ontologically defined, for example, how and where in general the being of animals is constituted by a "time," remains a problem for itself.

(c) The Temporality of Falling Prey[8]

The temporal interpretation of understanding and attunement not only came up against a *primary* ecstasy for each of these phenomena, but at the same time always came up against temporality as a *whole*. Just as the future primarily makes understanding possible, and having-been makes mood possible, the third constitutive factor of care, falling prey, has its existential meaning in the *present*. Our preparatory analysis of falling prey began with an interpretation of idle chatter, curiosity, and ambiguity.[9] Our temporal analysis of falling prey should follow the same path. We shall restrict our inquiry, however, to a consideration of curiosity because in it the specific temporality of falling prey is most easily seen. On the other hand, our analysis of idle chatter and ambiguity already presupposes a clarification of the temporal constitution of discourse and meaning (interpretation).

         Curiosity is an eminent tendency of Dasein's being in accordance with which Dasein takes care of a potentiality of seeing.[10] Like the

---

8. Cf. § 38.
9. Cf. §§ 35 et seq.
10. Cf. § 36.

concept of sight, "seeing" is not limited to perceiving with the "physical eyes." Perceiving [Vernehmen] in the broader sense lets what is at
hand and present [Vorhanden] be "bodily" encountered with regard
to their outward appearance. This letting something be encountered is
grounded in a present [Gegenwart]. This present provides the ecstatic
horizon in general within which being can be bodily *present* [*anwesend*].
Curiosity, however, does not make present what is objectively present
in order to *understand* it by lingering with it, but it seeks to see *only* in
order to see and have seen. As this making present that gets tangled up
in itself, curiosity has an ecstatic unity with a corresponding future and
having-been. Greed for the new indeed penetrates to something not   *347*
yet seen, but in such a way that making present attempts to withdraw
from awaiting. Curiosity is altogether inauthentically futural, in such
a way that it does not await a *possibility*, but in its greed only desires
possibility as something real. Curiosity is constituted by a dispersed
making present that, only making present, thus constantly tries to run
away from the awaiting in which it is nevertheless "held," although in
a dispersed way. The present "arises" from ["entspringt"] the awaiting that belongs to it in the sense of running away from it that we
emphasized. But the making present of curiosity that "arises" is so
little interested in the "matter in question" that, as soon as it catches
sight of it, it already is looking for the next thing. The making present
that "arises" from the awaiting of a definite, grasped possibility makes
possible ontologically the *not-lingering* that is distinctive of curiosity.
Making present does not "arise" from awaiting in such a way that it,
so to speak, disengages itself from awaiting and leaves it to itself, ontically understood. "Arising" is an ecstatic modification of awaiting in
such a way that awaiting *pursues* [*nachspringt*] making present. Awaiting gives itself up, so to speak; it no longer lets inauthentic possibilities
of taking care come toward it from what is taken care of, unless they
serve the purpose of an impatient making present. When awaiting is
ecstatically modified by a making present that no longer arises, but
pursues, this modification is the existential and temporal condition of
the possibility of distraction.

Making present is left more and more to itself as it is modified by
the awaiting that pursues. It makes present for the sake of the present.
Thus tangled up in itself, the dispersed not-lingering turns into *never
dwelling anywhere* [*Aufenthaltlosigkeit*]. This mode of the present is the
most extreme opposite phenomenon to the *Moment*. In this *never dwelling anywhere* Da-sein is everywhere and nowhere. *The Moment* brings
existence to the situation and discloses the authentic "there."

The more inauthentic the present is, that is, the more making
present comes to "itself," the more it flees from a definite potentiality-
of-being and closes it off. But then the future can hardly come back

at all to the being [Seiende] that has been thrown. In the "arising" of the present, one also forgets increasingly. The fact that curiosity always already keeps to what is nearest by, and has forgotten what went before, is not something resulting *from* curiosity, but the ontological condition for curiosity itself.

348

With regard to their temporal meaning, the characteristics of falling prey that we described—temptation, tranquillization, alienation, and self-entanglement—mean that the making present that "arises" seeks to temporalize itself out of itself in accordance with its ecstatic tendency. Dasein entangles itself, and this determination has an ecstatic meaning. The rapture of existence in making present does not mean that Dasein is separated from its I and its self. Even in the most extreme making present, it remains temporal, that is, awaiting and forgetting. Even in making present, Dasein still understands itself, although it is alienated from its ownmost potentiality-of-being that is primarily grounded in the authentic future and having-been. But since making present always offers something "new," it does not let Dasein come back to itself and constantly tranquillizes it anew. But this tranquillization again enforces the tendency toward arising. Curiosity is "brought about" not by the endless immensity of what has not yet been seen, but rather by the entangled kind of temporalizing of the arising present. Even if one has seen everything, curiosity *invents* new things.

The mode of temporalizing of the "arising" of the present is grounded in the essence of temporality, which is *finite*. Thrown into being-toward-death, Dasein initially and for the most part flees from this more or less explicitly revealed thrownness. The present arises from its authentic future and having-been, so that it lets Dasein come to authentic existence only by taking a detour through that present. The origin of the "arising" of the present, that is, of being entangled in lostness, is the primordial, authentic temporality itself that makes possible thrown being-toward-death.

The thrownness *before* which Dasein can indeed be brought *authentically* and in which it can authentically understand itself yet remains closed off from it with regard to where it comes from and how it comes ontically. But this closure is by no means only a factually existent lack of knowledge, but constitutes the facticity of Dasein. It also determines the *ecstatic* character of the abandonment of existence to the null ground of itself.

Initially, the throw [Wurf] of being-thrown into the world does not authentically get caught by Dasein. The "movement" in such a throw does not already come to a "stop" because Dasein "is there." Dasein is swept along in thrownness, that is, as something thrown into the world, it loses itself in the "world" in its being factically dependent on what is to be taken care of. The present, which constitutes the existential meaning of being swept along, never acquires another ecstatic horizon of its own accord, unless it is brought back from its lostness by a resolution so that both the actual situation and thus the primordial "limit-situation" of being-toward-death are disclosed as the Moment held on to.

(d) The Temporality of Discourse[11]

The complete disclosedness of the there constituted by understanding, attunement, and falling prey is articulated by discourse. Thus discourse does not temporalize itself primarily in a definite ecstasis. But since discourse is for the most part spoken in language and initially speaks by addressing the "surrounding world" in taking care of it and talking about it, *making present* has, of course, a *privileged* constitutive function.

Tenses, like the other temporal phenomena of language—"kinds of action" and "temporal stages"—do not originate from the fact that discourse "also" speaks about "temporal" processes, namely processes that are encountered "in time." Nor does the reason for this lie in the fact that speaking occurs "in psychical time." Discourse is *in itself* temporal, since all speaking about . . . , of . . . , or to . . . is grounded in the ecstatic unity of temporality. The *kinds of action* are rooted in the primordial temporality of taking care, whether it is related to things within time or not. With the help of the vulgar and traditional concept of time, which linguistics is forced to make use of, the problem of the existential and temporal structure of the kinds of action *cannot even be formulated.*[12] But because discourse is always talking about beings, although not primarily and predominantly in the sense of theoretical statements, our analysis of the temporal constitution of discourse and the explication of the temporal characteristics of language patterns can be tackled only if the problem of the fundamental connection between being and truth has been unfolded in terms of the problematic of temporality. Then the ontological meaning of the "is" can be defined, which a superficial theory of propositions and judgments has disfigured into the "copula." The "origination" of "significance" can be clarified and the possibility of the formulation of concepts can be made ontologically intelligible only in terms of the temporality of discourse, that is, of Dasein in general.

Understanding is grounded primarily in the future (anticipation or awaiting). Attunement temporalizes itself primarily in having-been (repetition or forgottenness). Falling prey is temporally rooted

---

11. Cf. § 34.
12. Cf. J. Wackernagel, *Vorlesungen über Syntax*, vol. 1 (1920), p. 15 and especially pp. 149–210. See also G. Herbig, "Aktionsart und Zeitstufe," *Indogermanische Forschung* 6 (1896): 167ff.

334 Being and Time

primarily in the present (making present or the Moment). And yet, understanding is always a present that "has-been." And yet, attunement temporalizes itself as a future that "makes present." And yet, the present "arises" from or is held by a future that has-been. From this it becomes evident that *temporality temporalizes itself completely in every ecstasis; that is, in the ecstatic unity of the actual, complete, temporalizing of temporality the wholeness of the structural whole of existence, facticity, and falling prey is grounded—that is the unity of the structure of care.*

Temporalizing does not mean a "succession" of the ecstasies. The future is *not later* than the having-been, and the having-been is *not earlier* than the present. Temporality temporalizes itself as a future that makes present, in the process of having-been.

The disclosedness of the there and the fundamental existentiell possibilities of Dasein, authenticity and inauthenticity, are founded in temporality. But disclosedness always pertains equiprimordially to the whole of being-in-the-world, to being-in as well as the world. In orientation toward the temporal constitution of disclosedness we must thus be able to demonstrate the ontological condition of the possibility that there can be beings that exist as being-in-the-world.

§ 69. *The Temporality of Being-in-the-World and the Problem of the Transcendence of the World*

The ecstatic unity of temporality—that is, the unity of the "outside-itself" in the raptures [Entrückungen] of the future, the having-been, and the present—is the condition of the possibility that there can be a being that exists as its "there." The being that bears the name Dasein is *"cleared."*[13] The light that constitutes this clearedness of Dasein is not a power or source, objectively present ontically, for a radiant brightness sometimes occurring in this being. What essentially clears this being, that is, makes it "open" as well as "bright" for itself, was defined as care, before any "temporal" interpretation. The full disclosedness of the there is grounded in care. This clearedness first makes possible any illumination or throwing light, any perceiving, "seeing," or having of something. We understand the light of this clearedness only if we do not look for an innate, objectively present power, but rather question the whole constitution of being of Dasein, care, as to the unified ground of its existential possibility. *Ecstatic temporality clears the there primordially.* It is the primary regulator of the possible unity of all the essential existential structures of Dasein.

Only in terms of the rootedness of Dasein in temporality, do we gain insight into the existential *possibility* of the phenomenon that we

351

---

13. Cf. § 28.

characterized at the beginning of our analytic of Dasein as its fundamental constitution; namely, *being-in-the-world*. At the beginning, it was
a matter of securing the unbreakable structural unity of this phenomenon. The question of the *ground of the possible unity* of this *articulated*
structure remained in the background. With the intention of protecting
the phenomenon from the most obvious and thus the most fatal tendencies to divide it up, the everyday mode of being-in-the-world closest to us—*taking care* of innerworldly beings at hand—was interpreted
more extensively. Now that *care* itself has been ontologically defined
and traced back to temporality as its existential ground, *taking care* in
its turn can be conceived *explicitly* in terms of care or temporality.

Our analysis of the temporality of taking care initially kept to
the mode of circumspectly having to do with things at hand. Then it
followed up the existential and temporal possibility that circumspect
taking care may be modified into a discovering of innerworldly beings
in the sense of certain possibilities of scientific investigation and discovering them by "only" looking at them. Our interpretation of the
temporality of the being together with *innerworldly* things at hand and
objectively present—being together circumspectly as well as theoretically taking care—shows at the same time how the same temporality is
already the advance condition of the possibility of being-in-the-world
in which being together with *innerworldly* beings is grounded as such.
The thematic analysis of the temporal constitution of being-in-the-
world led to the following questions: how is something like world
possible at all, in what sense *is* world, what and how does the world
transcend, how are "independent" innerworldly beings "connected"
with the transcending world? The *ontological exposition* of these questions does not already entail their answer. On the other hand, it does
bring about the clarification, previously necessary, of *the* structures with
reference to which the problem of transcendence is to be interrogated.
The existential and temporal interpretation of being-in-the-world will   *352*
consider three things: (a) the temporality of circumspect taking care; (b)
the temporal meaning of the modification of circumspect taking care
into theoretical knowledge of innerworldly things objectively present;
(c) the temporal problem of the transcendence of the world.

(a) The Temporality of Circumspect Taking Care

How are we to gain the perspective for an analysis of the temporality
of taking care? We called heedful being together with the "world" our
dealing in and with the surrounding world.[14] As examples of the phenomena of being together with . . . we chose the using, handling, and

---

14. Cf. § 15.

producing of things at hand and their deficient and undifferentiated modes, that is, the being together with things that belong to everyday need.[15] The authentic existence of Dasein also maintains itself in such taking care, even when it remains "indifferent" for it. Things at hand taken care of are not the cause of taking care, as if this were to arise only on the basis of the effects of innerworldly beings. Being together with things at hand can neither be explained ontically in terms of those things at hand nor can things at hand be derived from this kind of being. Taking care, as a kind of being of Dasein, and things taken care of, as innerworldly things at hand, are, however, not simply *present together*. Nevertheless, there is a "connection" between them. The properly understood 'whereby' of what is dealt with sheds light on heedful dealing itself. On the other hand, if we miss the phenomenal structure of the whereby of dealings, we fail to recognize the existential constitution of such dealings. It is indeed already an essential gain for the analysis of the beings encountered nearest to us if their specific handy character is not omitted. But we must understand further that heedful dealing never has to do with a single useful thing. Using and handling a definite useful thing remains as such oriented toward a context of useful things. If, for example, we look for a "misplaced" useful thing, we do not mean by this simply and primarily only what is looked for in an isolated "act," but the context of the whole of useful things has already been discovered. Any "going to work" and reaching for something does not bump up against a useful thing given in isolation out of nothing [Nichts], but in taking hold of a useful thing we come back to the useful thing grasped from the work-world that has always already been disclosed.

353   If in our analysis of dealing with things we aim at what we have to deal with, then our existent being together with beings taken care of must be given an orientation toward the totality of useful things, not toward an isolated useful thing at hand. Our reflection upon the eminent character of being of useful things at hand, *relevance*,[16] also forces us to this conception of what we have to do with. We understand the term relevance ontologically. To say that something has a relevance with and in something else is not supposed to ontically ascertain a fact, but to indicate the kind of being of things at hand. The relational character of relevance, with its . . . together with . . . points to the fact that *a* useful thing is ontologically impossible. Indeed, a single useful thing may be at hand while another is "missing." But here the fact makes itself known that one thing at hand belongs *to* the other. Heedful dealings can only let things at hand be encountered circumspectly

---

15. Cf. § 12.
16. Cf. § 18.

if it already understands something like a relevance in which things are. The being together with . . . that takes care in a way that circumspectly discovers amounts to letting something be in relevance, that is, to projecting relevance understandingly. *If letting things be relevant constitutes the existential structure of taking care, and if the latter as being together with . . . belongs to the essential constitution of care, and if care in its turn is grounded in temporality, then the existential condition of the possibility of letting something be relevant must be sought in a mode of the temporalizing of temporality.*

Letting something be relevant is found in the simplest handling of a useful thing. Relevance has an intentional character with reference to which the thing is usable or in use. Understanding the intention [Wozu] and context [Wobei] of relevance has the temporal structure of awaiting. Awaiting the intention, taking care can at the same time come back to something like relevance. *Awaiting [Gewärtigen]* the context [Wobei] and *retaining [Behalten]* the means of [Womit] relevance make possible in its ecstatic unity the specifically handy way in which the useful thing is made present.

Awaiting the towards-which [Wozu] is neither a reflection upon the "goal" nor an expectation of the imminent completion of the work to be produced. It does not have the nature of a thematic grasping at all. Nor does retaining what is relevant mean holding fast to it thematically. Handling things is no more related merely to what it handles than to what it uses in relevance. Rather, being relevant constitutes itself in the unity of awaiting and retaining in such a way that the making present arising from this makes the characteristic absorption in taking care in the world of its useful things possible. When one is "really" busy with something and totally immersed in it, one is neither only together with the work, nor with the tools, nor with both "together." Being in relevance, which is grounded in temporality, has already founded the unity of the relations in which taking care "moves" circumspectly.

354

A specific kind of *forgetting* is essential for the temporality that constitutes being in relevance. In order to be able to "really" get to work, "lost" in the world of tools and to handle them, the self must forget itself. But since an *awaiting* is always guiding in the unity of the temporalizing of taking care, the ownmost potentiality-of-being of Dasein, taking care is nevertheless placed in care, as we shall show.

The making present that awaits and retains constitutes the familiarity in accordance with which Dasein "knows its way around" as being-with-one-another in the public surrounding world. We understand letting things be in relevance existentially as letting-"be." On its basis, things at hand can be encountered by circumspection *as the beings* that they are. We can thus further clarify the temporality of taking care

if we pay attention to *the* modes of circumspectly letting something
be encountered that were characterized before[17] as conspicuousness,
obtrusiveness, and obstinacy. The useful thing at hand is precisely not
encountered with regard to its "true in-itself" by a thematic perception of things, but is encountered in the inconspicuousness of what is
found "obviously" and "objectively." But if something is conspicuous
in the totality of such beings, then this implies the possibility that the
totality of useful things as such also obtrudes itself. How must letting things be in relevance be structured existentially so that it can let
something conspicuous be encountered? This question is not aiming at
factical occasions that direct our attention to something already given,
but at the ontological meaning of this ability to direct our attention
as such.

Things that cannot be used, for example, a tool that will not
work, can be conspicuous only in and for someone using them. Even
the most sharp and persistent "perception" and "representation" of
things could never discover something like damage to the tool. The
355   using must be able to be hampered so that something unhandy can be
encountered. But what does this mean *ontologically*? The making present that awaits and retains is held up with regard to its absorption in
the relevant relations, and it is held up by something which afterwards
turns out to be a damage. The making present, which equiprimordially
awaits the what-for, is held fast with the tool used in such a way that
the what-for and the in-order-to are now explicitly encountered for
the first time. However, making present itself can only meet up with
something unsuited for . . . because it is already moving in an awaiting
retention of what is in relevance. Making present is "held up," that
is, in the unity of the awaiting that retains, it shifts more to itself and
thus constitutes the "inspection," checking, and removal of the disruption. If heedful dealings were simply a succession of "experiences"
occurring "in time," and even if these experiences "associated" with
each other as intimately as possible, letting a conspicuous, unusable
tool be encountered would be ontologically impossible. Whatever we
have made accessible in contexts of useful things, letting things be in
relevance as such must be grounded in the ecstatic unity of the making
present that awaits and retains.

And how it is possible to "ascertain" that something is missing,
that is, unhandy, and not just at hand in an unhandy way? Unhandy
things are not discovered circumspectly in *missing* something. "Finding
out" that something is not objectively present is based upon missing
something, and both have their own existential presuppositions. Missing something is by no means a not-making-present, but is a deficient

17. Cf. § 16.

mode of the present in the sense of the not-making-present of something expected or always already available. If circumspect letting things be relevant were not *awaiting* what is taken care of "from the very beginning," and if awaiting did not temporalize itself in the *unity with* a making present, Dasein could never "find out" that something is missing.

On the other hand, the possibility of *being surprised* lies in the fact that the making present that *awaits* a thing does *not await* something else that stands in a possible context of relevance with what one does await. The not awaiting of the making present that is lost first discloses the "horizontal" leeway in which something surprising can overcome Dasein.

That with which heedful dealings fail to cope, either by producing or procuring something, or even by turning away, keeping at a distance, protecting oneself from something, is revealed in its insurmountability. Taking care reconciles itself with this. But reconciling with something is a mode peculiar to circumspectly letting something be encountered. On the basis of this discovery, taking care can find what is inconvenient, disturbing, hindering, jeopardizing, or in general resistant in some way. The temporal structure of reconciling oneself with something lies in a *nonretention* that awaits and makes present. The making present that awaits does not, for example, count "on" something that is unsuitable, but is nevertheless available. Not counting on . . . is a mode of taking into account what one *cannot* hold on to. It is not forgotten, but retained so that it remains at hand precisely *in its unsuitability*. Things at hand like this belong to the everyday content of the factically disclosed surrounding world.

356

Only because things offering resistance are disclosed on the basis of the ecstatic temporality of taking care, can factical Dasein understand itself in its abandonment to a "world" of which it never becomes master. Even if taking care remains restricted to the urgency of everyday needs, it is never a pure making present, but arises from a retention that awaits; on the basis of such retention, or as such a "basis," Dasein exists in a world. For this reason factically existing Dasein in a way always already knows its way around, even in a strange "world."

When, in taking care, one lets something be in relevance, one's doing so is founded upon temporality and amounts to an altogether pre-ontological and unthematic way of understanding relevance and things at hand. In what follows, we shall show how the understanding of these determinations of being as such is, in the end, also founded in temporality. We must first demonstrate the temporality of being-in-the-world still more concretely. With this as our aim, we shall trace how the theoretical mode of behavior toward the "world" "arises" out of circumspect taking care of things at hand. The circumspect, as well as the theoretical discovery of innerworldly beings, is based upon being-in-the-world. The existential and temporal interpretation of these ways of discovering will prepare the temporal characterization of this fundamental constitution of Dasein.

(b) The Temporal Meaning of the Way in which Circumspect Taking Care Becomes Modified into the Theoretical Discovery of That Which is Present Within the World

357
When in the course of our *existential* and *ontological* analyses we ask about how *theoretical* discovery "arises" from *circumspect* taking care, this means that we are not making a problem out of the *ontic* history and development of science, or of its factical occasions or of its proximate goals. In searching for the *ontological genesis* of the theoretical mode of behavior, we are asking which of those conditions of possibility in the constitution of being of Dasein are existentially necessary for Dasein to be able to exist in the mode of scientific investigation? This question aims at an *existential concept of science*. This is distinct from the "logical" concept that understands science with regard to its results and defines it as a "context of causal relations of true, that is, valid propositions." The existential concept understands science as a mode of existence and thus a mode of being-in-the-world which discovers or discloses beings or being. However, a completely adequate existential interpretation of science cannot be carried out until the *meaning of being and the "connection" between being and truth*[18] *have been clarified* in terms of the temporality of existence. The following considerations are to prepare the understanding of *this central problematic* within which the idea of phenomenology can first be developed, as opposed to the preconception[19] indicated in an introductory fashion.

In accordance with the stage of our study attained up to now, a further restriction will be imposed upon our interpretation of the theoretical mode of behavior. We are only inquiring into the way in which circumspect taking care of things at hand changes over into the investigation of things objectively present found in the world, and we shall be guided by the aim of penetrating to the temporal constitution of being-in-the-world in general.

In characterizing the transformation [Umschlag] from "practically" circumspect handling and using and so on, to "theoretical" investigation, it would be easy to suggest that merely looking at beings is something that emerges when taking care *abstains* from any kind of use. Then what is decisive about the "origin" of theoretical behavior would lie in the *disappearance* of praxis. So if one posits "practical"

---

18. Cf. § 44.
19. Cf. § 7.

taking care as the primary and predominant kind of being of factical Dasein, the ontological possibility of "theory" will be due to the *absence* of praxis, that is, to a *privation*. But stopping a specific kind of use in heedful dealing does not simply leave its guiding circumspection behind as a remnant. Rather, taking care then transposes itself explic- itly into just-looking-around. But this is by no means the way in which the "theoretical" attitude of science is reached. On the contrary, the lingering [Verweilen] which comes about when busyness is abandoned can acquire the quality of a more precise kind of circumspection, such as "inspecting," checking what has been attained, as looking over the "operations" just now "at a standstill." To refrain from the use of tools is so far from "theory" that lingering, "reflecting" circumspection remains completely stuck in the tools at hand taken care of. "Practical" dealings have their *own* way of lingering. And just as praxis has its own specific sight ("theory"), theoretical research is not without its own praxis. Read- ing off the measurements that result from an experiment often requires a complicated "technological" set-up for the experimental arrangement. Observing with the microscope is dependent upon the production of "prepared slides." Archeological excavation that precedes any interpre- tation of the "findings" demands the most massive manual labor. But even the most "abstract" working out of problems and refining what has been gained, uses, for example, writing materials. As "uninterest- ing" and "obvious" as these components of scientific investigation may be, they are by no means ontologically indifferent. The explicit refer- ence to the fact that scientific behavior as a way of being-in-the-world is not only a "purely intellectual activity" might seem unnecessarily complicated and superfluous. If only it did not become clear from this triviality that it is by no means obvious where the ontological boundary between "theoretical" and "atheoretical" behavior really lies!

One will want to assert that all manipulation in the sciences is only in the service of pure observation, of the investigating discov- ery and disclosure of the "things themselves." Taken in its broadest sense, "seeing" regulates all "procedures" and retains its priority. "To whatever kind of objects one's knowledge may relate itself and by whatever means it may do so, still that through which it relates itself to them immediately, *and which all thinking as a means has as its goal* is *intuition.*"[20] The idea of the *intuitus* has guided all interpretation of knowledge ever since the beginning of Greek ontology up to today, whether that intuition is actually attainable or not. In accordance with the priority of "seeing," the demonstration of the existential genesis of science will have to start out by characterizing the circumspection that guides "practical" taking care of things.

20. Kant, *Kritik der reinen Vernunft*, B33 [Emphasis added by Heidegger]..

358

359   Circumspection moves in the relevant relations of the context of useful things at hand. It itself is again subject to the guidance of a more or less explicit view over [Übersicht] the totality of useful things in every world of tools and the public surrounding world belonging to it. This overview is not simply one that subsequently scrapes objectively present things together. What is essential in this overview is the primary understanding of the totality of relevance within which factical taking care always starts out. The overview, which illuminates taking care, gets its "light" from the potentiality-of-being of Dasein *for the sake of which* taking care exists as care. The circumspection of taking care that has this "overview" *brings* things at hand *nearer* to Dasein in its actual using and handling in the mode of interpreting what it has seen. We call the specific bringing near of what is taken care of by interpreting it circumspectly *deliberation* [Überlegung]. The schema peculiar to it is "if-then": if this or that is to be produced, put into use, or prevented, for example, then we need these or those means, ways, circumstances, or opportunities. Circumspect deliberation throws light on the actual factical position of Dasein in the surrounding world taken care of. Thus it never simply "confirms" the objective presence of a being or of its qualities. Deliberation can also come about without what is circumspectly approached itself being concretely at hand or present within reach [Sichtweite]. Bringing the surrounding world near in circumspect deliberation has the existential meaning of *making present* [*Gegenwärtigung*]. For visualization [*Vergegenwärtigung*] is only a mode of making present. In it, deliberation catches sight directly of what is needed but not at hand. Circumspection that visualizes [vergegenwärtigende Umsicht] does not relate itself to "mere representations" ["bloße Vorstellungen"].

         But circumspect making present is a phenomenon with more than one kind of foundation. First of all, it belongs to the full ecstatic unity of temporality. It is grounded in a *retention* of the context of useful things that Dasein takes care of in *awaiting* a possibility. What has already been disclosed in awaiting retention is brought nearer by one's deliberative making present or representing. But if deliberation is to be able to move in the scheme of "if-then," taking care must already understand a context of relevance in an "overview." What is addressed with the "if" must already be understood *as this and that*. For this, it is not necessary that the understanding of useful things be expressed predicatively. The scheme "something as something" is already prefigured in the structure of pre-predicative understanding. The as-structure is ontologically grounded in the temporality of understanding. Only because Dasein, awaiting a possibility (that is, here a *360*   what-for) has come back to a for-this (that is, retains a thing at hand), can *conversely* the making present that belongs to this awaiting retention start with this retention and *bring* it *explicitly nearer* in its reference to the what-for. The deliberation that brings near must, in the scheme of making present, adapt itself to the kind of being of what is to be brought near. The character of relevance of what is at hand is not first discovered by deliberation, but only gets brought near by it in such a way that it circumspectly lets that through which something is in relevance be seen *as* this.

The way the present is rooted in the future and in the having-been is the existential and temporal condition of the possibility that what is projected in circumspect understanding can be brought nearer in a making present in such a way that the present must adapt itself to what is encountered in the horizon of awaiting retention; that is, it must interpret itself in the schema of the as-structure. This answers our earlier question whether the as-structure is existentially and onto-logically connected[21] with the phenomenon of projecting. *Like understanding and interpretation in general, the "as" is grounded in the ecstatic and horizonal unity of temporality.* In our fundamental analysis of being, and indeed in connection with the interpretation of the "is" (which as a copula "expresses" the addressing of something as something), we must again make the as-phenomenon thematic and define the concept of the "schema" existentially.

However, what is the temporal characterization of circumspect deliberation and its schemata supposed to contribute to answering our current question of the genesis of the theoretical mode of behavior? Only enough to clarify the situation of Dasein in which a circumspect taking care changes over into theoretical discovery. We may then try to analyze this transformation itself following the guideline of an elemental statement of circumspect deliberation and its possible modifications.

In our circumspect use of tools, we can say that the hammer is too heavy or too light. Even the sentence that the hammer is heavy can express a heedful deliberation and mean that it is not light, that is, that it requires force to use it or it makes using it difficult. But the statement *can* also mean that the being before us, with which we are circumspectly familiar as a hammer, has a weight, namely, the "property" of heaviness. It exerts a pressure on what lies beneath it, and when that is removed, it falls. The discourse understood in this way is no longer in the horizon of the awaiting retention of a totality of useful things and its relations of relevance. What is said has been drawn from looking at what is appropriate for a being with "mass." What is now in view is appropriate for the hammer, not as a tool, but as a corporeal thing that is subject to the law of gravity. Circumspect talk about being "too heavy" or "too light" no longer has any "meaning";

361

---

21. Cf. § 32.

that is, the being now encountered of itself provides us with nothing in relation to which it could be "found" too heavy or too light.

Why does what we are talking about, the heavy hammer, show itself differently when our way of talking is modified? Not because we are keeping our distance from handling, nor because we are only looking *away* from the useful character of this being, but because we are looking *at* the thing at hand encountered in a "new" way, as something objectively present. *The understanding of being* guiding the heedful dealings with innerworldly beings *has been transformed.* But does this already constitute a scientific mode of behavior if we "comprehend" things at hand as something objectively present, instead of circumspectly deliberating about them? Moreover, even things at hand can be made a theme of scientific investigation and determination, for example, in examining someone's surrounding world, his milieu, in the context of a historiographical biography. The everyday context of useful things at hand, their historical origination and utilization, their factical role in Dasein—all these are the objects of the science of economics. Things at hand need not lose their character of being useful things in order to become the "object" of a science. A modification of our understanding of being seems not to be necessarily constitutive for the genesis of the theoretical mode of behavior "toward things." Certainly not, if this modification is supposed to mean a change of the kind of being which, in understanding the being in question, we understand it to possess.

In our first characterization of the genesis of the theoretical mode of behavior from circumspection, we have made basic a kind of theoretical grasping of innerworldly beings, a physical nature, in which the modification of our understanding of being amounts to a transformation. In the "physical" statement that "the hammer is heavy," we *overlook* not only the tool-character of the being encountered, but thus also that which belongs to every useful thing at hand: its place [Platz].

362 The place becomes indifferent. This does not mean that the objectively present thing loses its "location" ["Ort"] altogether. Its place becomes a position [Stelle] in space and time, a "world point," which is in no way distinguished from any other. This means that the multiplicity of places of useful things at hand within the confines of [umschränkte] the surrounding world is not simply modified to a mere multiplicity of positions; rather it means that the beings of the surrounding world are *released from this confinement* [entschränkt]. The totality of what is present becomes thematic.

In this case, a releasing of the surrounding world belongs to the modification of the understanding of being. Following the guideline of the understanding of being in the sense of objective presence, this release becomes at the same time a delimitation of the "region" of what is objectively present. The more appropriately the being of the beings to be investigated is understood in the guiding understanding of being and the more the totality of beings is articulated in its fundamental determinations as a possible area of subject-matter for a science, the more assured will be the actual perspective of methodological questioning.

The classic example for the historical development of a science, and even for its ontological genesis, is the origin of mathematical physics. What is decisive for its development lies neither in its higher evaluation of the observation of "facts," nor in the "application" of mathematics in determining events of nature, but the *mathematical projection of nature itself*. This project discovers in advance something constantly objectively present (matter) and opens the horizon for the guiding perspective on its quantitatively definable constitutive moments (motion, force, location, and time). Only "in the light of" a nature thus projected can something like a "fact" be found and be taken as a point of departure for an experiment defined and regulated in terms of this project. The "founding" of "factual science" was possible only because the researchers understood that there are in principle no "bare facts." What is decisive about the mathematical project of nature is again not primarily the mathematical element as such, but the fact that this project *discloses an a priori*. And thus the paradigm of the mathematical natural sciences does not consist in its specific exactitude and binding character for "everyone," but in the fact that in it the thematic beings are discovered in *the only* way that beings can be discovered: in the prior project of their constitution of being. When the basic concepts of the understanding of being by which we are guided have been worked out, the methods, the structure of conceptuality, the relevant possibility of truth and certainty, the kind of grounding and proof, the mode of being binding and the kind of communication—all these will be determined. The totality of these moments constitutes the complete existential concept of science.

363

The scientific projection of the beings somehow always already encountered lets their kind of being be explicitly understood in such a way that the possible ways of purely discovering innerworldly beings thus become evident. The articulation of the understanding of being, the definition of the subject-matter defined by that understanding, and the prefiguration of the concepts suitable to these beings, all belong to the totality of this projecting that we call *thematization*. It aims at freeing beings encountered within the world in such a way that they can "project" themselves back upon pure discovery, that is, they can become objects. Thematization objectifies. It does not first "posit" beings, but frees them in such a way that they become "objectively" subject to questioning and definition. The objectifying being together with innerworldly things objectively present has the character of a *distinctive kind of making present*.[22] It is above all distinguished from the present of circumspection by the fact that the discovering of the science in question solely awaits the discoveredness of what is present. This awaiting of discoveredness is grounded existentielly in a resoluteness of Dasein by means of which it projects itself upon its potentiality-of-being-in-the-"truth." This project is possible because being-in-the-truth constitutes a determination of the existence of Dasein. How science has its origin in authentic existence is not to be pursued here. It is simply a matter of understanding that and how the thematization of innerworldly beings presupposes being-in-the-world as the fundamental constitution of Dasein.

If the thematization of what is present—the scientific projection of nature—is to become possible, *Dasein must transcend* the beings thematized. Transcendence does not consist in objectification, rather objectification presupposes transcendence. But if the thematization of innerworldly beings objectively present is a transformation from taking care which circumspectly discovers, then a transcendence of Dasein must already underlie "practical" being together with things at hand.

Furthermore, if thematization modifies and articulates the understanding of being, insofar as Dasein, the being that thematizes, exists, it must already understand something like being. This understanding of being can remain neutral. Then handiness and objective presence are not differentiated, still less are they conceived ontologically. But for Dasein to be able to have something to do with a context of useful things, it must understand something like relevance, even if only unthematically. *A world must be disclosed to it.* The world is disclosed with the factical existence of Dasein, if indeed Dasein essentially exists as being-in-the-world. And if the being of Dasein is completely grounded in temporality, temporality must make possible being-in-the-world and thus the transcendence of Dasein, which in its turn supports the being together with innerworldly beings that takes care, whether theoretical or practical.

364

---

22. This thesis that all cognition aims at "intuition" has the temporal meaning that all cognition is a making present. Whether every science or even philosophical cognition aims at a making present must remain undecided here. Husserl uses the expression "making present" to characterize sense perception. Cf. *Logische Untersuchungen* (1901), vol. 2, pp. 588, 620. The *intentional* analysis of perception and intuition in general had to suggest this "temporal" characterization of the phenomenon. How the intentionality of "consciousness" is *grounded* in the ecstatic temporality of Dasein will be shown in the following division.

(c) The Temporal Problem of the Transcendence of the World

The understanding of a totality of relevance inherent in circumspect taking care is grounded in a previous understanding of the relations of in-order-to, what-for, for-that, and for-the-sake-of-which. We set forth the connection of these relations as significance.[23] Their unity constitutes what we call world. Now the question arises of how something like world in its unity with Dasein is ontologically possible? In what way must world *be,* such that Dasein can exist as being-in-the-world?

Dasein exists for the sake of a potentiality-of-being of itself. Existing, it is thrown, and as thrown, it is delivered over to beings that it needs *in order to* be able to be as it is, namely *for the sake of* itself. Since Dasein exists factically, it understands itself in this connection of the for-the-sake-of-itself in each instance with an in-order-to. That *within which* existing Dasein understands *itself is* "there" together with its factical existence. The wherein of primary self-understanding has the kind of being of Dasein. Existing, Dasein *is* its world.

We defined the being of Dasein as care. Its ontological meaning is temporality. We showed that and how temporality constitutes the disclosedness of the there. World is also disclosed in the disclosedness of the there. The unity of significance, that is, the ontological constitution of the world, must then also be grounded in temporality. *The existential and temporal condition of the possibility of the world lies in the fact that temporality, as an ecstatical unity, has something like a horizon.* The ecstasies are not simply raptures [Entrückungen] toward. . . . Rather, a "whereto" of being transported [Entrückung] belongs to each ecstasy. We call this whereto of the ecstasy the horizonal schema. The ecstatical horizon is different in each of the three ecstasies. The schema in which Dasein comes back to itself *futurally,* whether authentically or inauthentically, is the *for-the-sake-of-itself.* We call the schema in which Dasein is disclosed to itself in attunement as thrown, that *in the face of which* it has been thrown and that to which it has been delivered over. It characterizes the horizonal structure of the *having-been.* Existing for-the-sake-of-itself in being delivered over to itself as thrown, Dasein is at the same time making present as being together with. . . . The horizonal schema of the *present* is determined by the *in-order-to.*

The unity of the horizonal schemata of future, having-been, and present is grounded in the ecstatic unity of temporality. The horizon of the whole of temporality determines *that upon which* the being factically

365

---

23. Cf. § 18.

existing is essentially *disclosed*. With factical Dasein, a potentiality-of-being is always projected in the horizon of the future, "already being" is disclosed in the horizon of the having-been, and what is taken care of is discovered in the horizon of the present. The horizonal unity of the schemata of the ecstasies makes possible the primordial connection of the relations of the in-order-to with the for-the-sake-of-which. This means that on the basis of the horizonal constitution of the ecstatic unity of temporality, something like a disclosed world belongs to the being that is always its There.

Just as the present arises in the unity of the temporalizing of temporality from the future and the having-been, the horizon of a present temporalizes itself equiprimordially with those of the future and the having-been. Insofar as Dasein temporalizes itself, a world *is*, too. Temporalizing itself with regard to its being as temporality, Dasein *is* essentially "in a world" on the basis of the ecstatic and horizonal constitution of that temporality. The world is neither objectively present nor at hand, but temporalizes itself in temporality. It "is" "there" together with the outside-itself of the ecstasies. If no Dasein exists, no world is "there" either.

The world is already presupposed in one's being together with things at hand heedfully and factically, in one's thematization of what is present, and in one's objectifying discovery of the latter; that is, all these are possible only as modes of being-in-the-world. The world is transcendent, grounded in the horizonal unity of ecstatic temporality. It must already be ecstatically disclosed so that innerworldly beings can be encountered from it. Temporality already holds itself ecstatically in the horizons of its ecstasies and, temporalizing itself, comes back to the beings encountered in the there. With the factical existence of Dasein, innerworldly beings are also already encountered. That such beings are discovered in the there of its own existence is not under the control of Dasein. Only *what*, in *which* direction, *to what extent, and how* it actually discovers and discloses is a matter of freedom, although always within the limits of its thrownness.

The relations of significance that determine the structure of the world are thus not a network of forms that is imposed upon some material by a worldless subject. Rather, factical Dasein, ecstatically understanding itself and its world in the unity of the there, comes back from these horizons to the beings encountered in them. Coming back to these beings understandingly is the existential meaning of letting them be encountered in making them present; for this reason they are called innerworldly. The world is, so to speak, already "further outside" than any object could ever be. The "problem of transcendence" cannot be reduced to the question of how does a subject get outside to an object, whereby the totality of objects is identified with the idea of

366

the world. We must rather ask what makes it ontologically possible for beings to be encountered within the world and objectified as encountered beings? Going back to the ecstatically and horizonally founding transcendence of the world will give us the answer.

If the "subject" is conceived ontologically as existing Dasein, whose being [Sein] is grounded in temporality, we must say then that the world is "subjective." But this "subjective" world, as one that is temporally transcendent, is then "more objective" than any possible "object."

By tracing being-in-the-world back to the ecstatic and horizonal unity of temporality, we have made the existential and ontological possibility of this fundamental constitution of Dasein intelligible. At the same time it becomes clear that the concrete development of the structure of world in general and of its possible variations can be attempted only if an ontology of possible innerworldly beings is oriented toward a clarified idea of being in general with sufficient assurance. The possible interpretation of this idea requires that we set forth the temporality of Dasein beforehand; here our characterization of being-in-the-world will be of service.

§ 70. *The Temporality of the Spatiality Characteristic of Dasein*   367

Although the expression "temporality" does not mean what the talk about "space and time" understands by time, spatiality nevertheless seems to constitute another fundamental determination [Grundbestimmtheit] of Dasein corresponding to temporality. The existential and temporal analysis thus appears to reach a limit with the spatiality of Dasein, so that this being that we call Dasein must be addressed coordinately as "temporal" "and also" as spatial. Does the existential and temporal analysis of Dasein come to a halt on account of the phenomenon that we got to know as the spatiality of Dasein and that we showed to belong to being-in-the-world?[24]

If in the course of our existential interpretation we were to talk about the "spatio-temporal" determination of Dasein, then we could not mean that this being is present "in space and also in time"; this needs no further discussion. Temporality is the meaning of being of care. The constitution of Dasein and its modes of being are ontologically possible only on the basis of temporality, regardless of whether this being occurs "in time" or not. But then the specific spatiality of Dasein must be grounded in temporality. On the other hand, the demonstration that this spatiality is existentially possible only through temporality, cannot aim either at deducing space from time or at dissolving

---

24. Cf. §§ 22–24.

it into mere time. If the spatiality of Dasein is "embraced" by temporality in the sense of an existential foundation, this connection (which
is to be clarified in what follows) is also different from the priority of
time over space in Kant's sense. That the empirical representations of
what is objectively present "in space" occur as psychical events "in
time," so that the "physical" also occurs indirectly "in time," is not to
give an existential and ontological interpretation of space as a form
of sensibility, but rather to ascertain ontically that what is psychically
present runs its course "in time."

   We shall ask existentially and analytically about the temporal
conditions of the possibility of the spatiality of Dasein—the spatiality
that in turn founds the discovery of space within the world. Before
that we must remember in what way Dasein is spatial. Dasein can
be spatial only as care, in the sense of factically entangled existing.
Negatively this means that Dasein is never objectively present in space,
not even initially. Dasein does not fill out a piece of space as a real
thing or useful thing would do, so that the boundaries dividing it
from the surrounding space would themselves just define that space
spatially. In the literal sense, Dasein takes space in. It is by no means
merely objectively present in the piece of space that its corporeal body
[Leibkörper] fills out. Existing, it has always already made room for a
leeway [Spielraum]. It determines its own location in such a way that
it comes back from the space made room for to a "place" that it has
taken over. To be able to say that Dasein is present at a position in
space, we have to *conceive* this being beforehand in an ontologically
inappropriate way. Nor does the difference between the "spatiality" of
an extended thing and that of Dasein lie in the fact that Dasein *knows*
about space. Making room is so far from identical with the "representation" of something spatial that the latter presupposes the former.
Nor may the spatiality of Dasein be interpreted as a kind of imperfection that adheres to existence on account of the fatal "connection of
the spirit with a body." Rather, because Dasein is "spiritual," *and only
because it is spiritual*, can it be spatial in a way that essentially remains
impossible for an extended corporeal thing.

   The making room of Dasein is constituted by directionality and
de-distancing. How is something like this existentially possible on the
basis of the temporality of Dasein? The foundational function of temporality for the spatiality of Dasein is to be indicated briefly only to
the extent that it is necessary for later discussions of the ontological
meaning of the "coupling" of space and time. The directional discovery of something like a *region* belongs to the making room of Dasein.
With this expression we mean initially the whereto of the possible
belonging somewhere of useful things at hand in the surrounding
world. Whenever one comes across useful things, handles them, moves

368

them around, or out of the way, a region has already been discovered. Being-in-the-world that takes care is directed, directing itself. Belonging-somewhere has an essential relation to relevance. It is always factically determined in terms of the context of relevance of the useful things taken care of. The relevant relations are intelligible only in the horizon of a disclosed world. Their horizonal nature also first makes possible the specific horizon of the whereto of regional belonging. The self-directive discovering of a region is grounded in an ecstatically retentive awaiting of the possible hither and whither. As a directed awaiting of region, making room is equiprimordially a bringing-near [Nähern] (or de-distancing) of things at hand and objectively present. De-distancing, taking care comes back out of the previously discovered region to what is nearest [Nächste]. Bringing-near and the estimating and measurement of distances within what is present within the de-distanced world are grounded in a making-present that belongs to the unity of temporality in which directionality is possible, too.

369

Because Dasein as temporality is ecstatic and horizonal in its being, it can factically and constantly take along space for which it has made room. With regard to this space ecstatically made room for, the here of its actual factical location or situation never signifies a position in space, but the leeway of the range of the totality of useful things taken care of nearby—a leeway that has been opened in directionality and de-distancing.

In bringing-close [Näherung] that makes possible the handling and being occupied that is "absorbed in the matter," the essential structure of care—falling prey—makes itself known. Its existential and temporal constitution is distinguished by the fact that in falling prey, and thus also in the bringing near which is founded in "making present," the forgetting that awaits pursues the present. In the making present that brings something near from its wherefrom, making present loses itself in itself, and forgets the over there. For this reason, if the "observation" of innerworldly beings starts in such a making present, the illusion arises that "initially" only a thing is present, here indeed, but indeterminately, in a space in general.

Only on the basis of ecstatic and horizonal temporality is it possible for Dasein to break into space. The world is not objectively present in space; however, only within a world can space be discovered. The ecstatic temporality of the spatiality of Dasein makes the independence of space upon time comprehensible; but, on the other hand, this same temporality also makes intelligible the "dependency" of Dasein upon space—a dependency that makes itself manifest in the familiar phenomenon that both the self-interpretation of Dasein and the stock of meaning so pervasive in language in general are thoroughly dominated by "spatial representations." This priority of the spatial in the articulation of significations and concepts has its ground, not in some specific power of space, but rather in the kind of being of Dasein.* Essentially entangled, temporality loses itself in making present, and understands itself not only circumspectly in terms of the things at hand taken care of; rather, it takes its guidelines for articulating what is understood and can be interpreted in understanding in general from those spatial relations that making present constantly meets up with in what is at hand as present.

370   § 71. *The Temporal Meaning of the Everydayness of Dasein*

The analysis of the temporality of taking care showed that the essential structures of the constitution of being of Dasein that were interpreted *before* setting forth temporality, with the intention of arriving at temporality, must themselves be *taken back* existentially *into temporality*. At the very beginning of our analysis we did not choose a definite, eminent possibility-of-existence of Dasein as our theme, but our analytic was oriented toward the inconspicuous, average modes of existing. We called the kind of being in which Dasein holds itself initially and for the most part *everydayness*.[25]

What this expression basically means when it is ontologically defined remains obscure. At the beginning of our inquiry there was no way available of even making the existential and ontological meaning of everydayness a problem. But now the meaning of being of Dasein has been illuminated as temporality. Can there still be any doubt with regard to the existential and temporal significance of the term "everydayness"? Yet we are far removed from an ontological concept of this phenomenon. It even remains questionable whether the explication of temporality carried out up to now is adequate to explain the existential meaning of everydayness.

Everydayness evidently means *the* mode of existing in which Dasein hold itself "each day." And yet "each day" does not signify the sum of the "days" that are allotted to Dasein in its "lifetime." Although "each day" is not to be understood in the sense of the calendar, some such temporal determination still echoes in the significance of the "everyday." But the expression "everydayness" primarily signifies a certain *How* of existence that prevails in Dasein "as long as it lives." In our earlier analyses we often used the expression "initially and for the most part." "Initially" means the way in which Dasein is "manifest" in the being-with-one-another of publicness, even if it

---

* No opposition; both belong together.

25. Cf. § 9.

has "basically" "overcome" everydayness existentielly. "For the most part" signifies the way in which Dasein shows itself for everyone "as a rule," but not always.

Everydayness means the How in accordance with which Dasein "lives its day," whether in all of its modes of behavior or only in certain ways prefigured by being-with-one-another. Furthermore, being comfortable with what is customary belongs to this How, even if habit forces us to confront what is burdensome and "disagreeable." The tomorrow that everyday taking care waits for is the "eternal yesterday." The monotony of everydayness takes whatever the day happens to bring as variety. Everydayness determines Dasein even when it has not chosen the they as its "hero."   *371*

But these manifold qualities of everydayness by no means characterize it as the mere "aspect" that Dasein proffers when "one" "looks at" the things human beings do. Everydayness is a way *to be*—to which, of course, public manifestness belongs. But as a way of its own existing, everydayness is indeed more or less familiar to each and every "individual" Dasein through the attunement of the pallid lack of mood. Dasein can "suffer" dully from everydayness, sink into its dullness, and evade it by looking for new ways in which its dispersion in its affairs may be further dispersed. But existence can also master the everyday in the Moment, and of course often only "for the moment," but it can never extinguish it.

What is *ontically* so familiar in the factical interpretedness of Dasein that we do not even pay any attention to it, hides in itself enigma upon enigma existentially and ontologically. The "natural" horizon for starting the existential analytic of Dasein is *only seemingly obvious*.

But, after our earlier interpretation of temporality, are we in a more fruitful position with regard to the existential delimitation of the structure of everydayness? Or does this confusing phenomenon precisely make evident what is insufficient about our explication of temporality up to now? Have we not been constantly immobilizing Dasein in certain positions and situations, while "consistently" disregarding the fact that, in moving from day to day, it *stretches* itself *along* "temporally" in the succession of its days? The monotony, the habit, the "like yesterday, so today and tomorrow," and the "for the most part" cannot be grasped without recourse to the "temporal" stretching along of Dasein.

And is it not also a fact of existing Dasein that, passing its time, it takes "time" daily into account and regulates the "calculation" astronomically and with the calendar? Only if we bring the everyday "occurrence" of Dasein and the heedful calculation of "time" in this occurrence into the interpretation of the temporality of Dasein, will our orientation become comprehensive enough to enable us to make the ontological meaning of everydayness as such problematic. However, since basically nothing other is meant by the term everydayness than temporality, and since temporality makes the *being* of Dasein possible, an adequate conceptual delineation of everydayness can succeed only in the framework of a fundamental discussion of the meaning of being in general and its possible variations.

CHAPTER FIVE
Temporality and Historicity

§ 72. *The Existential and Ontological Exposition of the Problem of History*

All our efforts in the existential analytic are geared to the one goal of finding a possibility of answering the question of the *meaning of being* in general. The development of this *question* requires a delineation *of the* phenomenon in which something like being itself becomes accessible— the phenomenon of the *understanding of being*. But this phenomenon belongs to the constitution of being of Dasein. Only when this being [Seiende] has been interpreted beforehand in a sufficiently primordial way, can the understanding of being contained in its constitution of being itself be grasped, and only on that basis can we formulate the question of being understood in this understanding and the question of what such understanding "presupposes."

Although many structures of Dasein still remain in the dark with regard to particulars, it nonetheless seems that we have reached the requisite, primordial interpretation of Dasein with the clarification of temporality as the primordial condition of the possibility of *care*. Temporality was set forth with regard to the authentic potentiality-of-being-whole of Dasein. The temporal interpretation of care was then confirmed by demonstrating the temporality of heedful being-in-the-world. Our analysis of the authentic potentiality-of-being-whole revealed that an equiprimordial connection of death, guilt, and conscience is rooted in care. Can Dasein be understood still more primordially than in the project of its authentic existence?

Up to now we have not seen any possibility of a more radical starting point for our existential analytic. Nonetheless with regard to the above discussion of the ontological meaning of everydayness, a serious reservation comes to light: has indeed the whole of Dasein with respect to its authentic *being*-whole [Ganz*sein*] been captured in the fore-having of our existential analysis? It may be that the line of

373

questioning related to the wholeness of Dasein possesses a genuinely unequivocal character ontologically. The question itself may even have been answered with regard to *being-toward-the-end*. However, death is, after all, only the "end" of Dasein, and formally speaking, it is just *one* of the ends that embraces the totality of Dasein. But the other "end" is the "beginning," "birth." Only the being "between" birth and death presents the whole we are looking for. Accordingly, the previous orientation of our analytic would remain "one-sided," despite all its tendencies toward a consideration of *existing* being-whole and in spite of the genuineness with which authentic and inauthentic being-toward-death have been explicated. Dasein has been our theme only as to how it exists, more or less, "forward" and leaves everything that has been "behind." Not only did being-toward-the-beginning remain unnoticed, but so did, above all, the way Dasein *stretches along between* birth and death. Precisely the "connection of life," in which, after all, Dasein constantly somehow holds itself, was overlooked in our analysis of being-whole.

Must we not take back our point of departure of temporality as the meaning of being of the totality of Dasein, even though what we addressed as the "connection" between birth and death is ontologically completely obscure? Or does *temporality*, as we set it forth, first give the *foundation* [*Boden*] on which to provide an unequivocal direction for the existential and ontological question of that "connection"? Perhaps it is already a gain in the field of this inquiry if we learn not to take these problems too lightly.

What seems "more simple" than the nature of the "connection of life" between birth and death? It *consists of* a succession of experiences "in time." If we pursue this characterization of the connection in question and above all of the ontological assumption behind it in a more penetrating way, something remarkable happens. In this succession of experiences only the experience that is present "in the actual now" is "really" ["eigentlich"] "real" ["wirklich"]. The experiences past and just coming, on the other hand, are no longer or not yet "real." Dasein traverses the time-span allotted to it between the two boundaries in such a way that it is "real" only in the now and hops, so to speak, through the succession of nows of its "time." For this reason one says that Dasein is "temporal." The self maintains itself in a certain sameness throughout this constant change of experiences. Opinions diverge as to how this persistent self is to be defined and how one is to determine what relation it may possibly have to the changing experiences. The being of this persistently changing connection of experiences remains undetermined. At bottom, however, and whether one admits it or not, something objectively present "in time," but of course "unthinglike," has been posited in this characterization of the connection of life.

With regard to what was developed as the meaning of being of   *374*
care under the rubric of temporality, we found that by following the
guideline of the vulgar interpretation of Dasein, which within its own
limits is justified and adequate, we could not carry through a genuine
ontological analysis of the way Dasein *stretches along* between birth and
death; indeed, if we take this interpretation as our guideline, we could
not even establish such an analysis as a problem.

Dasein does not exist as the sum of the momentary realities of
experiences that succeed each other and disappear. Nor does this succession gradually fill up a framework. For how should that framework
be present, when it is always only the experience that one is having
"right now" that is "real," and when the boundaries of the framework—birth that is past and death that is yet to come—are lacking
reality. At bottom, even the vulgar interpretation of the "connectedness
of life" does not think of a framework spanned "outside" of Dasein
and embracing it, but correctly looks for it in Dasein itself. When, however, one tacitly regards this being ontologically as something present
"in time," an attempt at any ontological characterization of the being
"between" birth and death gets stranded.

Dasein does not first fill up an objectively present path or stretch
"of life" through the phases of its momentary realities, but stretches
itself along in such a way that its own being is constituted beforehand
as this stretching along. The "between" of birth and death already lies
in the being of Dasein. On the other hand, it is by no means the case
that Dasein "*is*" real in a point of time, and that, in addition, it is then
"surrounded" by the nonreality of its birth and its death. Understood
existentially, birth [Geburt] is never something past in the sense of what
is no longer present, and death is just as far from having the kind of
being of something outstanding that is not yet present but will come.
Factical Dasein exists as being born [existiert gebürtig]*, and in being
born it is also already dying [gebürtig stirbt es] in the sense of being-
toward-death. Both "ends" and their "between" *are* as long as Dasein
factically exists, and they *are* in the sole way possible on the basis of the
being of Dasein as *care*. In the unity of thrownness and the fleeting, or
else anticipatory, being-toward-death, birth and death "are connected"
in the way appropriate to Dasein. As care, Dasein *is* the "between."

But the constitutional totality of care has the possible *ground* of its
unity in temporality. The ontological clarification of the "connectedness

---

* "Gebürtig" [being born] needs to be distinguished from "Geburt" [born]: the former
refers to something that is continuous, not to an event that is "past"; the latter refers
to the singular event of one's birth. The parallel here is to "Sterben" [dying] and "Tod"
[death]: one continually "dies" so long as one is ["gebürtig stirbt es"], while "death"
refers to an event still outstanding. Being born and dying are the existential meaning
of birth and death. "Being born" is thus the name for the natality of Dasein that needs
to be existentially understood in conjunction with Dasein's mortality which Heidegger
characterizes as "dying" or as "being-toward-death." [TR]

of life," that is, of the specific way of stretching along, movement, and persistence of Dasein, must accordingly be approached in the horizon *375* of the temporal constitution of this being. The movement of existence is not the motion of something objectively present. It is determined from the stretching along of Dasein. The specific movement of the *stretched out stretching itself along*, we call the *occurrence* [*Geschehen*] of Dasein. The question of the "connectedness" of Dasein is the ontological problem of its occurrence. To expose the *structure of occurrence* and the existential and temporal conditions of its possibility means to gain an *ontological* understanding of *historicity* [*Geschichtlichkeit*].

With the analysis of the specific movement and persistence appropriate to the occurrence of Dasein, our inquiry returns to the problem that was touched upon right before the exposition of temporality: to the question of the constancy of the self that we determined as the who of Dasein.[1] Self-constancy is a mode of being of Dasein and is thus grounded in a specific temporalizing of temporality. The analysis of occurrence introduces the problems found in a thematic investigation into temporalization as such.

If the question of historicity leads back to those "origins," then the *place* of the problem of history has thus already been decided upon. We must not search in historiography as the science of history. Even if the scientific and theoretical kind of treatment of the problem of "history" does not just aim at an "epistemological" (Simmel) clarification of historiographical comprehension, or at the logic of the concept formation of historiographical presentation (Rickert), but is rather oriented toward the "objective side," history is still only accessible in this line of questioning as the *object* of a science. The basic phenomenon of history, which is prior to and underlies the possibility of making something thematic by historiography, is thus irrevocably set aside. How history can become a possible *object* for historiography can be gathered only from the kind of being of what is historical, from historicity and its rootedness in temporality.

If historicity itself is to be illuminated in terms of temporality, and primordially in terms of *authentic* temporality, then it is essential to this task that it can only be carried out by structuring it phenomenologi-*376* cally.*[2] The existential and ontological constitution of historicity must be mastered in *opposition to* the vulgar interpretation of the history of Dasein that covers over. The existential construction of historicity has its definite supports in the vulgar understanding of Dasein and is guided by those existential structures attained so far.

---

* Project.

1. Cf. § 64.
2. Cf. § 63.

We shall first describe the vulgar concept of history, so that we may give our investigation an orientation as to the factors which are generally held to be essential for history. Here it must become clear what is primordially considered as historical. Thus the entry point for the exposition of the ontological problem of historicity has been designated.

Our interpretation of the authentic potentiality-of-being-whole of Dasein and our analysis of care as temporality arising from that interpretation offer the guideline for the existential construction of historicity. The existential project for the historicity of Dasein only reveals what already lies enveloped in the temporalizing of temporality. Corresponding to the rootedness of historicity in care, Dasein always exists as authentically or inauthentically historical. What we had in view under the rubric of everydayness for the existential analytic of Dasein as the closest horizon gets clarified as the inauthentic historicity of Dasein.

Disclosure and interpretation belong essentially to the occurrence of Dasein. From the kind of being of this being that exists historically, there arises the existentiell possibility of an explicit disclosure and conception of history. Making it thematic, that is, the *historiographical* disclosure of history, is the presupposition for the possibility of "the foundation of the historical world in the human sciences." The existential interpretation of historiography as a science aims solely at a demonstration of its ontological provenance from the historicity of Dasein. Only from here are the boundaries to be staked out within which a theory of science oriented toward the factical workings of science may expose itself to the chance elements of its line of questioning.

The analysis of the historicity of Dasein attempted to show that this being is not "temporal" because it "is in history," but that, on the contrary, it exists and can exist historically only because it is temporal in the ground of its being.

Nevertheless, Dasein must also be called "temporal" in the sense of its being "in time." Factical Dasein needs and uses the calendar and the clock even without a developed historiography. What occurs "with it," it experiences as occurring "in time." In the same way, the processes of nature, whether living or lifeless, are encountered "in time." They are within-time. So while our analysis of how the "time" of within-time-ness has its source in temporality will be deferred until the next chapter,[3] it would be easy to put this before the discussion of the connection between historicity and temporality. What is historical is ordinarily characterized with the aid of the time of within-time-ness. But if this vulgar characterization is to be stripped of its seeming self-evidence and exclusiveness, historicity is to be "deduced" beforehand purely from the primordial temporality of Dasein. This

377

3. Cf. § 80.

is required by the way these are "objectively" connected. But since time as within-time-ness also "stems" from the temporality of Dasein, historicity and within-time-ness turn out to be equiprimordial. The vulgar interpretation of the temporal character of history is thus justified within its limits.

After this first characterization of the course of the ontological exposition of historicity in terms of temporality, do we still need explicit assurance that the following inquiry does not believe that the problem of history can be solved by a sleight of hand? The paucity of the available "categorial" means and the uncertainty of the primary ontological horizons become all the more obtrusive, the more the problem of history is traced to its *primordial rootedness*. In the following reflections, we shall content ourselves with indicating the ontological place of the problem of historicity. Basically, the following analysis is solely concerned with furthering the pioneering investigations of Dilthey. Today's present generation has not as yet made them its own.

Our exposition of the existential problem of historicity—an exposition, moreover, that is of necessity limited by our fundamental and ontological aim—is divided up as follows: the vulgar understanding of history and the occurrence of Dasein (§ 73); the fundamental constitution of historicity (§ 74); the historicity of Dasein and world history (§ 75); the existential origin of historiography from the historicity of Dasein (§ 76); the connection of the previous exposition of the problem of historicity with the investigations of Dilthey and the ideas of Count Yorck (§ 77).

378

§ 73. *The Vulgar Understanding of History and the Occurrence of Dasein*

Our next goal is to find the entry point for the primordial question of the essence of history, that is, for the existential construction of historicity. This point is designated by what is primarily historical. Thus our reflections begin with a characterization of what is meant by the expressions "history" ["Geschichte"] and "historical" ["geschichtlich"] in the vulgar interpretation of Dasein. They are ambiguous.

The most obvious ambiguity of the term "history" has often been noted and it is by no means "vague." It makes itself known in the fact that it means "historical reality" as well as the possibility of a science of it. We shall provisionally discard the meaning of "history" in the sense of a science of history (historiography).

Among the meanings of the expression "history" that refer neither to the science of history nor history as an object, but rather to this being [Seiende] itself which has not necessarily been objectified, the one in which this being is understood as something *past* claims a preeminent use. This significance makes itself known in talk such as "this or that already belongs to history." Here "past" means on the one hand no longer present [vorhanden], or else indeed still present, but without "effect" ["Wirkung"] on the "present" ["Gegenwart"]. However, what is historical as what is past also has the opposite significance when we say that one cannot evade history. Here history means what is past,* but is nevertheless still consequential [Nachwirkende]. However, what is historical, as what is past, is understood in a positively or privatively effective relation to the "present" in the sense of what is real "now" and "today." "The past" has a remarkable double meaning here. Here "the past" belongs irrevocably to an earlier time; it belonged to former events and can yet still be present "now"—for example, the remains of a Greek temple. A "bit of the past" is still "present" in it.

Thus history does not so much mean the "past" in the sense of what is past, but the *derivation* [*Herkunft*]† from it. Whatever "has a history" is in the nexus of a becoming. In such becoming the "development" is sometimes a rise, sometimes a fall. Whatever "has a history" in this way can at the same time "make" history. "Epoch making," it "presently" defines a "future" ["Zukunft"]. Here history means "an event nexus that is a 'productive nexus' " that moves through the "past," the "present," and the "future." Here the past has no special priority.

379

Furthermore, history signifies the whole of beings that change "in time," the transformations and destinies of humankind, human institutions and their "cultures," as distinct from nature that similarly moves "in time." In this instance, history does not so much mean the kind of being [Sein], the occurrence, as the region of beings that one distinguishes from nature with regard to the essential determination of the existence of human being as "spirit" and "culture," although nature, too, belongs in a way to history thus understood.

And finally, what has been handed down as such is taken to be "historical," whether it be known historiographically or taken over as being self-evident and concealed in its derivation.

If we consider the four meanings together, we find that history is the specific occurrence of existing Dasein happening in time, in such a way that the occurrence—which in being-with-one-another is "past" and, at the same time, "handed down" [überlieferte"] and still having an effect—is taken to be history in the sense emphasized.

The four meanings have a connection in that they are related to human being as the "subject" of events. How is the kind of occurrence of these events to be determined? Is the occurrence a succession of processes, a changing appearance and disappearance of events? In

---

* What preceded beforehand and now remaining behind.
† "Herkunft" [derivation] has the sense as well as "ancestry," "source," or "birth" (as a reference to parentage). It echoes the word "Zukunft" [future] which is used later in the same paragraph. [TR]

what way does this occurrence of history belong to Dasein? Is Dasein factically already "objectively present" beforehand, and then at times gets into "a history"? Does Dasein first *become* historical through a concatenation of circumstances and events? Or is the being of Dasein first constituted by occurrence, so that *only because Dasein is historical in its being* are anything like circumstances, events, and destinies ontologically possible? Why does the function of the past get emphasized in the "temporal" characterization of Dasein occurring "in time"?

If history belongs to the being of Dasein, and if this being [Sein] is grounded in temporality, it seems logical to begin the existential analysis of historicity with *those* characteristics of what is historical that evidently have a temporal meaning. Thus a more precise characterization of the remarkable priority of the "past" in the concept of history should prepare the exposition of the fundamental constitution of historicity.

380

The "antiquities" preserved in museums (for example, household things) belong to a "time past," and are yet still objectively present in the "present." How are these useful things historical when they are, after all, *not yet* past? Only because they became an *object* of historiographical interest, of the cultivation of antiquity and national lore? But such useful things can only, after all, be *historiographical objects* because they are somehow in themselves *historical*. We repeat the question: with what justification do we call these beings historical when they are not yet past? Or do these "things" "in themselves" yet have "something past" about them although they are still objectively present today? *Are* these objectively present things then still what they were? Evidently these "things" have changed. The tools have become fragile and worm-eaten "in the course of time." But *the* specific character of the past that makes them something historical does not lie in this transience that continues even during their objective presence in the museum. But then what is past about the useful thing? What *were* the "things" that they no longer are today? They are still definite useful things, but out of use. However, if they were still in use, like many heirlooms in the household, would they then not be historical? Whether in use or out of use, they are no longer what they were. What is "past"? Nothing other than the *world* within which they were encountered as things at hand belonging to a context of useful things and used by heedful Dasein existing-in-the-world. That *world* is no longer. But what was previously *innerworldly* in that world is still objectively present. As useful things belonging to that world, what is *now* still objectively present can nevertheless belong to the *"past."* But what does it mean that the world no-longer-is? World *is* only in the mode of *existing* Dasein, which, as being-in-the-world, is *factical*.

The historical character of antiquities that have been preserved is thus grounded in the "past" of that Dasein to whose world that past belongs. According to this, only "past" Dasein would be historical, but not "present" Dasein. However, can Dasein be *past* at all, if we define "past" as now *"no longer present or at hand"*? Obviously Dasein can *never* be past, not because it is imperishable, but because it can essentially *never* be *objectively present*. Rather, if it is, it *exists*. But a Dasein that no longer exists is not past in the ontologically strict sense; it is rather *having-been-there* [*da-gewesen*]. The antiquities still objectively present have a "past" and a character of history because they belong to useful things and originate from a world that has-been—the world of a Dasein that has-been-there. Dasein is what is primarily historical. But does Dasein first *become* historical by no longer being there? Or *is* it historical precisely as factically existing? *Is Dasein something that has-been only in the sense of having-been-there, or has it been as something making present and futural, that is, in the temporalizing of its temporality?*

381

From this preliminary analysis of the useful things belonging to history that are still objectively present and yet somehow "past," it becomes clear that this kind of being [Seiendes] is historical only on the basis of its belonging to the world. But the world has a historical kind of being because it constitutes an ontological determination of Dasein. Furthermore, we can see that when one designates a time as "the past," the meaning of this is not unequivocal, but the "past" ["Vergangenheit"] is obviously distinct from *having-been* [*Gewesenheit*], which we came to know as a constituent of the ecstatic unity of the temporality of Dasein. But saying this only makes the enigma more acute: why is it that precisely the "past" or, more appropriately, the having-been *predominately* determines what is historical when, after all, having-been temporalizes itself equiprimordially with present and future?

We asserted that Dasein is what is *primarily* historical. But *secondarily* historical is what is encountered within the world, not only useful things at hand in the broadest sense, but also *nature* in the surrounding world as the "historical ground." We call beings unlike Dasein that are historical by reason of their belonging to the world that which is world-historical. It can be shown that the vulgar concept of "world history" arises precisely from our orientation toward what is secondarily historical. What is world-historical is not first historical on the basis of a historiographical objectification, but rather *as the being* that is in itself encountered in the world.

The analysis of the historical character of a useful thing still objectively present not only led us back to Dasein as what is primarily historical, but at the same time made it dubious whether the temporal characteristics of what is historical should be primarily oriented toward the being-in-time of something objectively present at all. Beings do not become "more historical" as we go on to a past ever farther away, so that what is most ancient would be the most

382   authentically historical. However, the "temporal" distance from now and today has no primarily constitutive significance for the historicity of authentically historical beings, not because they are not "in time" or are timeless, but rather because they *primordially* exist *temporally in a way that* nothing objectively present "in time," whether passing away or coming into being, could ever, by its ontological essence, be temporal in such a way.

It will be said that these are overly complicated remarks. No one denies that human existence is basically the primary "subject" of history, and the vulgar concept of history cited says this clearly enough. But the thesis that "Dasein is historical" not only refers to the ontic fact that human being represents a more or less important "atom" in the mechanism of world history, and remains the plaything of circumstances and events, but poses the problem *why, and on the basis of what ontological conditions, does historicity belong to the subjectivity of the "historical" subject as its essential constitution?*

§ 74. *The Essential Constitution of Historicity*

Factically, Dasein always has its "history," and it can have something of the sort because the being of this being is constituted by historicity. We want to justify this thesis with the intention of setting forth the *ontological* problem of history as an existential one. The being of Dasein was defined as care. Care is grounded in temporality. Within the scope of temporality we must accordingly search for an occurrence that determines existence as historical. Thus the interpretation of the historicity of Dasein turns out to be basically just a more concrete elaboration of temporality. We revealed temporality initially with regard to the mode of authentic existing that we characterized as anticipatory resolution. To what extent does this involve an authentic occurrence of Dasein?

We determined resoluteness as self-projection upon one's own being guilty[4] that is reticent and ready for anxiety. It attains its authenticity as *anticipatory* resoluteness.[5] In this, Dasein understands itself with regard to its potentiality-of-being in a way that confronts death in order to take over completely the being [Seiende] that it itself is in its thrownness. Resolutely taking over one's own factical "there" implies
383   at the same time resolve in the situation. In the existential analytic we cannot, on principle, discuss what Dasein *factically* resolves upon. Our present inquiry excludes even the existential project of factical possibilities of existence. Nevertheless, we must ask whence *in general* can the possibilities be drawn upon which Dasein factically projects

---

4. Cf. § 60.
5. Cf. § 62.

itself? Anticipatory self-projection upon the insuperable possibility of existence—death—guarantees only the totality and authenticity of resoluteness. But the factically disclosed possibilities of existence are not to be learned from death. All the less so since anticipation of that possibility is not a speculation about it, but rather precisely means coming back to the factical there. Is taking over the thrownness of the self into its world supposed to disclose a horizon from which existence seizes its factical possibilities? Did we not moreover say that Dasein never gets behind its thrownness?[6] Before we rashly decide whether Dasein draws its authentic possibilities of existence from thrownness or not, we must assure ourselves that we have a complete conception of this fundamental determination of care.

To be sure, as thrown, Dasein is delivered over to itself and its potentiality-of-being, *but as being-in-the-world*. As thrown, it is dependent upon a "world," and exists factically with others. Initially and for the most part, the self is lost in the they. It understands itself in terms of the possibilities of existence that "circulate" in the present day "average" public interpretedness of Dasein. Mostly these possibilities are made unrecognizable by ambiguity, but they are still familiar. Authentic existentiell understanding is so far from extricating itself from traditional interpretedness that it always seizes its chosen possibility in resolution from out of and in opposition to that interpretedness, and yet again for it.

The resoluteness in which Dasein comes back to itself discloses the actual factical possibilities of authentic existing *in terms of the heritage* which that resoluteness *takes over* as thrown. Resolute coming back to thrownness involves *handing oneself over* to traditional possibilities, although not necessarily *as* traditional ones. If everything "good" is a matter of heritage and if the character of "goodness" lies in making authentic existence possible, then handing down a heritage is always constituted in resoluteness. The more authentically Dasein resolves itself, that is, understands itself unambiguously in terms of its ownmost eminent possibility in anticipating death, the more unequivocal and the less haphazard is the choice in finding the possibility of its existence. Only the anticipation of death drives every random and "preliminary" possibility out. Only being free *for* death gives Dasein its absolute goal and pushes existence into its finitude. The finitude of existence thus seized upon tears one back out of endless multiplicity of closest possibilities offering themselves—those of comfort, shirking and taking things easy—and brings Dasein to the simplicity of its *fate* [*Schicksals*]. This is how we designate the primordial occurrence of Dasein that lies in authentic resoluteness in which it *hands itself*

384

---

6. Cf. § 58.

down to itself, free for death, in a possibility that it inherited and yet has chosen.

Dasein can only be reached by the blows of fate because in the basis of its being it *is* fate in the sense described. Existing fatefully in resoluteness handing itself down, Dasein is disclosed as being-in-the-world for the "coming" of "fortunate" circumstances and for the cruelty of chance. Fate does not first originate with the collision of circumstances and events. Even an irresolute person is driven by them, more so than someone who has chosen, and yet he can "have" no fate.

If Dasein, anticipating, lets death become powerful in itself, then, as free for death, it understands itself in its own *higher power* of its finite freedom. In this way it takes over the *powerlessness* of being abandoned to itself in that freedom, which always only *is* in having chosen the choice, and it becomes clear about the chance elements in the situation disclosed. But if fateful Dasein essentially exists as being-in-the-world in being-with others, then its occurrence is an occurrence-with and is determined as *destiny* [*Geschick*]. With this term, we designate the occurrence of the community of a people. Destiny is not composed of individual fates, nor can being-with-one-another be conceived of as the mutual occurrence of several subjects.[7] These fates are already guided beforehand in being-with-one-another in the same world and in the resoluteness for definite possibilities. In communication and in struggle the power of destiny first becomes free. The fateful destiny of Dasein in and with its "generation"[8] constitutes the complete, authentic occurrence of Dasein.

385

Fate is the powerless higher power making itself available for adversities, the power of reticent self-projection, ready for anxiety, upon one's own being-guilty. As such, fate requires the constitution of being of care, that is, temporality, as the ontological condition of its possibility. Only if death, guilt, conscience, freedom, and finitude live together equiprimordially in the being of a being, as they do in care, can that being exist in the mode of fate, that is, be historical in the ground of its existence.

Only a being that is essentially futural in its being so that it can let itself be thrown back upon its factical there, free for its death and shattering itself on it, that is, only a being that, as futural, is equiprimordially **having-been,** *can hand down to itself its inherited possibility, take over its own thrownness and be* **in the Moment** *for "its time." Only authentic temporality that is at the same time finite makes something like fate, that is, authentic historicity, possible.*

It is not necessary that resoluteness *explicitly* know the provenance of the possibilities upon which it projects itself. However, in the temporality of Dasein, and only in it, lies the possibility of *explicitly*

---

7. Cf. § 26.
8. On the concept of "generation," cf. W. Dilthey, "Über das Studium der Geschichte der Wissenschaften vom Menschen, der Gesellschaft und dem Staat" (1875), *Gesammelte Schriften*, vol. 5 (1924), pp. 36–41.

fetching [holen] from the traditional [überlieferten] understanding of Dasein the existentiell potentiality-of-being upon which it projects itself. Resoluteness that returns to itself and hands itself down then becomes the *repetition* [*Wiederholung*] of a possibility of existence that has been handed down. *Repetition is explicitly handing down* [*Überlieferung*], that is, going back to the possibilities of the Dasein that has been there. The authentic repetition of a possibility of existence that has been—the possibility that Dasein may choose its hero—is grounded existentially in anticipatory resoluteness; for in resoluteness the choice is first chosen that makes one free for the struggle over what is to follow [kämpfende Nachfolge]* and fidelity [Treue] to what can be repeated. The handing down of a possibility that has been in repeating it, does not, however, disclose the Dasein that has been there in order to actualize it again. The repetition of what is possible neither brings back "what is past," nor does it bind the "present" back to what is "outdated." Arising from a resolute self-projection, repetition is not convinced by "something past," in just letting it come back as what was once real. Rather, repetition *responds* to the possibility of existence that has been-there. But responding [Erwiderung] to this possibility in a resolution is at the same time, *as a response belonging to the Moment*, the *renunciation* [*Widerruf*] of that which is working itself out in the today as "past." Repetition neither abandons itself to the past, nor does it aim at progress. In the Moment, authentic existence is indifferent to both of these alternatives.

386

We characterize repetition as the mode of resolution handing itself down, by which Dasein exists explicitly as fate. But if fate constitutes the primordial historicity of Dasein, history has its essential weight neither in what is past nor in the today and its "connection" with what is past, but in the authentic occurrence of existence that arises from the *future* of Dasein. As a mode of being of Dasein, history has its roots so essentially in the future that death, as the possibility of Dasein we characterized, throws anticipatory existence back upon its *factical* thrownness and thus first gives to *having-been* its unique priority in what is historical. *Authentic being-toward-death, that is, the finitude of temporality, is the concealed ground of the historicity of Dasein.* Dasein does not first become historical in repetition, but rather because as temporal it is historical, it can take itself over in its history, retrieving itself. For this, historiography is still not needed.

---

* "Nachfolge" has several senses here. It refers not only to what follows, to a succession, but also to a sort of imitation. "Nachfolge Christi," the "imitation of Christ," is a phrase that comes to mind in this passage. There are a number of words in this paragraph that echo one another—holen-Wiederholen [fetch-repetition], Erwiderung-Widerruf [response-renounce]—and there are a number of words that refer to a movement in some direction—Rückgang, Fortschritt, Zurückbinden, Wiederholung, Wiederbringen, vorlaufen, überliefern, überkommen, Nachfolge. At the center of these words is the word "Wiederholung" ["repetition"], which refers back to the title opening section (§ 1) of *Being and Time* with its announcement of the "The Necessity of an Explicit Repetition of the Question of Being." [TR]

We call fate the anticipatory handing oneself down to the there of the Moment that lies in resoluteness. In it destiny is also grounded, by which we understand the occurrence of Dasein in being-with-others. Fateful destiny can be explicitly disclosed in repetition with regard to its being bound up with the heritage handed down to it. Repetition first makes manifest to Dasein its own history. The occurrence itself and the disclosedness belonging to it, or the appropriation of it, is existentially grounded in the fact that Dasein is ecstatically open as a temporal being.

What we characterized up to now as historicity, in conformity with the occurrence lying in anticipatory resoluteness, we shall now call the *authentic* historicity of Dasein. It became clear in terms of the phenomena of handing down and repetition, rooted in the future, why the occurrence of authentic history has its emphasis in having-been.

387 However, it remains all the more enigmatic how this occurrence, as fate, is to constitute the whole "connection" of Dasein from its birth to its death. What can going back to resoluteness add to this by way of clarification? Is not each resolution just *one* more single "experience" ["Erlebnis"] in the succession of the whole connection of experiences? Is the "connection" of authentic occurrence supposed to consist of an seamless succession of resolutions? Why does the question of the constitution of the "connectedness of life" not find an adequate and satisfactory answer? Is our investigation overhasty so that, in the end, it clings too much to the answer, without having tested the *question* beforehand as to its legitimacy? Nothing became more clear from the course of the existential analytic thus far than the fact that the ontology of Dasein falls back into the temptations of the vulgar understanding of being again and again. We can cope with this methodologically only by pursuing the *origin* of the question of the constitution of the connection of Dasein, no matter how "self-evident" this question may be, and by determining in what ontological horizon it moves.

If historicity belongs to the being of Dasein, then even inauthentic existence must be historical. What if the *inauthentic* historicity of Dasein determined our line of questioning about a "connectedness of life" and blocked access to authentic historicity and the "connection" peculiar to it? However that may be, if the exposition of the ontological problem of history is supposed to be adequate and complete, we cannot escape considering the inauthentic historicity of Dasein.

§ 75. *The Historicity of Dasein and World History*

Initially and for the most part, Dasein understands itself in terms of what it encounters in the surrounding world and what it circumspectly takes care of. This understanding is no mere taking cognizance of itself which simply accompanies all the modes of behavior of Dasein. Under-

standing signifies self-projection upon the actual possibility of being-
in-the-world, that is, existing as this possibility. Thus understanding,
as common sense [Verständigkeit], also constitutes the inauthentic
existence of the they. What everyday taking care of things encounters
in public being-with-one-another is not just useful things and works,
but at the same time what "is going on" with them: "occupations,"   388
undertakings, incidents, mishaps. The "world" belongs to everyday
trade and traffic as the soil from which they have grown and the stage
where they are displayed. In public being-with-one-another others are
encountered in the activities in which "one" "gets into the swim of
things" [mitschwimmt] "oneself." One always knows about it, talks
about it, furthers it, fights it, retains it, and forgets it primarily with
regard to *what* is being done and *what* will "come out of it." We initially
calculate the progress, arrest, adjustment, and "output" of individual
Dasein in terms of the course, status, change, and availability of what is
taken care of. As trivial as the reference to the understanding of Dasein
of everyday common sense may be, ontologically this understanding
is by no means transparent. But then why should the "connectedness"
of Dasein not be determined in terms of what is taken care of and
"experienced"? Do not useful things and works and everything that
Dasein spends time with also belong to "history"? Is the occurrence
of history then only the isolated course of "streams of experience" in
individual subjects?

Indeed, history is neither the connectedness of movements in the
alteration of objects, nor the free-floating succession of experiences of
"subjects." Does that occurrence of history then pertain to the "linking" of subject and object? Even if one assigns the occurrence to the
subject-object relationship, one must still ask about the kind of being of
this linking as such if this linking is what basically "occurs." The thesis
of the historicity of Dasein does not say that the worldless subject is
historical, but that what is historical is the being that exists as being-in-
the-world. *The occurrence of history is the occurrence of being-in-the-world.*
The historicity of Dasein is essentially the historicity of the world which,
on the basis of its ecstatic and horizonal temporality, belongs to the
temporalizing of that temporality. In so far as Dasein factically exists,
it already encounters that which has been discovered within the world.
*With the existence of historical being-in-the-world, things at hand and pres-
ent have always already been drawn into the history of the world.* Tools and
works, for example books, have their "fates"; buildings and institutions
have their history. And even nature is historical. It is precisely *not* historical to the extent that we speak about "natural history,"[9] but nature is

---

9. For the question of ontologically differentiating "natural occurrence" from the movement of history, cf. F. Gottl, *Die Grenzen der Geschichte* (1904). These reflections have not
been appreciated sufficiently at all.

389 historical as a countryside, as areas that have been inhabited or exploit-
ed, as battlefields and cultic sites. These innerworldly beings as such *are*
historical, and their history does not signify something "external" that
simply accompanies the "inner" history of the "soul." We shall call these
beings *world-historical.* Here we must note that the expression "world
history" that we have chosen, and that is here understood ontologically,
has a double meaning. On the one hand, it signifies the occurrence of
world in its essential existent [existenten] unity with Dasein. But at the
same time it means the innerworldly "occurrence" of what is at hand
and objectively present, since innerworldly beings are always discovered with the factically existent world. The historical world is factically
only as the world of innerworldly beings. What "occurs" with tools and
works as such has its own character of motion, and this character has
been completely obscure up to now. For example, a ring that is "presented" and "worn" does not simply undergo a change of location in
its being. The movement of occurrence in which "something happens
to it" cannot be grasped at all in terms of motion as change of location.
That is true of all world-historical "processes" and events, and in a way
even of "natural catastrophes." Quite apart from the fact that we would
necessarily go beyond the limits of our theme if we were to pursue the
problem of the ontological structure of world-historical occurrence, we
cannot do this because the intention of this exposition is to lead us to
the ontological enigma of the movement of occurrence in general.

We only want to delimit *that* range of phenomena that we also
necessarily have in mind ontologically when we speak about the historicity of Dasein. On the basis of the temporally founded transcendence
of the world, what is world-historical is always already "objectively"
there in the occurrence of existing being-in-the-world, *without being
grasped historiographically.* And since factical Dasein is absorbed and
entangled in what it takes care of, it initially understands its history
as world history. And since, furthermore, the vulgar understanding of
being understands "being" as objective presence without further differentiation, the being of what is world-historical is experienced and
interpreted in the sense of objective presence that arrives, is present,
and disappears. And finally since the meaning of being in general is
taken to be what is absolutely self-evident, the question of the kind
of being of what is world-historical and of the movement of occurrence in general is "after all really" only unfruitful and unnecessarily
complicated verbal sophistry.

390 Everyday Dasein is dispersed in the multiplicity of what "happens" daily. The opportunities and circumstances that taking care
keeps "tactically" awaiting in advance result in "fate." Inauthentically
existing Dasein first calculates its history in terms of what it takes care
of. In so doing, it is driven about by its "affairs." So if Dasein wants to

come to itself, it must first *pull itself together* [*Zusammenholen*] from the *dispersion* and the *disconnectedness* of what has just "happened," and, because of this, there at last arises from the horizon of the understanding of inauthentic historicity the *question* of how one is to establish Dasein's "connectedness" in the sense of the experiences of the subject which are "also" present. The possibility that this horizon for the question should be the dominant one is grounded in irresoluteness that constitutes the essence of the in-constancy of the self.

We have thus pointed out the *origin* of the question of Dasein's "connectedness" in the sense of the unity with which experiences are linked together between birth and death. At the same time, the provenance of this question betrays its inappropriateness with regard to a primordial existential interpretation of the totality of occurrence of Dasein. But, on the other hand, with the predominance of this "natural" horizon of questioning, it becomes explicable why precisely the authentic historicity of Dasein—fate and repetition—looks as if it, least of all, could provide the phenomenal basis for bringing into the form of an ontologically founded problem what is basically intended with the question of the "connectedness of life."

This question cannot ask: how does Dasein acquire such a unity of connection [Zusammenhang] so that it can subsequently link together the succession of "experiences" that has ensued and is still ensuing; rather, it asks in which of its own kinds of being *does it lose itself in such a way that it must, as it were, pull itself together only subsequently out of its dispersion, and invent [erdenken] for itself a unity in which this together [Zusammen] is embraced?* Lostness in the they and in world history revealed itself earlier as a flight from death. This flight from . . . reveals being-*toward*-death as a fundamental determination of care. Anticipatory resoluteness brings this being-toward-death to authentic existence. But we interpreted the occurrence of this resoluteness, which anticipates and hands down and repeats the heritage of possibilities as authentic historicity. Does perhaps the primordial stretching along of the whole of existence, which is not lost and does not need a connection, lie in historicity? The resoluteness of the self against the inconstancy of dispersion is in itself a *steadiness that has been stretched along*—the steadiness in which Dasein as fate "incorporates" into its existence birth and death and their "between" in such a way that in such constancy it is in the Moment for what is world-historical in its actual situation. In the fateful repetition of possibilities that have-been, Dasein brings itself back "immediately," that is, temporally and ecstatically, to what has already been before it. But then, when its heritage is thus handed down to itself, "birth," in coming back from the insuperable possibility of death, is *taken into existence*, so that existence may accept the thrownness of its own there freer from illusion.

391

Resoluteness constitutes the *fidelity* of existence to its own self. As resoluteness ready for *anxiety*, fidelity is at the same time a possible reverence for the sole authority that a free existence can have, for the possibilities of existence that can be repeated. Resoluteness would be misunderstood ontologically if one thought that it *is* real as "experience" only as long as the "act" of resolution "lasts." In resoluteness lies the existentiell constancy which, in keeping with its essence, has already anticipated [vorweggenommen] every possible Moment arising from it. As fate, resoluteness is freedom to *give up* a definite resolution, as may be required in the situation. Thus the constancy of existence is not interrupted, but precisely confirmed in the Moment. Constancy is not first formed either through or by "Moments" adjoining each other, but rather the Moments arise from the temporality, *already stretched along*, of that repetition which is futurally in the process of having-been.

On the other hand, in inauthentic historicity the primordial stretching along of fate is concealed. With the inconstancy of the they-self, Dasein makes present its "today." Awaiting the next new thing, it has already forgotten what is old. The they evades choice. Blind toward possibilities, it is incapable of retrieving what has been, but only retains what is and receives the "real" that has been left over of the world-historical that has been: the remnants, and the information about them that is available. Lost in the making present of the today, it understands the "past" in terms of the "present." In contrast, the temporality of authentic historicity, as the Moment that anticipates and retrieves, *undoes* the character of *making-present* of the today and weans one off of the conventions of the they. Inauthentic historical existence, on the other hand, burdened with the legacy of a "past" that has become unrecognizable to it, looks for what is modern. Authentic historicity understands history as the "return" [Wiederkehr"] of what is possible and knows that a possibility returns only when existence is open for it fatefully, in the Moment, in resolute repetition.

The existential interpretation of the historicity of Dasein constantly gets caught up [gerät] unexpectedly in shadows. The obscurities are all the more difficult to dispel when the possible dimensions of appropriate questioning are not disentangled and when everything is haunted by the *enigma* of *being* and, as has now become clear, of *movement*. Nevertheless, we may venture an outline of the ontological genesis of historiography as a science in terms of the historicity of Dasein. It should serve as a preparation for the clarification of the task of a historical destruction of the history of philosophy[10] to be carried out in what follows.

392

---

10. Cf. § 6.

§ 76. The Existential Origin of Historiography from the Historicity of Dasein

That historiography, like every science as a mode of being of Dasein, is factically and actually "dependent" upon the "dominant worldview" needs no discussion. However, beyond this fact, we must inquire into the ontological possibility of the origin of the sciences from the constitution of being of Dasein. This origin is still not very transparent. In the context to be addressed our analysis will acquaint us in outline with the existential origin of historiography only to the extent that it will shed more light upon the historicity of Dasein, and its roots in temporality.

If the being of Dasein is fundamentally historical, then every factical science evidently remains bound to this occurrence. But historiography presupposes the historicity of Dasein in its own distinctive way.

At first one wants to clarify this by referring to the fact that historiography, as a science of the history of Dasein, must "presuppose" the being that is primordially historical as its possible "object." But history must not only *be* in order for a historiographical object to be accessible; and historiographical cognition, as an actual mode of behavior of Dasein, is not only historical. Rather, the *historiographical disclosure of history is in itself rooted in the historicity of Dasein in accordance with its ontological structure*, whether it is factically carried out or not. This connection is what the talk about the existential origin of historiography from the historicity of Dasein means. To throw light on this connection means methodologically to project ontologically the *idea* of historiography in terms of the historicity of Dasein. On the other hand, it is not a matter of "abstracting" the concept of historiography from some factical procedure of the sciences today nor of assimilating it to that procedure. For what guarantee do we have in principle that this factical procedure indeed represents historiography's primordial and authentic possibilities? And even if that is true (we shall refrain from any decision about this) the concept could still be "discovered" in fact only if guided by the idea of historiography that had already been understood. On the other hand, however, the existential idea of historiography is not accorded a greater legitimacy if the historian's factical mode of behavior confirms it by agreeing with it. Nor does the idea become "false" if the historian contests any such agreement.

The idea of historiography as a science implies that it has grasped the *disclosure* of historical beings as its own task. Every science is primarily constituted by thematization. What is known prescientifically in Dasein as disclosed being-in-the-world is projected upon its specific being. The region of beings is limited by this project. The accesses to it contain their methodological "directive," and the conceptual structure of interpretation is prefigured. If we may postpone the question of whether a "history of the present" is possible and assign to histo-riography the task of disclosing the "past," the historiographical the-matization of history is possible only if the "past" has always already been disclosed in general. Quite apart from the question of whether sufficient sources are available for a historiographical envisagement of the past, the *way to it* must be *open* in general for the historiographical return to it. That something like this is true and how it is possible is by no means obvious.

But since the being of Dasein is historical, that is, since it is open in its character of having-been on the basis of ecstatical and horizonal temporality, the way is in general freed for such thematization of the "past" as can be carried out in existence. And because Dasein *and only* Dasein is primordially historical, what historiographical thematization presents as the possible object of its investigation must have the kind of being of Dasein *that has-been-there*. Together with factical Dasein as being-in-the-world, there *is* also always world history. If Dasein is no longer there, then the world, too, is something that has-been-there. This is not in conflict with the fact that what was formerly at hand within the world does not yet pass away, but is available "historiographically" for the present as something that has not passed away and belongs to the world that has-been-there.

Remains, monuments, and records that are still objectively pres-ent are *possible* "material" for the concrete disclosure of Dasein that has-been-there. These things *can* become *historiographical* material only because they have a *world-historical* character in accordance with their own kind of being. And they *become* such material only by being under-stood from the outset with regard to their innerworldliness. The world already projected is determined by way of an interpretation of the world-historical material that has been "preserved." The acquisition, sifting, and securing of such material does not first bring about a return to the "past," but rather already presupposes *historical being toward* the Dasein that has-been-there, that is, the historicity of the historian's exis-tence. This existence existentially grounds historiography as science, down to the most inconspicuous, "mechanical" procedures.[11]

If historiography is rooted in historicity in this way, then we should also be able to determine from there what the *object* of historiography "really" is. The delimitation of the primordial theme of historiography must be carried out in conformity with authentic historicity and its disclosure of what-has-been-there, of repetition. Repetition understands Dasein that has-been-there in its authentic possibility that has-been. The

394

---

11. Concerning the constitution of historiographical understanding, cf. E. Spranger, "Zur Theorie des Verstehens und zur geisteswissenschaftlichen Psychologie," *Festschrift für Johannes Volkelt* (1918), pp. 357ff.

"birth" of historiography from authentic historicity thus means that the primary thematization of the object of historiography projects Dasein that has-been-there upon its ownmost potentiality-of-existence. Does historiography thus have *what is possible* as its theme? Does not its whole "meaning" lie in "facts," in what has factually been?

However, what does it mean that Dasein "factually" ["tatsächlich"] is? If Dasein is "really" actual only in existence, its "factuality" is, after all, constituted precisely by its resolute self-projection upon a chosen potentiality-of-being. What has "factually" really been there, however, is then the existentiell possibility in which fate, destiny, and world history are factically determined. Because existence always is only as factically thrown, historiography will disclose the silent power of the possible with greater penetration the more simply and concretely it understands having-been-in-the-world in terms of its possibility, and it "only" presents it as such.

If historiography, which itself arises from authentic historicity, reveals by repetition the Dasein that has-been-there in its possibility, it has also already made the "universal" manifest in what is unique. The question of whether historiography only has as its object a series of unique, "individual" events, or whether it also has "laws," is radically mistaken. Its theme is neither only a singular occurrence, nor something universal floating above these occurrences, it is rather the possibility that has been factically existent. This possibility is not repeated as such, that is, authentically understood historiographically, if it is distorted into the pallor of a supratemporal pattern. Only factically authentic historicity, as resolute fate, can disclose the history that has-been-there in such a way that in repetition the "power" of the possible breaks into factical existence, that is, comes toward it in its futurality. Historiography by no means takes its point of departure from the "present" and what is "real" only today—any more than does the historicity of unhistorical Dasein—in order to then grope its way back from there to a past. Rather, even *historiographical* disclosure temporalizes itself *out of the future*. The "*selection*" of what is to become a possible object for historiography *has already been made* in the factical existentiell *choice* of the historicity of Dasein, in which historiography first arises and in which it uniquely *is*.

The historiographical disclosure of the "past" is grounded in fateful repetition and is so far from being "subjective" that it alone guarantees the "objectivity" of historiography. For the objectivity of a science chiefly regulates itself by its capacity to *bring* the being that is its theme, uncovered in the primordiality of its being [Sein], *to* the understanding. In no science are the "universal validity" of standards and the claims to "universality" that are demanded by the they and its common sense *less* possible criteria of "truth" than in authentic historiography.

Only because the central theme of historiography is always the *possibility* of existence that has-been-there, and because the latter always factically exists in a world-historical way, can historiography demand of itself a relentless orientation toward "facts." For this reason factical research has many branches and makes the history of useful things, works, culture, spirit, and ideas its object. At the same time history, handing itself down, is in itself always in an interpretedness that belongs to it, and that has a history of its own; so that for the most part it is only through traditional history that historiography penetrates into what has-been-there itself. This explains why concrete historiographical research can always keep to its authentic theme in varying degrees of nearness. The historian who from the outset has "thrown" himself into the "worldview" of an era has not yet proven that he understands his subject-matter authentically and historically, and not just "aesthetically." On the other hand, the existence of a historian who "only" edits sources may be determined by an authentic historicity.

Thus, even the prevalence of a differentiated historiographical interest in the most remote and primitive cultures is in itself no proof of the authentic historicity of an "age." Ultimately, the emergence of the problem of "historicism" is the clearest indication that historiography strives to alienate Dasein from its authentic historicity. Historicity does not necessarily need historiography. Unhistoriographical ages are, as such, not also automatically unhistorical.

The possibility that historiography in general can be either an "advantage" or a "disadvantage" "for life" is based on the fact that life is historical in the roots of its being and has thus, factically existing, always already decided upon authentic or inauthentic historicity. Nietzsche recognized what is essential about "advantage and disadvantage of historiography for life" in the second of his *Untimely Mediations* (1874) and stated it unequivocally and penetratingly. He distinguishes three kinds of historiography: the monumental, the antiquarian, and the critical, without demonstrating explicitly the necessity of this triad and the ground of its unity. *The threefold character of historiography is prefigured in the historicity of Dasein.* At the same time historicity enables us to understand why authentic historiography must be the factical and concrete unity of these three possibilities. Nietzsche's division is not accidental. The beginning of his *Untimely Meditations* makes us suspect that he understood more than he made known.

As historical, Dasein is possible only on the basis of temporality. Temporality temporalizes itself in the ecstatic-horizonal unity of its raptures. Dasein exists as futural authentically in the resolute disclosure of a chosen possibility. Resolutely coming back to itself, it is open, in repetition, for the "monumental" possibilities of human existence. The historiography arising from this historicity is "monumental." As hav-

396

ing-been, Dasein is delivered over to [überantwortet] its thrownness. In appropriating the possible in repetition, there is prefigured at the same time the possibility of reverently preserving the existence that has-been-there, in which the possibility taken up became manifest. As monumental, authentic historiography is thus "antiquarian." Dasein temporalizes itself in the unity of future and the having-been as the present. The present, as the Moment, discloses the today authentically. But insofar as the today is interpreted on the basis of the futurally repetitive understanding of a possibility taken up from existence, authentic historiography ceases to make the today present; that is, it becomes the painful way of detaching itself from the entangled publicness of the today. As authentic, monumental-antiquarian historiography is necessarily a critique of the "present." Authentic historicity is the foundation of the possible unity of the three kinds of historiography. But the *ground* on which authentic historiography is founded is *temporality* as the existential meaning of being of care.

The existential and historical origin of historiography may be presented concretely by analyzing the thematization that constitutes this science. Historiographical thematization centers on developing the hermeneutical situation that is opened up—once historically existing Dasein has made its resolution—to the disclosure in repetition of what has-been-there. The possibility and the structure of *historiographical truth* are to be set forth in terms of the *authentic disclosedness* ("truth") *of historical existence*. But since the fundamental concepts of the historiographical sciences—whether they pertain to the objects of these sciences or to the way these are treated—are concepts of existence, the theory of the human science presupposes a thematic and existential interpretation of the *historicity* of Dasein. Such an interpretation is the constant goal that Dilthey's investigations attempt to approach and that is illuminated more penetratingly by the ideas of Count Yorck von Wartenburg.

§ 77. *The Connection of the Foregoing Exposition of the Problem of Historicity with the Investigations of Dilthey and the Ideas of Count Yorck*

The confrontation with the problem of history grew out of an appropriation of Dilthey's work. It was corroborated, and at the same time strengthened, by Count Yorck's theses that are scattered throughout his letters to Dilthey.[12]

---

12. Cf. *Briefwechsel zwischen Wilhelm Dilthey und dem Grafen Paul Yorck von Wartenburg, 1877–1897* (Halle an der Salle, 1923).

398

The image of Dilthey still prevalent today is that of the "sensi-tive" interpreter of the history of the spirit, especially the history of literature, who "also" concerned himself with the distinction between the natural and the human sciences, attributing a distinctive role to the history of these sciences and also to "psychology," then letting the whole merge into a relativistic "philosophy of life." For a superficial consideration, this sketch is "correct." But it misses the "substance." It covers over more than it reveals.

Dilthey's investigations can be divided schematically into three areas: studies regarding the theory of the human sciences [Geisteswis-senschaften] and the distinction between these and the natural sci-ences; investigations into the history of the sciences of human being, society, and the state; and endeavors toward a psychology in which the "whole fact of being human" is to be presented. Investigations in scientific theory, the history of science, and hermeneutical psychology constantly interpenetrate and overlap each other. When one direction predominates, the others are motives and means. What appears to be disunity and uncertain, chance "attempts," is an elemental restlessness, of which the one goal is to understand "life" philosophically and to secure for this understanding a hermeneutical foundation in terms of "life itself." Everything is centered in the "psychology" that is supposed to understand "life" in the historical context of its development and its effects, as the *way* in which human being *is*, as the possible *object* of the human sciences and especially the *root* of these sciences. Hermeneutics is the self-clarification of this understanding; it is also the methodology of historiography, though only in a derivative form.

In contemporaneous discussions, Dilthey's own investigations for laying the foundations for the human sciences were forced one-sid-edly into the field of a theory of science; and it was with a regard for such discussions that his publications were often oriented in this direction. The "logic of the human sciences" was by no means central for him—no more than he was striving in his "psychology" merely to improve the positivistic science of the psychical.

Dilthey's friend, Count Yorck, gives unambiguous expression to Dilthey's ownmost philosophical tendency when he refers to *"our com-mon interest in understanding historicity."*[13] Dilthey's investigations are only now becoming accessible in their complete scope; if we are to make them our own we need the constancy and concretion of coming to terms with them in principle. This is not the place[14] for a detailed discussion of the problems that motivated him or how they moti-

399

---

13. *Briefwechsel*, p. 185 [Heidegger's italics].
14. This is not necessary since we have G. Misch to thank for a concrete presentation of Dilthey that aims at the central tendencies essential to any discussion of his work. Cf. W. Dilthey, *Gesammelte Schriften*, vol. 5 (1924), *Vorbericht*, pp. vii–cxvii.

vated him. We shall, however, describe in a provisional way some of
Count Yorck's central ideas by selecting characteristic passages from
the letters.

In these communications Yorck's own tendency is brought to life
by Dilthey's questions and work, and this tendency can be seen in his
stance with regard to the tasks of the fundamental discipline—analytical psychology. About Dilthey's Academy treatise, "Ideas on a Descriptive and Analytic Psychology" (1894), he writes:

> It is definitely established that self-reflection is the primary means
> of knowing, and that the primary procedure of knowing is analy-
> sis. From this standpoint principles are formulated that are verified
> by their own findings. No progress is made toward critically break-
> ing down constructive psychology and its assumptions, explaining
> it and thus refuting it from within. . . . Your resistance to breaking
> things down critically (i.e., for demonstrating their provenance psy-
> chologically and carrying this out trenchantly in detail) is, in my
> opinion, connected with your concept of the theory of knowledge
> and the position you assign to it. (*Briefwechsel*, p. 177)

> The *explanation* of inapplicability—the fact is set up and made
> clear—is given only by a theory of knowledge. It has to give jus-
> tifications for the adequacy of scientific methods; it has to pro-
> vide the grounds for a doctrine of method, instead of having its
> methods taken from individual areas, rather haphazardly I must
> say. (p. 179)

At bottom Yorck is demanding a logic preceding the sciences and
guiding them as did Platonic and Aristotelian logic, and this demand
includes the task of developing, positively and radically, the various
categorial structures of the being that is nature and the being that
is history (Dasein). Yorck finds that Dilthey's investigations *"too little
emphasize the generic difference between the ontic and the historical"* (p.
191, my emphasis).

> In particular, the procedure of comparison is claimed as the meth-
> od of the human sciences. Here I disagree with you. . . . Compari-
> son is always aesthetic and is bound to form. Windelband assigns
> forms to history. Your concept of the type is an entirely inward   *400*
> one. Here it is a matter of characteristics, not forms. For Wind-
> elband, history is a series of pictures, individual forms, aesthetic
> demands. For the natural scientist, besides science, only aesthetic
> pleasure remains as a kind of human tranquillizer. Your concept
> of history is, after all, that of a nexus of forces, unities of force, to

which the category of form should be applied only in a symbolic sense. (p. 193)

In terms of his certain instinct for the "difference between the ontic and the historical," Yorck knew how strongly traditional historical investigation is geared to "purely ocular determinations" (p. 192) that aim at what has body and form.

Ranke is a great ocularist, for whom things that have vanished can never become *realities*. . . . Ranke's whole manner can be explained in terms of his limiting the stuff of history to the political. Only the political is dramatic. (p. 60)

The modifications that have come in the course of time seem inessential to me, and here I probably judge differently from you. For example, I think that the so-called historical school is merely a side-current within the same riverbed, and represents only one branch of an old and thorough-going opposition. The name has something deceptive about it. *That school was not historical at all* [Heidegger's italics], but rather an antiquarian one, construing things aesthetically, whereas the great, dominating movement was one of mechanical construction. Thus what it added methodologically to the method of rationality was only a general feeling. (pp. 68f.)

The genuine philologist conceives of historiography as a cabinet of antiquities. Where nothing is palpable—where only a living psychical transposition guides us—these gentlemen never get there. At heart they are natural scientists, and they become skeptics all the more because experimentation is lacking. One must keep completely away from the petty details, for example, how often Plato was in Magna Graecia or Syracuse. There is no vitality in that. This superficial affectation that I have seen through critically finally boils down to a large question mark and is put to shame by the great realities of Plato, Homer, and the New Testament. Everything actually real becomes a schema when it is considered as a "thing in itself," when it is not experienced. (p. 61)

The 'scientists' confront the powers of the times in a way similar to the over-refined French society of the revolutionary period. Here as there, formalism, the cult of form, the defining of relationship is the final word of wisdom. Of course, this direction of thought has its own history that, I believe, is not yet written. The groundlessness of this thinking and of the faith in it (and

401

such thinking is, epistemologically considered, a metaphysical attitude) is a historical product. (p. 39)

The groundswells evoked by the principle of eccentricity that led to a new time more than four hundred years ago seem to me to have become exceedingly broad and flat; knowledge has progressed to the point of negating itself; man has become so far removed from himself that he has lost sight of himself. "Modern man," i.e., man since the Renaissance, is ready to be buried. (p. 83)

In contrast, "all written history that is alive and not just depicting life is critique" (p. 19). "But historical knowledge is, for the best part, knowledge of hidden sources" (p. 109). "With history, what creates a spectacle and catches the eye is not the main thing. The nerves are invisible, as is the essential in general. And as it is said that 'if you were quiet, you would be strong,' the variation of this is also true: if you are quiet, you will perceive, i.e., understand" (p. 26). "And then I enjoy the silent soliloquy and commerce with the spirit of history. This spirit did not appear to Faust in his study, nor to Master Goethe either. They would not have flinched from it in alarm, no matter how serious and compelling the apparition was. For it is brotherly and akin to us in a sense deeper and other than the inhabitants of bush and field. These exertions are like Jacob's wrestling, a sure gain for the wrestler himself. That is what matters first of all" (p. 133).

Yorck gets his clear insight into the fundamental character of history as "virtuality" from his knowledge of the characteristics of the being of human existence itself, thus precisely not in a theoretical and scientific way oriented to the object of historical observation: "That the whole psychophysical datum *is* not [being = the being-present of nature], but rather lives, is the core of historicity. And if self-reflection is directed not to an abstract ego, but to the fullness of my self, it will find me historically determined, just as physics knows me as cosmically determined. I am history as well as nature ..." (p. 71). And Yorck, who saw through all false "relational definitions" and "groundless" relativisms, did not hesitate to draw the final conclusion from his insight into the historicity of Dasein.

But, on the other hand, a systematic that is separated from history is methodologically inadequate for the inner historicity of self-consciousness. Just as physiology cannot be studied in abstraction from physics, neither can philosophy from historicity, especially if it is critical. . . . Self-relation and historicity are like breathing and atmospheric pressure and, although this sounds rather paradoxical, it seems to me methodologically like a residue of metaphysics not to historicize philosophizing. (p. 69)

Since philosophizing is living, there is (do not be alarmed), in my opinion, a philosophy of history—but who could write it! Certainly not in the way it has been interpreted and attempted up to now, and you have declared yourself irrefutably against this. The line of questioning up to now was simply false, even impossible, but it is not the only one. Thus there is no longer any real philosophizing that is not historical. The separation between systematic philosophy, and historical presentation is essentially wrong. (p. 251)

Being able to become practical is now, of course, the real basis for the justification of any science. But the mathematical *praxis* is not the only one. The practical aim of our standpoint is the pedagogical one in the broadest and deepest sense of the word. It is the soul of all true philosophy, and the truth of Plato and Aristotle. (pp. 42f).

You know what I think about the possibility of ethics as a science. Nevertheless, it can always be improved. Who are such books really for? Registries about registries! The only thing worthy of notice here is the drive from physics to ethics. (p. 73)

If one conceives of philosophy as a manifestation of life and not as the expectoration of a groundless thinking (and such thinking appears groundless because one's look is turned away from the ground of consciousness), the task is as meager in its results as it is complicated and laborious in arriving at them. Freedom from prejudice is what it presupposes, and that is difficult to obtain. (p. 250)

That Yorck set out to grasp the historical categorially, as opposed to the ontic (ocular), and to elevate "life" into its appropriate scientific understanding, becomes clear from his reference to the kind of difficulty of such investigations. The aesthetic and mechanistic kind of thinking

403

is more easily expressed verbally than an analysis that goes behind intuition, and this can be explained by the wide extent to which words have their provenance in the ocular. . . . On the other hand, what penetrates to the ground of vitality eludes an exoteric presentation; hence all its terminology is not intelligible to all, it is symbolic and inevitable. From the special kind of philosophical thinking follows the special character of its linguistic expression. (pp. 70f.)

But you are familiar with my predilection for paradox, which I justify by the fact that paradox is a mark of truth, that the *communis
opinio* is surely never in the truth, but is like an elemental precipi-
tate of a halfway understanding that makes generalizations; in its
relationship to truth it is like the sulphurous fumes that lightning
leaves behind. Truth is never an element. To dissolve elemental
public opinion and, if possible, to make possible the shaping of
individuality in seeing and regarding, would be a pedagogical
task for the state. Then instead of a so-called public conscience—
instead of this radical externalization—individual conscience, i.e.,
conscience, would again become powerful. (pp. 249f.)

If one has an interest in understanding historicity, one is brought
to the task of developing the "generic difference between the ontic
and the historical." Thus we have ascertained the *fundamental goal of
the "philosophy of life."* Still, our line of questioning needs a more *fundamental* radicalization. How else is historicity to be philosophically
grasped and "categorially" conceived in its difference from the ontic
than by bringing the "ontic" as well as the "historiographical" into a
more primordial unity so that they can be compared and distinguished?
But that is possible only if we attain the following insights:

1. The question of historicity is an *ontological* question about the constitution of being of historical beings.
2. The question of the ontic is the *ontological* question of the constitution of being of beings unlike Dasein, of what is present in the
   broadest sense.
3. The ontic is only *one* area of beings.

The idea of being encompasses the "ontic" *and* the "historiographical."
This idea is what must be "generically differentiated."
    It is not by chance that Yorck calls nonhistorical beings simply the
ontic. That just reflects the unbroken dominance of traditional ontology that, coming from the *ancient* questioning of being, holds fast to
the ontological problematic and fundamentally narrows it down. The
problem of the difference between the ontic and the historiographical can be worked out as a problem to be investigated only if it has
made sure of its guideline beforehand by clarifying, through fundamental
ontology, the question of the meaning of being in general.[15] Thus it   *404*
becomes clear in what sense the preparatory existential and temporal
analytic of Dasein is resolved to cultivate the spirit of Count Yorck in
the service of Dilthey's work.

---

15. Cf. §§ 5 and 6.

CHAPTER SIX

Temporality and Within-Timeness as the Origin of the Vulgar Concept of Time

§ 78. *The Incompleteness of the Foregoing Temporal Analysis of Dasein*

To demonstrate that and how temporality constitutes the being of Dasein, we showed that historicity, as the constitution of being of existence, is "at bottom" temporality. Our interpretation of the temporal nature of history was carried out without regard to the "fact" that every occurrence runs its course "in time." In the course of the existential and temporal analysis of historicity, there was no room for the everyday understanding of Dasein that factically knows all history only as an occurrence "within time." If the existential analytic is to make Dasein ontologically transparent in its very facticity, the factical, "ontic-temporal" interpretation of history must also *explicitly* be given its due. It is all the more necessary that the time "in which" beings are encountered be given a *fundamental* analysis, since not only history, but natural processes, too, are determined "by time." However, more elemental than the circumstance that the "time factor" occurs in the *sciences* of history and nature, is the fact that, before all thematic investigation, Dasein "reckons with time" and orients itself *according to it*. And here again *the* "reckoning" of Dasein "with its time" remains decisive, the reckoning that precedes any use of instruments that are geared to determining time. This reckoning is prior to such instruments, and first makes possible something like the use of clocks.

Factically existing, actual Dasein either "has time" or it "has none." It either "takes time" or "cannot take time." Why does Dasein take "time" and why can it "lose" it? From where does it take time? How is this time related to the temporality of Dasein?

Factical Dasein takes account of time without existentially understanding temporality. The elemental mode of behavior of reckoning with time must be clarified before we turn to the question of what it means that beings are "in time." All the modes of behavior of Dasein

405   are to be interpreted in terms of its being, that is, in terms of temporality. We must show how Dasein *as* temporality temporalizes a mode
of behavior that is related in *such* a way to time that it takes account
of it. Our characterization of temporality up to now is thus not only
generally incomplete, since we did not pay heed to all the dimensions
of the phenomenon, but it has fundamental gaps in it because something like world time belongs to temporality itself in the strict sense of
the existential and temporal concept of world. We wish to understand
how that is possible and why it is necessary. In this way we can throw
light on the vulgarly familiar "time" "in which" beings occur, as well
as the within-timeness of these beings.

Everyday Dasein taking time initially finds time in things at hand
and objectively present encountered within the world. It understands
time thus "experienced" in the horizon of the understanding of being
that is closest to it, that is, as something that is itself somehow present.
We must clarify how and why the development of the vulgar concept
of time comes about in terms of the temporally grounded constitution
of being of Dasein taking care of time. The vulgar concept of time owes
its provenance to a leveling down of primordial time. By demonstrating
that this is the source of the vulgar concept of time, we shall justify our
earlier interpretation of temporality as *primordial time.*

In the development of the vulgar concept of time, there is a remarkable vacillation as to whether a "subjective" or an "objective" character
should be attributed to time. When one conceives it as being in itself, it
is attributed primarily to the "soul." And when it has the character of
"belonging to consciousness" it still functions "objectively." In Hegel's
interpretation of time both possibilities are in a way elevated to a higher
unity. Hegel attempts to determine the connection between "time" and
"spirit" in order to make it intelligible why spirit, as history, "falls into
time." In its *results*, the foregoing interpretation of the temporality of
Dasein and the way world time belongs to it seems to agree with Hegel.
But since our analysis of time is already distinguished from the outset
in principle from that of Hegel, and since its orientation is precisely the
opposite of his in that it aims at fundamental ontology, a short presentation of Hegel's interpretation of the relation between time and spirit can
help to clarify indirectly our existential and ontological interpretation
of the temporality of Dasein, world time, and the origin of the vulgar
concept of time, and it will conclude our discussion for now.

406   The question as to whether and how time has any "being"
["Sein"], why and in what sense we designate it as "existing"
["seiend"], cannot be answered until we have shown how temporality itself, in the whole of its temporalizing, makes possible something
like an understanding of being and addressing of beings. Our chapter
will be divided as follows: the temporality of Dasein and taking care

of time (§ 79); time taken care of and within-timeness (§ 80); within-
timeness and the genesis of the vulgar concept of time (§ 81); a comparison of the existential and ontological connection of temporality,
Dasein and world time with Hegel's interpretation of the relation of
time and spirit (§ 82); the existential and temporal analytic of Dasein
and the fundamental ontological question of the meaning of being in
general (§ 83).

§ 79. *The Temporality of Dasein and Taking Care of Time*

Dasein exists as a being that, in its being, is concerned *about* that being
itself. Essentially ahead of itself, it has projected itself upon its potentiality-
of-being *before* going on to any mere consideration of itself. In its project
it is revealed as something thrown. Thrown and abandoned to the world,
it falls prey to it in taking care of it. As care, that is, as existing in the
unity of the entangled, thrown project, this being [Seiende] is disclosed
as there. Being-together with others [Mitseiend mit Anderen], it keeps
itself in an average interpretedness that is articulated in discourse and
expressed in language. Being-in-the-world has always already expressed
itself, and *as being-together-with* beings encountered within the world, it
constantly expresses *itself* in addressing and talking over what is taken
care of. The circumspect taking care of common sense is grounded in
temporality, in the mode of making present that awaits and retains. As
taking care in calculating, planning, preparing ahead, and preventing, it
always already says, whether audibly or not: "then" . . . that will happen,
"*before*" . . . that will get settled, "*now*" . . . that will be made up for, that
which "*on that former occasion*" failed or eluded us.

   In the "then," taking care expresses itself in awaiting, retaining
in the "on that former occasion," and making present in the "now."
In the "then"—but for the most part inexplicitly—lies the "now not
yet," that is, it is spoken in a making present that awaitingly retains or
forgets. The "on that former occasion" contains the "now no longer."
With it, retaining expresses itself as a making present that awaits. The
"then" and the "on that former occasion" are understood with regard
to a "now," that is, making present has a peculiar weight. Indeed, it   407
always temporalizes itself in a unity with awaiting and retaining, even
if these are modified into a forgetting that does not await. In this mode
of forgetting, temporality gets caught in the present that primarily says
"now-now" in making present. What taking care awaits as what is
nearest to it is addressed in the "right away"; what has been made
initially available or lost is addressed in the "just now." The horizon
of retaining that expresses itself in the "on that former occasion" is the
"*earlier*," the horizon for the "then" is the "*later on*" ("in the future"),
the horizon for the "now" is the "*today*."

But every "then" is, *as such,* a "then, when . . . ,"; every "on that
former occasion" is an "on that former occasion when . . ." every "now"
is a "now that . . .". We shall call this seemingly self-evident relational
structure of the "now," "on that former occasion," and "then" *datability.*
We completely leave aside the question whether this datability is factically carried out with regard to a "date" on the calendar. Even without
such "dates," the "now" and "then" and "on that former occasion" are
more or less dated in a definite way. If the dating is not made more
definite, that does not mean that the structure of datability is lacking
or is a matter of chance.

*What is that to which such datability essentially belongs and what is dat-
ability based upon?* But can a more superfluous question be raised than
this one? With the "now that . . .", we mean, after all, a "point in time,"
"as one knows." The "now" is time. Undeniably we understand the
"now . . . that," "then . . . when," "on that former occasion . . . when" in
a way as also being connected with "time." That all this means "time"
itself, how that is possible, and what "time" signifies—these are matters of which we have no conception in our "natural" understanding of
the "now" and so on. Is it then self-evident that we "right away understand" and "naturally" express something like "now," "then," and "on
that former occasion"? Where do we get this "now . . . that . . ."? Did
we find something like this among innerworldly beings, among those
that are objectively present? Obviously not. Have we found it at all?
Have we ever started to look for it and ascertain it? It is "always"
available to us without our ever having explicitly taken it over, and
we make constant use of it, although not always in verbal expression.
Even in the most trivial, offhand kind of everyday talk (for example,
"it is cold") we also have in mind a "now that . . .". Why does Dasein
express a "now that . . .", "then when . . .", "on that former occasion
when . . ." in addressing what it takes care of, although mostly without
verbalizing it? First, because in addressing itself to something inter-
pretively, it expresses *itself* too; that is, it expresses its circumspect and
408   understanding *being together with* things at hand that lets them be discovered and encountered. And secondly because this addressing and
discussing that also interprets *itself* is grounded in a *making present,*
and is possible only as this.[1]

The making present that awaits and retains interprets *itself.*
And that is again possible only because, in itself ecstatically open,
it is always already disclosed to itself and can be articulated in the
interpretation that understands and speaks. *Since temporality is ecstatically and horizonally constitutive of the clearedness of the there, it is already*

---

1. Cf. § 33.

always interpreted primordially in the there and is thus familiar. The making present that interprets itself, that is, what has been interpreted and addressed in the "now," is what we call "time." What is made known in this is simply that temporality, recognizable as ecstatically open, is initially and for the most part known only in this interpretedness that takes care. But while time is "immediately" intelligible and recognizable, this does not, however, preclude the possibility that primordial temporality as such, as well as the origin of expressed time temporalizing itself in it, may remain unknown and unconceived.

The fact that the structure of datability belongs essentially to what is interpreted with the "now," "then," and "on that former occasion" becomes the most elemental proof that what has been interpreted originates from temporality interpreting itself. Saying "now," we always already also understand a "now that . . ." without actually saying it. Why? Because the "now" interprets a *making present of* beings. In that "now that . . ." lies the *ecstatic* nature of the present. The *datability* of the "now," "then," and "on that former occasion" is the *reflex* [*Widerschein*] of the *ecstatic* constitution of temporality, and is *thus* essential for time itself that has been expressed. The structure of the datability of the "now," "then," and "on that former occasion" is evidence for the fact that they *stem from temporality and are themselves time.* The interpretive expression of "now," "then," and "on that former occasion" is the most primordial* way of *giving the time.* In the *ecstatic* unity of temporality that is understood along with datability, but unthematically and unrecognizable as such, Dasein has always already been disclosed to itself as being-in-the-world, and innerworldly beings have been discovered along with it; because of this, interpreted time too always already has a date-stamp [Datierung] on the basis of the beings encountered in the disclosedness of the there; now that . . . the door slams; now that . . . my book is missing, etc.

Because they have the same origin from *ecstatic* temporality, the horizons that belong to the "now," "then," and "on that former occasion" also have the character of datability as "today when . . .", "later on when . . . ", and "earlier when . . .".   *409*

If awaiting, understanding itself in the "then," interprets itself and in so doing, as a making present, understands what it is awaiting in terms of its "now," then the "and now not yet" already lies in the "specification" of the "then." The awaiting that makes present understands the "until then." Interpretation articulates this "until then"—namely, "it has its time"—as the *in-between* that also has a relation of datability. This relation is expressed in the "meanwhile." Taking care can again articulate in awaiting the "during" itself by specifying

---

* nearest

further "thens." The "until then" is subdivided by a number of "from
then . . . until thens" that, however, have been "embraced" beforehand
in the awaiting project of the primary "then." The "lasting" is articulated in the understanding of the "during" that awaits and makes present. This duration is again the time revealed in the *self*-interpretation
of temporality, a time that is thus actually, but unthematically, understood in taking care as a "span." The making present that awaits and
retains interprets a "during" with a "span," only because in so doing
it is disclosed to *itself* as being ecstatically *stretched along* in historical
temporality, even though it does not know itself as this. But here a
further peculiarity of time which has been "specified" shows itself. Not
only does the "during" have a span, but every "now," "then," and "on
that former occasion" is always spanned with the structure oí datability, with a changing span: "now" in the intermission, at dinner, in the
evening, in summer; "then" at breakfast, while climbing, and so on.

The taking care that awaits, retains, and makes present, "allows
itself" time in this or that way and gives this time to itself in taking
care, even without determining the time by any specific reckoning,
and before any such reckoning has been done. Here time dates itself
in one's actual mode of allowing oneself time heedfully; it does this
on the basis of what is actually disclosed in what is taken care of in
the surrounding world and in attuned understanding, that is, on the
basis of what one does "all day long." The more Dasein is absorbed
in awaiting what is taken care of and, not awaiting itself, forgets itself,
the more its time that it "allows" itself is *covered over* by this mode of
"allowing." In the everyday "just passing through life" that takes care,
Dasein never understands itself as running along in a continuously
enduring succession of sheer "nows." On the basis of this covering
over, the time that Dasein allows itself has gaps in it, so to speak. We
410   often cannot reconstruct a "day" when we come back to the time that
we have "used." Yet the time that has gaps in it does not go to pieces
in this lack of togetherness; it is rather a mode of temporality that is
always already disclosed and ecstatically *stretched along*. The mode in
which time that is "allowed" "elapses," and the way in which taking
care gives that time to itself more or less explicitly, can be phenomenally
explicated appropriately only if, on the one hand, we avoid the theoretical "representation" of a continuous stream of nows, and if, on the
other hand, the possible modes in which Dasein gives and allows itself
time are to be conceived of as primarily determined in terms of *how it
"has" its time in a manner corresponding to its actual existence.*

In an earlier passage, authentic and inauthentic existing were
characterized with regard to the modes of the temporalizing of temporality upon which such existing is founded. Accordingly, the irresoluteness of inauthentic existence temporalizes itself in the mode of a

making present that does not await but forgets. The irresolute person understands himself in terms of the events and accidents nearest by that are encountered in such making present and urge themselves upon him in changing ways. Busily losing *himself* in what is taken care of, the irresolute person *loses his time* in them, too. Hence his characteristic way of talking: "I have no time." Just as the person who exists inauthentically constantly loses time and never "has" any, it is the distinction of the temporality of authentic existence that in resoluteness it never loses time and "always has time." For the temporality of resoluteness has, in regard to its present, the character of the *Moment*. The Moment's authentic making present of the situation does not itself take the lead, but is *maintained* in the future that has-been. Existence defined by the Moment [augenblickliche Existenz] temporalizes itself as fatefully whole, stretching along in the sense of the authentic, historical *constancy* of the self. This kind of temporal existence "constantly" has its time *for* that which the situation requires of it. But resoluteness discloses the there in this way only as situation. Thus the resolute person can never encounter what is disclosed in such a way that he could lose his time on it in an irresolute way.

Factically thrown Dasein can "take" and lose time for itself only because a "time" is allotted to it as temporality ecstatically stretched along with the disclosedness of the there grounded in that temporality.

As disclosed, Dasein exists factically in the mode of *being-with* with the others. It keeps itself in a public, average intelligibility. The "now that...", "then when..." interpreted and expressed in everyday being-with-one-another, are understood in principle, although they are unequivocally dated only within limits. In the "nearest" being-with-one-another, several people can say *"now"* together, and each can date the "now" in a different way: now that this or that happens. The "now" expressed is spoken by each one in the publicness of being-with-one-another-in-the-world. The time interpreted and expressed by any particular Dasein is thus also always already *made public* as such on the basis of its ecstatic being-in-the-world. Since everyday taking care understands itself in terms of the "world" taken care of, it knows the "time" that it takes for itself *not as its own*, but rather heedfully *exploits* the time that "there is," the time with which the *they* reckons. But the publicness of "time" is all the more compelling the more factical Dasein *explicitly takes care* of time by expressly taking it into account.

411

§ 80. *Time Taken Care of and Within-Timeness*

Thus far we have only had to understand how Dasein, grounded in temporality, takes care of time in existing, and how time makes itself public for being-in-the-world in the taking care that interprets. We did not determine at all in what sense the public time expressed "*is*," or whether it can be addressed as *existing* [*seiend*] at all. Before any decision as to whether public time is "merely subjective" or whether it is "objectively real," or neither of the two, the phenomenal character of public time must first be determined more precisely.

Making time public does not occur occasionally and subsequently. Rather, since as ecstatic and temporal Dasein *is* always already disclosed, and because understanding and interpretation belong to existence, time has also already made itself public in taking care. One orients oneself *according to it*, so that it must somehow be available for everyone.

Although taking care of time can be carried out in the mode of dating that we characterized as based on events in the surrounding world, this always occurs in the horizon of a taking care of time that we know as astronomical and calendrical *time-reckoning*. This reckoning is not a matter of chance, but has its existential and ontological necessity in the fundamental constitution of Dasein as care. Since Dasein essentially exists as thrown and entangled, it interprets its time heedfully by way of a reckoning with time. *In this reckoning*, the "real" *making public* of time temporalizes itself so that we must say that *the thrownness of Dasein is the reason "there is" public time.* If we are to demonstrate that public time has its origin in factical temporality, and if we are to assure ourselves that this demonstration is as intelligible as possible, we must characterize beforehand the time interpreted in general in the temporality of taking care, if only to make clear that the essence of taking care of time does *not* lie in the application of numerical procedures in dating. What is existentially and ontologically decisive about *reckoning* with time must not be seen in the quantification of time, but must be more primordially conceived in terms of the temporality of Dasein reckoning with time.

"Public time" turns out to be *the* time "in which" innerworldly things at hand and objectively present are encountered. This requires that we call these beings unlike Dasein beings *within-time*. The interpretation of within-timeness gives us a more primordial insight into the essence of "public time" and at the same time makes it possible to define its "being."

The being of Dasein is care. This being exists entangled as thrown. Delivered over to the "world" discovered with its factical there and dependent upon it in taking care, Dasein awaits its potentiality-of-being-in-the-world in such a way that it reckons *with* and *on* whatever is in eminent *relevance* for the sake of its potentiality-of-being. Everyday *circumspect* being-in-the-world needs the *possibility of sight*, that is, brightness, if it is to take care of things at hand within what is present. With the factical disclosedness of world, nature has been discovered for Dasein. In its thrownness Dasein is subject to the changes of day

412

and night. Day with its brightness gives it the possibility of sight, night takes it away.

Awaiting the possibility of sight while circumspectly taking care, Dasein, understanding itself in terms of its daily work, gives itself its time with the "then when it dawns." The "then" taken care of is dated in terms of what is in the proximate context of relevance of the surrounding world, with getting light, the sunrise. Then, when the sun rises, it is *time for*. . . . Thus Dasein dates the time that it must take for itself in terms of what is encountered within the world and in the horizon of being delivered over to the world as something for which there is an eminent relevance for the circumspect potentiality-of-being-in-the-world. Taking care makes use of the "handiness" of the sun giving forth light and warmth. The sun dates the time interpreted in taking care. From this dating arises the "most natural" measure of time, the day. And since the temporality of Dasein that must take its time is finite, its days are also already numbered. "While it is day" gives to awaiting that takes care the possibility of determining in a precautionary way the "thens" of what is to be taken care of, that is, of dividing up the day. This dividing up in its turn is carried out with regard to what dates time: the moving sun. Like sunrise, sunset and noon are distinctive "places" that this heavenly body occupies. Dasein, thrown into the world, temporalizing, and giving itself time, takes account of its regular recurring passage. The occurrence of Dasein is a *daily one* by reason of interpreting time by dating it—a way that is prefigured in its thrownness into the there.

This dating of things in terms of the heavenly body giving forth light and warmth, and in terms of its distinctive "places" in the sky, is a way of giving time which can be done in our being-with-one-another "under the same sky," and which can be done for "everyone" at any time in the same way so that within certain limits everyone is initially agreed upon it. That which dates is available in the surrounding world and yet not restricted to the actual world of useful things taken care of. Rather, in that world, the natural environment [Umweltnatur] and the public surrounding world [öffentliche Umwelt] are always discovered along with it.[2] At the same time everyone can "count on" this public dating in which everyone gives himself his time. It makes use of a *measure* that is available to the public. This dating reckons with time in the sense of *time measurement* that needs something to measure time, that is, a clock. *This means that with the temporality of Dasein as thrown, delivered over to the world, and giving itself time, something like a "clock" is also discovered, that is, a handy thing that has become accessible in its regular recurrence in a making present that awaits.* Thrown being-together-with things at hand is grounded in temporality. Temporality is the reason for the clock. As the condition of the possibility of the factical necessity of

413

---

2. Cf. § 15.

the clock, temporality is at the same time the condition for its discoverability. For while the course of the sun is encountered along with the discoveredness of innerworldly beings, it is only by making it present in awaitingly retaining, and by doing so in a way that interprets itself, that dating in terms of things at hand in the public surrounding world is made possible and is also required.

414   The "natural" clock, which is always already discovered with the factical thrownness of Dasein grounded in temporality, first motivates and at the same time makes possible the production and use of still more handy clocks. It does this in such a way that these "artificial" clocks must be "adjusted" to the "natural" one if they are to make the time primarily discovered in the natural clock accessible in its turn.

Before we characterize the main traits in the development of reckoning with time and the use of clocks in their existential and ontological meaning, we first want to characterize more completely the time taken care of in the measurement of time. If the time we take care of is "really" made public only when it gets measured, then public time is to be accessible in a way that has been phenomenally unveiled. We must have access to it by following up the way that which has been dated shows itself when dated in this "reckoning" way.

When the "then" that interprets itself in heedful awaiting gets dated, this dating includes some such statement as: then—when it dawns—it is *time for* the day's work. The time interpreted in taking care is always already understood as time for. . . . The actual "now that so and so" is as *such* either *appropriate* or *inappropriate*. The "now"—and thus every mode of interpreted time—is not only a "now that . . ." that is essentially datable, but is at the same time essentially determined by the structure of appropriateness or inappropriateness. Interpreted time has by its very nature the character of "time for . . ." or "not the time for . . .". The making present that awaits and retains in taking care understands time in its relation to a what-for, that is in turn ultimately anchored in a for-the-sake-of-which of the potentiality-of-being of Dasein. With this relation of in-order-to, time made public reveals *the* structure that we got to know earlier[3] as *significance*. It constitutes the worldliness of the world. As time-for . . . , the time that has been made public essentially has the character of world. Thus we shall call the time making itself public in the temporalizing of temporality *world time*. And we shall designate it thus not because it is *present* as an *innerworldly* being (that it can never be), but because it belongs *to the world* in the existentially and ontologically interpreted sense. How the essential relations of the structure of the world (for example, the *"in-order-to"*) are connected with public time (for example, the *"then-when"*) on the basis of the ecstatic and horizonal constitution of temporality must be

---

3. Cf. §§ 18 and 69c.

shown in what follows. At any rate, only now can time taken care of
be completely characterized as to its structure: it is datable, spanned,
and public and, as having this structure, it belongs to the world itself.
Every "now," for example, that is expressed in a natural, everyday way,
has this structure and is understood as such—even if unthematically   *415*
and preconceptually—when Dasein allows itself time in taking care.

The disclosedness of the natural clock belongs to Dasein that
exists as thrown and entangled, and in this disclosedness factical
Dasein has at the same time already brought about an eminent making
public of the time taken care of. This making public is enhanced and
strengthened as time reckoning is perfected and the use of the clock
becomes more refined. The historical development of time reckoning
and the use of the clock is not to be presented here historiographically
with all its possible variations. Rather, we want to ask existentially and
ontologically what mode of temporalizing of the temporality of Dasein
is made manifest in the *direction* that reckoning with time and the use
of the clock have developed. The answer to this question must further
a more primordial understanding of the fact that *time measured*—that is,
at the same time the explicit making public of time taken care of—is
grounded *in the temporality* of Dasein and indeed in a quite definite
temporalizing of that temporality.

If we compare "primitive" Dasein, which we used as the basis
for the analysis of "natural" time-reckoning, with more "advanced"
Dasein, we find that for more advanced Dasein, day and the presence
of sunlight no longer possess an eminent function because this Dasein
has the "advantage" of even being able to turn the night into day.
Similarly, we no longer need to glance explicitly and directly at the
sun and its position to ascertain the time. The manufacture and use
of one's own measuring instruments permits us to read off the time
directly by clocks explicitly produced for this purpose. The what hour
is it?, is the "what time is it?" Because the clock—in the sense of what
makes possible a public reckoning of time—must be regulated by the
"natural" clock, even the use of clocks is grounded in the temporality
of Dasein that, with the disclosedness of the there, first makes possible
a dating of the time taken care of. This is a fact, even if it is covered
over when the time is read off. Our understanding of the *natural* clock
that develops with the progressive discovery of *nature* directs us to new
possibilities for time measurement that are relatively independent of
the day and of any explicit observation of the sky.

But in a way even "primitive" Dasein makes itself independent
of reading off the time directly from the sky, since it does not ascertain the position of the sun in the sky, but measures the shadow cast
by some being available at any time. That can happen at first in the   *416*
most simple form of the ancient "peasant's clock." In the shadow that
constantly accompanies everyone, we encounter the sun with respect

to its changing presence at different places. The various lengths of the shadow during the day can be paced off "at any time." Even if the lengths of the individual's body and feet are different, still the *relation* of the two remain constant within certain limits. Thus, for example, when one takes care of making an appointment, one designates the time publicly by saying: "When the shadow is so many footsteps long, then we will meet each other over there." Here in being-with-one-another in the more narrow limits of a surrounding world nearest to us, we tacitly presuppose that the "locations" at which the shadow is paced off are at the same latitude. Dasein does not even need to wear this clock, in a certain sense it is this clock itself.

The public sundial, in which the line of a shadow is counterposed to the course of the sun and moves along a graduated dial, needs no further description. But why do we find something like time at the position that the shadow occupies on the dial? Neither the shadow nor the graduated dial is time itself, nor is the spatial relation between them. Where, then, is the time that we read off directly not only on the "sundial" but also on every pocketwatch?

What does reading off the time signify? "To look at the clock" cannot simply mean to contemplate the tool at hand in its changes and to follow the positions of the pointer. Ascertaining what time it is in using the clock, *we say*, whether explicitly or not, *now* it is such an hour and so many minutes, *now* it is time to . . . , or there is still time . . . , namely *now* until. . . . Looking at the clock is grounded in and guided by a taking-time-for-oneself. What already showed itself in the most elemental reckoning of time becomes clearer here: looking at the clock and orienting oneself *according to time* is essentially a *now-saying*. Here the now is always already understood and *interpreted* in its complete structural content of datability, spannedness, publicness, and worldliness. This is so "obvious" that we do not take any notice of it at all; still less do we know anything about it explicitly.

But now-saying is the discoursing articulation of a *making present* that temporalizes itself in unity with an awaiting that retains. The dating carried out in the use of the clock turns out to be the eminent making present of something present. Dating does not simply take up a relation with something present, but taking up a relation itself has the character of *measuring*. Of course, the number of the measurement can be read off directly. However, this means that when a length is to be measured, we understand that our standard is contained in it, that is, we determine the frequency of its *presence* [*Anwesenheit*] in that length. Measuring is constituted temporally when a present standard is made present in a present length. The idea of a standard implies unchangingness; this means that it must be present [vorhanden] in its constancy at every time for everyone. When the time taken care of is

417

dated by *measuring*, one interprets that time by looking at something present [Vorhandenes] and making it present—something that would not become accessible as a standard or as something measured except by our eminent making present. Since the making present of what is present [das Gegenwärtigen von Anwesendem] has special priority in the dating that measures, when one measures and reads off the time by the clock, one also expresses the now with special emphasis. Thus in *measuring time*, time gets *made public* in such a way that it is encountered in each case and at each time for everyone as "now and now and now." This time "universally" accessible in clocks is found as an *objectively present multiplicity of nows*, so to speak, though time measurement is not directed thematically toward time as such.

The temporality of factical being-in-the-world is what primordially makes the opening up of space [Raumerschließung] possible; and spatial Dasein has always been referred to a here of the character of Dasein out of an over there that has been discovered. Because of all this, the time taken care of in the temporality of Dasein is always bound up with some location of Dasein with regard to its datability. It is not that time is tied to a location, but rather temporality is the condition of the possibility that dating may be bound up with the spatially-local in such a way that the latter is binding for everyone as a measure. Time is not first coupled with space, but the "space" that is supposedly to be coupled with it is encountered only on the basis of temporality taking care of time. Inasmuch as both the clock and time-reckoning are grounded in the *temporality* of Dasein, which constitutes this being as historical, we can show ontologically how the use of the clock is itself historical and how every clock as such "has a history."[4]

The time made public in our measurement of it by no means   *418* turns into space because we date it in terms of spatial relations of measurement. Nor is what is existentially and ontologically essential in time *measurement* to be sought in the fact that dated "time" is determined numerically in terms of *spatial* distances and changes in the *location* of some spatial thing. Rather, what is ontologically decisive lies in the specific *making present* that makes measurement possible. Dating in terms of what is present "spatially" is so far from a spatialization of time that this supposed spatialization signifies nothing other than that a being that is present [vorhandenen Seienden] for everyone

---

4. We shall not go into the problem of *time measurement* in the theory of relativity here. The illumination of the ontological foundations of this measurement already presupposes a clarification of world time and within-timeness on the basis of the temporality of Dasein and the explication of the existential and temporal constitution of the discovery of nature and the temporal meaning of measurement in general as well. An axiomatics of the technique of physical measurement is *based* on these investigations and can never from its perspective explicate the problem of time as such.

in every now is made present [Gegenwärtigen] in its own presence [Anwesenheit]. Measuring time is essentially such that it is necessary to say now, but in obtaining the measurement we, as it were, forget what has been measured as such so that nothing is to be found except distance and number.

The less time Dasein has to lose while taking care of time, the more "precious" time becomes and the *handier* [*handlicher*] the clock must be. Not only must time be able to be given "more precisely," but the determination of time itself should require the least time possible, though it must still agree with the time given by others.

For the time being, we only wanted to point out the general "connection" of the use of the clock with temporality that takes time for itself. Just as the concrete analysis of the astronomical time-reckoning in its full development belongs to the existential and ontological interpretation of the discovery of nature, the foundations of calendrical and historiographical "chronology" can also be set forth only in the scope of tasks of an existential analysis of historiographical cognition.[5]

419   The measurement of time brings about a making public of time, so that only in this way does what we usually call "time" become familiar. In taking care, "its time" is attributed to every thing. It "has" it, and like every innerworldly being, it can "have" it only because it is "in time" in general. The time "in which" innerworldly beings are encountered we know as world time. On the basis of the ecstatic and horizonal constitution of temporality to which it belongs, world time has *the same* transcendence as the world. With the disclosedness of world, world time is made public, so that every being-together-with *innerworldly* beings that temporally takes care understands those beings as circumspectly encountered "in time."

The time "in which" what is present moves or is at rest is *not* "*objective*," if by this is meant the being-present-in-itself [An-sich-vorhanden-sein] of beings encountered within the world. But time is

---

5. For a first attempt at the interpretation of chronological time and "historical numeration," cf. the Freiburg habilitation lecture of the author (SS 1915): "Der Zeitbegriff in der Geschichtswissenschaft," published in the *Zeitschrift für Philosophie und philosophische Kritik* vol. 161 (1916): 173ff. The connections between historical numeration, astronomically calculated world time, and the temporality and historicity of Dasein need further investigation. Cf. also G. Simmel, "Das Problem der historischen Zeit," in *Philos. Vorträge*, published by the Kantgesellschaft, no. 12 (1916). The two fundamental works on the development of historiographical chronology are: Josephus Justus Scaliger, *De emendatione temporum*, 1583, and Dionysius Petavius, S.J., *Opus de doctrina temporum*, 1627. For the ancient time reckoning, cf. G. Bilfinger, *Die antiken Stundenangaben* (1888); *Der bürgerliche Tag: Untersuchungen über den Beginn des Kalendertages im klassischen Altertum und im christlichen Mittelalter* (1888); H. Diels, *Antike Technik*, 2nd ed. (1920), pp. 155–232, "Die antike Uhr." On more recent chronology, see Fr. Rühl, *Chronologie des Mittelalters und der Neuzeit* (1897).

just as little "subjective," if we understand by that being present and occurring in a "subject." *World time is "more objective" than any possible object because, with the disclosedness of the world, it always already becomes ecstatically and horizonally "objectified" ["objiciert"] as the condition of the possibility of innerworldly beings.* Thus, contrary to Kant's opinion, world time is found *just as directly* in what is physical as in what is psychical, and not just by way of a detour through the psychical. Initially "time" shows itself in the sky, that is, precisely where one finds it in the natural orientation *toward it*, so that "time" is even identified with the sky.

But world time is also "more subjective" than any possible subject because it first makes possible the being [Sein] of the factically existing self, that being [Sein] which, as is now well understood, is the meaning of care. "Time" is present neither in the "subject" nor in the "object," neither "inside" nor "outside," and it "is" "prior" to every subjectivity and objectivity, because it presents the condition of the very possibility of this "prior." Does it then have a "being" ["Sein"] at all? And, if not, is it then a phantom or is it "more existent" ["seiender"] than any possible being ["Sein"]? Any investigation that goes further in the direction of these questions will bump into the same "limit" that already posed itself for our provisional discussion of the connection between truth and being.[6] In whatever way, these questions are to be answered in what follows—or are to be asked primordially—we must first understand that temporality, as ecstatic and horizonal, first temporalizes something like *world* time that constitutes a within-timeness of things at hand and present. But then these beings can never be called "temporal" in the strict sense. Like every being unlike Dasein, they are atemporal, whether they occur, arise, and pass away as something real, or subsist "ideally."

420

If world time thus belongs to the temporalizing of temporality, it can neither be volatized "subjectivistically" nor be "reified" in a bad "objectification." These two possibilities can be avoided with clear insight—not just by vacillating insecurely between them—only if we understand how everyday Dasein theoretically conceives "time" in terms of its nearest understanding of time, and only if we understand how this concept of time and its dominance blocks the possibility of understanding what it means in terms of primordial time, that is, as *temporality*. Everyday taking care that gives itself time finds "time" in innerworldly beings that are encountered "in time." Thus our illumination of the genesis of the vulgar concept of time must take its point of departure from within-timeness.

---

6. Cf. § 44c.

§ 81. *Within-Timeness and the Genesis of the Vulgar Concept of Time*

How does something like "time" initially show itself for everyday, circumspect taking care? In what mode of taking care and using tools does
it become *explicitly* accessible? If time has been made public with the
disclosedness of world, if it has always already been taken care of with
the discoveredness of innerworldly beings belonging to the disclosedness of world since Dasein calculates time reckoning with *itself*, then
the mode of behavior in which "one" orients oneself explicitly *toward
time* lies in the use of the clock. The existential and temporal meaning
of the clock turns out to be making present of the moving pointer. By
following the positions of the pointer in a way that makes present, one
counts them. This making present temporalizes itself in the ecstatic unity
of a retaining that awaits. To *retain* the "on that former occasion" in
making present means that in saying-now one is open for the horizon
of the earlier, that is, the now-no-longer. To *await* the "then" in *making
present* means that in saying-now one is open for the horizon of the
later, that is, the now-not-yet. *What shows itself in this making present
is time.* Then how are we to define the *time* manifest in the horizon of
the use of the clock that is circumspect and takes time for itself in taking care? *This time is what is counted, showing itself in following, making
present, and counting the moving pointer in such a way that making present
temporalizes itself in ecstatic unity with retaining and awaiting horizonally
open according to the earlier and later.* But that is nothing more than an
existential and ontological interpretation of the definition that Aristotle
gave of time: τοῦτο γάρ ἐστιν ὁ χρόνος, ἀριθμὸς κινήσεως κατὰ τὸ
πρότερον καὶ ὕστερον. "This, namely, is time: that which is counted in
the motion encountered in the horizon of the earlier and the later."[7] As
strange as this definition may appear at first glance, it is "self-evident"
and genuinely drawn if the existential and ontological horizon from
which Aristotle took it is delimited. The origin of time thus revealed is
not a problem for Aristotle. His interpretation of time rather moves in
the direction of the "natural" understanding of being. However, since
that understanding and the being understood in it have been made a
problem in principle in our present inquiry, the Aristotelian analysis
of time can be thematically interpreted only *after* the solution to the
question of being, and indeed in such a way that that analysis gains a
fundamental significance for a positive appropriation of the critically
limited line of questioning of ancient ontology in general.[8]

All subsequent discussion of the concept of time *fundamentally*
holds itself to the Aristotelian definition, that is, it makes time thematic

421 (margin)

7. Cf. *Physics*, IV 11, 219b1 et seq.
8. Cf. § 6.

in the way that it shows itself in circumspect taking care. Time is "what is counted," that is, it is what is expressed and what is meant, although unthematically, in the making present of the *moving* pointer (or shadow). In making present what is moved in its motion, one says "now here, now here, and so on." What is counted are the nows. And they show themselves "in every now" as "right-away-no-longer-now" and "just-now-not-yet." The world time "caught sight of" in this way in the use of the clock we shall call *now-time*.

The more "naturally" the taking care of time that gives itself   422
time reckons with time, the less it dwells together with the expressed time as such. Rather, it is lost in the useful things taken care of that always have their time. The more "naturally" taking care determines and specifies time—that is, the less it is directed toward treating time as such thematically—all the more does the being-together-with what is taken care of (the being-together making present and falling prey) say unhesitatingly (whether with or without utterance): now, then, on that former occasion. And thus time shows itself for the vulgar understanding as a succession of constantly "present" ["vorhanden"] nows that pass away and arrive at the same time. Time is understood as a sequence, as the "flux" of nows, as the "course of time." *What is implied by this interpretation of world time taken care of*?

We can answer this if we go back to the *complete* essential structure of world time and compare this with that with which the vulgar understanding of time is acquainted. We set forth *datability* as the first essential factor of time taken care of. It is grounded in the ecstatic constitution of temporality. The "now" is essentially a now-that. . . . The datable now that is understood in taking care, although not grasped as such, is always appropriate or inappropriate. *Significance* belongs to the now-structure. Thus we called time taken care of *world* time. In the vulgar interpretation of time as a succession of nows, both datability and significance are *lacking*. The characterization of time as pure sequence does *not* let these two structures "appear." The vulgar interpretation of time *covers* them *over*. The ecstatic and horizonal constitution of temporality, in which the datability and significance of the now are grounded, is *leveled down* by this covering over. The nows are cut off from these relations, so to speak, and, as thus cut off, they simply range themselves along after one another so as to constitute the succession.

This covering over and leveling down of world time that is carried out by the vulgar understanding of time is no accident. Rather, precisely *because* the everyday interpretation of time keeps itself solely in the perspective of commonsense taking care, and because it understands only what "shows" itself in this horizon, these structures must escape it. What is counted in the measurement of time taken care of, the now, is also understood in taking care of things at hand and present [Vorhanden]. Since *this* taking care of time comes back to the time itself that has also been understood and "contemplates" it, it sees the nows (that are also somehow "there") in the horizon of *the* understanding of being by which this taking care is itself constantly guided.[9] The *nows* are thus in a way *also present*, that is, beings are encountered *and also* the now. Although it is not explicitly stated that the nows are present like things, still they are "seen" ontologically in the horizon of the idea of objective presence [Vorhandenheit]. The nows *pass away*, and the past ones constitute the past. The nows *arrive*, and the future ones define the "future." The vulgar interpretation of world time as now-time does not have the horizon available at all by which such things as world, significance, and datability can be made more accessible. These structures remain necessarily covered over, all the more so since the vulgar interpretation of time consolidates this covering over by the way in which it conceptually develops its characterization of time.

423

The succession of nows is interpreted as something somehow objectively present; for it itself moves "in time." We say that *in* every now it is now, *in* every now it already disappears. The now is now in *every* now, thus constantly present *as the same*, even if in every now another may be disappearing as it arrives. Yet it does show at the same time the constant presence of itself as *this* changing thing. Thus even Plato, who had this perspective of time as a succession of nows that come into being and pass away, had to call time the image of eternity: εἰκὼ δ' ἐπενόει κινητόν τινα αἰῶνος ποιῆσαι, καὶ διακοσμῶν ἅμα οὐρανὸν ποιεῖ μένοντος αἰῶνος ἐν ἑνὶ κατ' ἀριθμὸν ἰοῦσαν αἰώνιον εἰκόνα, τοῦτον ὃν δὴ χρόνον ὠνομάκαμεν.[10]

The succession of nows is uninterrupted and has no gaps. No matter how "far" we penetrate in "dividing" the now, it is still always now. One regards the continuity of time in the horizon of something indissolubly present. Ontologically oriented toward something constantly present, one either looks for the problem of the continuity of time, or one leaves the *aporia* alone. Here the specific structure of world time must remain *covered over*, since it is *spanned* together with ecstatically founded datability. The spannedness of time is not understood in terms of the horizonal *being stretched along* of the ecstatic unity of temporality that has made itself public in taking care of time. The fact that it is *always already* now in every now, no matter how

---

9. Cf. § 21.
10. Cf. *Timaeus* 37d ["But he decided to make a kind of moving image of the eternal; and while setting the heavens in order, he made an eternal image, moving according to number—an image of that eternity which abides in oneness. It is to this image that we have given the name of 'time.' "]

momentary, must be conceived in terms of what is *still* "earlier" out of which every now arises; that is, it must be conceived in terms of the ecstatically being stretched along of the temporality that is foreign to any continuity of something present, but that in turn presents the condition of the possibility of the access to something continuous and present.

424

The main thesis of the vulgar interpretation of time—namely, that time is "infinite"—reveals most penetratingly the leveling down and covering over of world time and thus of temporality in general belonging to this interpretation. Initially, time presents itself as an uninterrupted succession of nows. Every now is already either a just now or a right-away. If the characterization of time keeps primarily and exclusively *to this succession* then, in principle, no beginning and no end can be found in it as such. Every last now, *as a now*, is always *already* a right-away that is no longer, thus it is time in the sense of the no-longer-now, of the past. Every first now is always a just-now-not-yet, thus it is time in the sense of the not-yet-now, the "future." Time is thus endless "in both directions." This thesis about time is possible only on the basis of an orientation *toward a free-floating in-itself of a course of nows objectively present*, whereby the complete phenomenon of the now is covered over with regard to the datability, worldliness, spannedness, and publicness of Dasein, so that it has dwindled to an unrecognizable fragment. If "one thinks" the succession of nows "to the end" either from the perspective of objective presence or the lack of it, an end can never be found. In *this way of thinking* time through to the end, one *must* always *think* more time; from this one concludes that time *is* infinite.

But in what is this leveling down of world time and covering over of temporality grounded? In the being of Dasein itself that we interpreted as *care* in a preparatory way.[11] Thrown and entangled, Dasein is initially and for the most part lost in what it takes care of. But in this lostness, the flight of Dasein from its authentic existence that we characterized as anticipatory resoluteness makes itself known, and this is a flight that covers over. In such heedful fleeing lies the flight *from* death, that is, a looking away *from* the end of being-in-the-world.[12] This looking away from . . . is in itself a mode of the ecstatic, *futural* being *toward* the end. Looking away from finitude, the inauthentic temporality of entangled everyday Dasein must fail to recognize authentic futurality and thus temporality in general. And if the vulgar understanding of Dasein is guided by the they, then the self-forgetful "representation" of the "infinity" of public time can first anchor itself. The they never dies

---

11. Cf. § 41.
12. Cf. § 51.

because it is *unable* to die, since death is always my own and is understood authentically only in anticipatory resoluteness in an existentiell
way. The they, which never dies and misunderstands being-toward-
the-end, nonetheless interprets the flight from death in a characteristic
way. Up to the end "it always has more time." Here a way of having
time makes itself known in the sense of being able to lose it: "right
now this, then that . . .". Here it is not as if the finitude of time were
understood, but quite the opposite. Taking care is out to snatch as
much as possible from time that is still coming and "goes on." Publicly,
time is something that everyone can and does take. The leveled-down
succession of nows remains completely unrecognizable with regard to
its provenance from the temporality of individual Dasein in everyday
being-with-one-another. How is the course of "time" to be affected at
all if a human being, present "in time," no longer existed? Time goes
on as it already "was," after all, when a human being "entered life."
One knows only public time that, leveled down, belongs to everyone,
and that means to no one.

However, just as one who flees death is pursued by it even as
one evades it, and just as in turning away from it one has to see it
nonetheless, the harmless endless succession of nows that just runs on
imposes itself "on" Dasein in a remarkably enigmatic way. Why do
we say that time *passes away* when we do not emphasize *just as much*
how it comes into being? With regard to the pure succession of nows,
both could, after all, be said with equal justification. In talking about
time's *passing away*, Dasein ultimately understands more about time
than it would like to admit, that is, the *temporality* in which world time
temporalizes itself is *not completely closed off* despite all covering over.
Talking about time's passing away gives expression to the "experience" that time cannot be halted. This "experience" is again possible
only on the basis of wanting to halt time. Herein lies an inauthentic
awaiting of "moments" that already *forgets* the moments as they slip by.
The awaiting of inauthentic existence that makes present and forgets
is the condition of the possibility of the vulgar experience of time's
passing away. Since Dasein is futural in being ahead-of-itself, it must,
in awaiting, understand the succession of nows as one that *slips away*
and passes away. *Dasein knows fleeting time from the "fleeting" knowledge
of its death.* In the kind of talk that emphasizes time's passing away,
the *finite futurality* of the temporality of Dasein is publicly reflected.
And since even in the talk about time's passing away death can remain
covered over, time shows itself as a passing away "in itself."

But even in this pure succession of nows passing away in itself,
primordial time reveals itself in spite of all leveling down and covering over. The vulgar interpretation determines the flux of time as an
irreversible succession. Why can time not be reversed? Especially when

one looks exclusively at the flux of nows, it is incomprehensible in itself why the sequence of nows should not accommodate itself to the reverse direction. The impossibility of this reversal has its basis in the provenance of public time in temporality, whose temporalizing, primarily futural, "goes" ecstatically toward its end in such a way that it "is" already toward its end.

The vulgar characterization of time as an endless, irreversible succession of nows passing away arises from the temporality of entangled Dasein. *The vulgar representation of time has its natural justification.* It belongs to the everyday kind of being of Dasein and to the understanding of being initially dominant. Thus even *history* is initially and for the most part understood *publicly* as an occurrence *within time.* This interpretation of time loses its exclusive and distinctive justification only if it claims to convey the "true" concept of time and to be able to sketch out the sole possible horizon for the interpretation of time. Rather, we found that only from the temporality of Dasein and its temporalizing does it become intelligible *why and how world time belongs to it.* This interpretation of the complete structure of world time is drawn from temporality and gives us guidelines for "seeing" the covering over contained in the vulgar concept of time, and for estimating how far the ecstatic and horizonal constitution of temporality has been leveled down. This orientation toward the temporality of Dasein, however, at the same time makes it possible to demonstrate the provenance and the factical necessity of this covering over that levels down, and to examine the arguments for the vulgar theses on time.

On the other hand, from the *reverse* direction [*umgekehrt*] temporality remains *inaccessible* in the horizon of the vulgar understanding of time. Not only must now-time be oriented primarily toward temporality in the order of possible interpretation, but it temporalizes itself only in the inauthentic temporality of Dasein; so if we pay attention to the derivation of now-time from temporality we are justified in addressing temporality as *primordial time.*

Ecstatic and horizonal temporality temporalizes itself *primarily* out of the *future.* However, the vulgar understanding of time sees the fundamental phenomenon of time in the *now,* and indeed in the sheer now, cut off in its complete structure, that is called the "present." One can gather from this that there is in principle no prospect of explaining or even deriving the ecstatic and horizonal phenomenon of the *Moment* that belongs to authentic temporality *from this now.* Thus the ecstatically understood future—the datable, significant "then"—does not coincide with the vulgar concept of the "future" in the sense of the sheer nows that have not yet arrived and are only arriving. Nor does the ecstatic having-been, the datable, significant "on that former occasion," coincide with the concept of the past in the sense of the past

427

sheer nows. The now is not pregnant with the not-yet-now, but rather the present arises from the future in the primordial, ecstatic unity of the temporalizing of temporality.[13]

Although, initially and for the most part, the vulgar experience of time knows only "world time," it nonetheless also always accords world time an *eminent* relation to "soul" and "spirit." And it does this even when an explicit and primary orientation toward philosophical questioning of the "subject" is absent. Two characteristic passages will suffice as evidence for this. Aristotle says: εἰ δὲ μηδὲν ἄλλο πέφυκεν ἀριθμεῖν ἢ ψυχὴ καὶ ψυχῆς νοῦς, ἀδύνατον εἶναι χρόνον ψυχῆς μὴ οὔσης.[14] And Augustine writes: *inde mihi visum est, nihil esse aliud tempus quam distentionem; sed cuius rei nescio; et mirum si non ipsius animi.*[15] Thus in principle the interpretation of Dasein as temporality does not lie beyond the horizon of the vulgar concept of time. And Hegel made an explicit attempt to point out the way in which time, understood in the vulgar sense, is connected with spirit. For Kant, on the other hand, time is indeed "subjective," but stands unconnected "next to" the "I think."[16] The grounds that Hegel explicitly gave for the connection between time and spirit are well suited for clarifying indirectly the foregoing interpretation of Dasein as temporality and our exhibition of the origin of world time from it.

428

§ 82. *The Contrast of the Existential and Ontological Connection of Temporality, Dasein, and World Time with Hegel's Conception of the Relation between Time and Spirit*

History, which is essentially the history of spirit, runs its course "in time." Thus "the development of history falls into time."[17] But Hegel is not satisfied with establishing the within-timeness of spirit as a fact, rather he attempts to understand how it is *possible* for spirit to fall into

---

13. We do not need to discuss in detail the fact that the traditional concept of eternity in the significance of the "standing now" (*nunc stans*) is drawn from the vulgar understanding of time and defined in orientation toward the idea of "constant" objective presence. If the eternity of *God* could be philosophically "constructed," it could be understood only as more primordial and "infinite" temporality. Whether or not the *via negationis et eminentiae* could offer a possible way remains an open question.

14. *Physics*, 14, 223a25; cf. 11, 218b29–219a1, 219a4–6. ["But if nothing other than the soul or the soul's mind were naturally equipped for numbering, then if there were no soul, time would be impossible."]

15. *Confessions* XI.26. ["Hence it seemed to me that time is nothing else than an extendedness; but of what sort of thing it is an extendedness I do not know; and it would be surprising if it were not an extendedness of the soul itself."]

16. On the other hand, how a more radical understanding of time emerges in Kant than in Hegel will be shown in the first section of the second part of this treatise.

17. Hegel, *Die Vernunft in der Geschichte. Einleitung in die Philosophie der Weltgeschichte*, ed. G. Lasson (1917), p. 133.

time, which is the "completely abstract, the sensuous."[18] Time must be able to receive spirit, as it were. And spirit must in turn be related to time and its essence. Thus we must discuss two things: (1) how does Hegel define the essence of time? (2) what belongs to the essence of spirit that makes it possible for it to "fall into time"? Our answer to these two questions will serve merely to *elucidate* our interpretation of Dasein as temporality and to do so by way of a comparison. We shall make no claim to give even a relatively complete treatment of the allied problems in Hegel, especially since we have no intention of "criticizing" Hegel. Because Hegel's concept of time presents the most radical way in which the vulgar understanding of time has been given form conceptually, and one that has received too little attention, the contrast of this concept with the idea of temporality that we have expounded is one that especially suggests itself.

(a) Hegel's Concept of Time

The "systematic place" in which a philosophical interpretation of time is carried out can serve as a criterion for understanding the basic guideline which leads the fundamental conception of time. The first traditional, thematically detailed interpretation of the vulgar understanding of time is to be found in Aristotle's *Physics*, in the context of an ontology of *nature*. "Time" is connected with "location" and "motion." True to the tradition, Hegel's analysis of time has its place in the second part of his *Encyclopedia of the Philosophical Sciences* bearing the title: "The Philosophy of Nature." The first division treats mechanics. Its first section is dedicated to a discussion of "space and time." They are the "abstract outside-of-one-another."[19]

429

Although Hegel puts space and time together, this does not amount simply to juxtaposing them externally: space "and also time." "Philosophy fights against this 'also'." The transition from space to time does not mean that they are treated in adjoining paragraphs, but "space itself goes over." Space "is" time, that is, time is the "truth" of space.[20] According to Hegel, if space is *thought* dialectically *in what it is*, this being of space reveals itself as time. How must space be thought?

Space is the unmediated indifference of nature's being-outside-itself.[21] That means that space is the abstract multiplicity of the points distinguishable in it. Space is not interrupted by these points, but

---

18. Ibid.
19. Cf. Hegel, *Encyklopädie der philosophischen Wissenschaften im Grundrisse*, ed. G. Bolland (Leiden, 1906), §§ 254 et seq. This edition also has the "Zusätze" from Hegel's lectures.
20. Ibid., § 257, *Zusatz*.
21. Ibid., § 254.

neither does it first arise from them by way of joining them together. Space, differentiated by the differentiable points that are themselves in space, remains undifferentiated. The differentiations themselves have the nature of what they differentiate. Nevertheless, the point is a *negation* of space in that it differentiates something in space, though in such a way that it itself remains in space as this negation (the point is, after all, space). The point does not lift itself out of space as something other than space. Space is the undifferentiated outside-one-another of the multiplicity of points. But space is not a point; it is rather, as Hegel says, "punctuality."[22] This is the basis of the statement in which Hegel thinks space in its truth, that is, as time:

430
"Negativity, which related itself as point to space and in which it develops its determinations as line and plane, is, however, in the sphere of self-externality [Außersichsein] equally *for itself* and its determinations; but, at the same time, as positing in the sphere of self-externality, it appears as indifferent to that which is quietly side by side. Thus posited for itself, it is time."[23]

If space is represented, that is, immediately intuited [unmittelbar angeschaut] in the indifferent subsistence of its distinctions, the negations are, so to speak, simply given. But this representing does not yet grasp space in its being. That is possible only in thought—as the synthesis that goes through thesis and antithesis and sublates them. Space is *thought* and thus grasped in its being only if the negations do not simply subsist in their indifference, but are sublated, that is, themselves negated. In the negation of negation (that is, punctuality) the point posits itself *for itself* and thus emerges from the indifference of subsistence. Posited for itself, it distinguishes itself from this or that point; it is *no longer* this one and *not yet* that one. In positing itself for itself, it posits the succession in which it stands, the sphere of being-outside-of-itself [Außersichsein] that is now the negated negation. The sublation of punctuality as indifference signifies that it can no longer lie quietly in the "paralyzed stillness of space." The point "rebels" against all the other points. According to Hegel, this negation of negation as punctuality is time. If this discussion has any demonstrable meaning at all, it can mean nothing other than that the positing of itself for itself of each point is a now-here, now-here, and so on. Every point "is" posited for itself as a now-point. "Thus the point has actuality in time." *That through which* the point can posit itself for itself, always as this point, is always a now. The condition of the *possibility* of the point's positing itself for itself is the now. This condition of possibility constitutes the *being* of the point, and being is at the same time having been thought [Gedachtheit]. Thus, since the pure thinking of punctuality, that is, of space, always "thinks" the now and the being-outside-itself of the nows, space "is" time. How is time itself defined?

---

22. Ibid., § 254, *Zusatz*.
23. Cf. Hegel, *Encyklopädie*, critical edition of Hoffmeister (1949), § 257.

"As the negative unity of being-outside-itself, time is similarly something absolutely abstract and ideal. It is the being that, in being, is not, and, in not being, is: it is intuited becoming. This means that the absolutely momentary distinctions that immediately sublate themselves are determined as external, but external to themselves."[24] Time reveals itself for this interpretation as "intuited becoming." According to Hegel, this signifies a transition from being to nothingness, or from nothingness to being.[25] Becoming is coming into being as well as passing away. Being, or nonbeing, "goes over." What does this mean with regard to time? The being of time is the now. But since every now either "now" is-*no*-longer, or now is-*not*-yet, it can also be grasped as nonbeing. Time is "*intuited*" becoming, that is, the transition that is not thought, but simply presents itself in the succession of nows. If the essence of time is determined as "intuited becoming," this reveals the fact that time is understood primarily in terms of the now, in the way that such a now can be found by sheer intuition.

431

We do not need any complicated discussion to make it clear that in his interpretation of time, Hegel is wholly moving in the direction of the vulgar understanding of time. Hegel's characterization of time in terms of the now presupposes that the now remains covered over and leveled down in its full structure, so that it can be intuited [angeschaut] as something present, even if it is present only "ideally."

The fact that Hegel interprets time in terms of this primary orientation toward the now that has been leveled down is proved by the following statements: "The now is enormously privileged—it 'is' nothing but the individual now, but this now that is so proudly exclusive is dissolved, diffused, and turned to dust by my expressing it."[26] "Moreover, in nature where time is now, no '*stable*' difference between those dimensions (past and future) ever comes about."[27] "In the positive sense of time one can thus say that only the present is, the before and after are not; but the concrete present is the result of the past and pregnant with the future. Thus the true present is eternity."[28]

If Hegel calls time "intuited becoming" ["angeschaute Werden"], neither coming into being nor passing away has priority in it. Nevertheless, on occasion he characterizes time as the "abstraction of consuming," and thus formulates the vulgar experience and interpretation of time in the most radical way.[29] On the other hand, when Hegel really defines time, he is consistent enough to grant no such priority to consuming

432

---

24. Ibid., § 258.
25. Cf. Hegel, *Wissenschaft der Logik*, Bk. I, Div., 1, Chap. 1 (ed. G. Lasson, 1923), pp. 66ff.
26. Cf. *Encyklopädie*, § 258, *Zusatz*.
27. Ibid., § 259.
28. Ibid., § 259, *Zusatz*.
29. Ibid., § 258, *Zusatz*.

and passing away as that to which the everyday experience of time rightly adheres; for Hegel can no more provide dialectical grounds for this priority than for the "circumstance" (that he introduces as self-evident) that precisely when the point posits itself for itself, the now turns up. So even when he characterizes time as becoming, Hegel understands this becoming in an "abstract" sense that goes beyond the representation of the "flux" of time. The most appropriate expression for Hegel's conception of time thus lies in the determination of time as the *negation of negation* (that is, of punctuality). Here the succession of nows is formalized in the most extreme sense and leveled down to an unprecedented degree.[30] It is only in terms of this formal and dia-

---

30. In terms of the priority of the now leveled down, it becomes clear that Hegel's conceptual determination of time also follows the course of the *vulgar* understanding of time, and that means at the same time the *traditional* concept of time. We can show that Hegel's concept of time is even drawn *directly* from Aristotle's *Physics*. In the *Jenenser Logik* (cf. G. Lasson's edition, 1923) that was outlined at the time of Hegel's habilitation, the analysis of time of the *Enzyklopädie* is already developed in all its essential constituents. The section on time (pp. 202ff.) reveals itself to even the crudest examination as a *paraphrase* of the *Aristotelian* treatise on time. Already in the *Jenenser Logik*, Hegel develops his conception of time in the framework of the philosophy of nature (p. 186); its first part is entitled "The System of the Sun" (p. 186). Following the conceptual determination of the ether and motion, Hegel discusses the concept of time. The analysis of space is still subordinate here. Although the dialectic already shows through, it does not have the later, rigid, schematic form, but still makes possible a more flexible understanding of the phenomena. On the way from Kant to Hegel's developed system, a decisive incursion of Aristotelian ontology and logic comes about once more. As a fact this has been long familiar. But the way, manner, and limits of this influence are just as obscure today. A *concrete*, comparative, *philosophical* interpretation of Hegel's *Jenenser Logik* and Aristotle's *Physics* and *Metaphysics* will shed new light on this. A few rough references will suffice for the above reflections.

Aristotle sees the essence of time in the νῦν; Hegel in the now. Aristotle conceives the νῦν as ὅρος; Hegel interprets the now as "limit." Aristotle understands the νῦν as στιγμή; Hegel interprets the now as point. Aristotle characterizes the νῦν as τόδε τι; Hegel calls the now the "absolute this." Aristotle connects χρόνος with σφαῖρα, in accordance with the tradition; Hegel emphasizes the "circular course" of time. Of course, Hegel misses the central tendency of Aristotle's analysis of time which exposes a foundational connection (ἀκολουθεῖν) between the νῦν, ὅρος, στιγμή, and τόδε τι.

Despite all differences in justification, Bergson's conception agrees with Hegel's thesis that space "is" time. Bergson just turns it around: Time (*temps*) is space. Bergson's interpretation of time, too, obviously grew out of an interpretation of Aristotle's treatise on time. It is not just a matter of an external literary connection that simultaneously with Bergson's *Essai sur les données immédiates de la conscience*, where the problem of *temps* and *durée* is expounded, a treatise of Bergson's appeared with the title: *Quid Aristoteles de loco senserit*. With regard to the Aristotelian definition of time as ἀριθμὸς κινήσεως, Bergson analyzes *number* before analyzing time. Time as space (cf. *Essai*, p. 69) is *quantitative* succession. Duration is described on the basis of a counter-orientation toward *this* concept of time as *qualitative* succession. This is not the place for a critical discussion of Bergson's concept of time and other present-day interpretations of time. To the extent that anything essential has been gained at all beyond Aristotle and Kant, the concern is more with grasping time and "time consciousness." By referring to the direct connection of Hegel's concept of time and Aristotle's analysis of time, we do not intend to charge Hegel with being "dependent," but to point out the *fundamental ontological import of this affiliation* for Hegel's *Logic*. On "Aristotle and Hegel," cf. the essay with this title of Nicolai Hartmann in *Beiträge zur Philosophie des deutschen Idealismus*, vol. 3 (1923), pp. 1–36.

lectical concept of time that Hegel can produce a connection between time and spirit.

(b) Hegel's Interpretation of the Connection between Time and Spirit   433

In what way has spirit itself been understood such that Hegel can say it is in accordance with spirit that it is actualized to fall into time, with time defined as the negation of a negation? The essence of spirit is the *concept*. By this Hegel understands not the universal that is intuited in a genus as the form of what is thought, but the form of the thinking that thinks itself: conceiving of *oneself—as grasping* of the non-I. Since grasping the *non*-I presents a differentiation, there lies in the pure concept, as the grasping of *this* differentiation, a differentiation of the difference. Thus Hegel can define the essence of spirit formally and apophantically as the negation of a negation. This "absolute negativity" gives a logically formalized interpretation of Descartes' *cogito me cogitare rem* in which he sees the essence of *conscientia*.

Accordingly, the concept is the self-conceiving conception of the self; as this, the self is authentic, that is *free*. "The *I* is the pure concept itself that has come to *existence* as the concept."[31] "But the I is this *first* pure unity relating itself to itself, not directly, but rather, in abstracting from all determinateness and content and going back to the freedom   434 of the limitless identity with itself."[32] Thus the I is "*universality*," but it is "individuality" *just as* immediately.

This negating of negation is both the "absolute unrest" of spirit and also its *self-revelation*, which belongs to its essence. The "progression" of spirit actualizing itself in history contains a "principle of exclusion."[33] However, in this exclusion what is excluded does not get detached from the spirit, it gets *surmounted*. Making itself free in overcoming and, at the same time, supporting, characterizes the freedom of spirit. Thus "progress" never means a quantitative more, but is essentially qualitative, and indeed has the quality of spirit. "Progression" is known and knowing itself in its goal. In every step of its "progress," spirit has to overcome "itself" as the truly inimical hindrance of its aim.[34] The goal of the development of spirit is "to attain its own concept."[35] The development of itself is "a hard, infinite struggle against itself."[36]

---

31. Cf. Hegel, *Wissenschaft der Logik*, vol. 2 (ed. Lasson, 1923), part 2, p. 220.
32. Ibid.
33. Cf. Hegel, *Die Vernunft in der Geschichte. Einleitung in die Philosophie der Weltgeschichte*, ed. G. Lasson (1917), p. 130.
34. Ibid., p. 132.
35. Ibid.
36. Ibid.

Since the restlessness of the development of *spirit* bringing itself to its concept is the *negation of a negation*, it is in accord with its self-actualization to fall "into *time*" as the immediate *negation of a negation*. For "time is the *concept* itself that *is there*, and represents itself to consciousness as empty intuition. For this reason spirit necessarily appears in time, and it appears in time so long as it has not *grasped* its pure concept, that is, has not annulled time. Time is the pure self that is *externally* intuited and *not grasped* by the self, the concept merely intuited."[37] Thus spirit, *in accordance with its essence*, necessarily appears in time. "Thus world history in general is the interpretation of spirit in time, just as the idea interprets itself in nature as space."[38] The "excluding" that belongs to the movement of development contains a relation to nonbeing. That is time, understood in terms of the self-insistence [sich aufspreizenden] of the now.

435   Time is "abstract" negativity. As "intuited becoming," it is the differentiated self-differentiation that is directly to be found, the concept that "is there," that is, present [vorhanden]. As something present and thus external to spirit, time has no power over the concept, but the concept is rather "the power of time."[39]

Hegel shows the possibility of the historical actualization of spirit "in time" by going back to *the sameness of the formal structure of spirit and time as the negation of a negation*. The most empty, formal-ontological and formal-apophantical abstraction into which spirit and time are externalized makes possible the production of a kinship of the two. But since, at the same time, time is yet conceived in the sense of world time that has been absolutely leveled down, so that its provenance thus remains completely covered over, it simply confronts spirit as something present [Vorhandenes]. For this reason spirit *must first* fall "into time." It remains obscure what indeed is signified ontologically by this "falling" and the "actualization" of spirit that has power over time and really "exists" outside of it. Just as Hegel throws little light on the origin of time that has been leveled down, he leaves totally unexamined the question of whether the essential constitution of spirit *as* the negating of negation is possible at all in any other way than on the basis of primordial temporality.

Whether Hegel's interpretation of time and spirit and their connection is correct and has an ontologically primordial basis at all cannot be discussed now. However, *that* the formal and dialectical "construction" of the connection of spirit and time can be ventured *at all* reveals the primordial kinship of both. Hegel's "construction" was prompted

---

37. Cf. *Phänomenologie des Geistes, Werke,* vol. 2, p. 604.
38. Cf. *Die Vernunft in der Geschichte,* p. 134.
39. Cf. *Encyklopädie,* § 258.

by his arduous struggle to conceive the "concretion" of spirit. The following statement from the concluding chapter of his *Phenomenology of Spirit* makes that known: "Thus time appears as the very fate and necessity of spirit when it is not in itself complete—the necessity of its giving self-consciousness a richer share in consciousness, of setting in motion the *immediacy of the in-itself* (the form in which substance is in consciousness), or conversely, of its realizing and revealing the in-itself taken as what is *inward* (and this is what is first *inward*), that is, vindicating it for the certainty of itself."[40]

Our existential analytic of Dasein, on the other hand, begins with the "concretion" of factically thrown existence itself in order to reveal temporality as what makes such existence primordially possible. "Spirit" does not first fall into time, but *exists as* the primordial *temporalizing* of temporality. Temporality temporalizes world time, in whose horizon "history" can "appear" as an occurrence within time. Spirit does not fall *into* time, but factical existence "falls," in falling prey, *out of* primordial, authentic temporality. This "falling," however, itself has its existential possibility in a mode of temporalizing that belongs to temporality.

§ 83. *The Existential and Temporal Analytic of Dasein and the Fundamental Ontological Question of the Meaning of Being in General*

The task of the foregoing considerations was to interpret the *primordial totality* of factical Dasein with regard to its possibilities of authentic and inauthentic existing, and to do so existentially and ontologically *in terms of its ground. Temporality* revealed itself as this ground and with this it revealed itself as the meaning of being of care. Thus what the *preparatory* existential analytic of Dasein contributed *prior* to setting forth temporality has now been *taken back* into temporality as the primordial structure of the totality of being of Dasein. In terms of the possible ways in which primordial time can temporalize itself, we have provided the grounds for those structures that we only "indicated" earlier. Setting forth the constitution of being of Dasein, however, still remains only *one way* that we may take. Our *goal* is to work out the question of being in general. Our *thematic* analytic of existence needs in its turn the light from a previously clarified idea of being in general. That is especially true if the statement expressed in our introduction is retained as a standard for every philosophical investigation: philosophy is universal phenomenological ontology, beginning with a hermeneutic of Dasein which, as an analytic of *existence*, has made

436

40. Cf. *Phänomenologie des Geistes*, p. 605.

fast the guideline for all philosophical questioning at the point where it *arises* and into which it is *folded back.**[41] Of course, this thesis must not be taken dogmatically, but as a formulation of the fundamental problem still "veiled": can ontology be grounded *ontologically* or does it also need for this an *ontic* foundation, and *which* being must take over the function of this foundation?

437
The distinction between the being of existing Dasein and the being of beings unlike Dasein (for example, objective presence) may seem to be illuminating, but it is only the *point of departure* for the ontological problematic; it is nothing with which philosophy can rest and be satisfied. We have long known that ancient ontology deals with "reified concepts" and that the danger exists of "reifying consciousness." But what does reifying mean? Where does it arise from? Why is being "initially" "conceived" in terms of what is objectively present, *and not* in terms of things at hand that do, after all, lie *still nearer* to us? *Why* does this reification come to dominate again and again? How is the being of "consciousness" *positively* structured so that reification remains inappropriate to it? Is the "distinction" between "consciousness" and "thing" sufficient at all for a primordial unfolding of the ontological problematic? Do the answers to these questions lie along our way? And can the answer even be *searched for* as long as the *question* of the meaning of being in general remains unasked and unclarified?

We can never inquire into the origin and the possibility of the "idea" of being in general with the means of formal and logical "abstraction," that is, not without a secure horizon for questions and answers. We must look for a *way*† to illuminate the fundamental ontological question and *follow* it. Whether that way is at all the *only* one or even the *right* one can be decided only after we have *followed* it. The *conflict* with respect to the interpretation of being cannot be settled *because it has not yet even been kindled*. In the end, one cannot just "rush into" this conflict; rather, igniting this conflict already requires a preparation. It cannot be "jumped into," but the beginning of the strife already needs preparation. This investigation is solely *underway* to that. Where does it stand?

Something like "being" ["Sein"] has been disclosed in the understanding of being that belongs to existing Dasein as a way in which it understands. The preliminary disclosure of being, although it is unconceptual, makes it possible for Dasein as existing being-in-the-world to be related *to beings*, to those it encounters in the world as well as to

---

41. Cf. § 7.

* Thus not philosophy of existence [Existenzphilosophie]
† Not "the" sole way.

itself in existing. *How is the disclosive understanding of being belonging to Dasein possible at all?* Can the question be answered by going back to the *primordial constitution of being* of Dasein that understands being? The existential and ontological constitution of the totality of Dasein is grounded in temporality. Accordingly, a primordial mode of temporalizing of ecstatic temporality itself must make the ecstatic project of being in general possible. How is this mode of the temporalizing of temporality to be interpreted? Is there a way leading from primordial *time* to the meaning of *being*? Does *time* itself reveal itself as the horizon of *being*?

LEXICON

On the Composition of the Lexicon

The following Lexicon has been composed with the help of the following sources, each of which is here gratefully acknowledged as a powerful tool, each in its own way, in the labor of the composition:

Hildegard Feick, *Index zu Heideggers 'Sein und Zeit'*. Fourth newly revised edition by Susanne Ziegler. Tübingen: Niemeyer, 1961, ⁴1991.

Rainer A. Bast/Heinrich P. Delfosse, *Handbuch zum Textstudium von Martin Heideggers 'Sein und Zeit'*. Vol. 1: *Stellenindizes; Philologisch-kritscher Apparat*. Stuttgart-Bad Cannstatt: frommann-holzboog, 1980.

John Macquarrie & Edward Robinson, translators of Martin Heidegger, *Being and Time*. New York/San Francisco: Harper-Collins, 1962. "Index of English Expressions (Latin Expressions; Greek Expressions; Proper Names)," pp. 524–589.

All three sources, including the English index by Macquarrie and Robinson, refer to the pagination of the original *German* edition of *Sein und Zeit* (Tübingen: Max Niemeyer Verlag, ⁷1953, ¹⁶1986), pp. 1–437, which since the 14ᵗʰ edition (1977) also lists Heidegger's later marginal remarks in an appendix, "Randbemerkungen aus dem Handexemplar des Autors," pp. 439–445. Following this convention, the following Lexicon also lists the pagination of the original German edition, which is to be found in the margins of the pages of this translation.

Since this translation presents Heidegger's later marginal remarks as footnotes to the appropriate pages within the body of the translation itself, these footnotes will be identified in the Lexicon by the designation "fn" preceded by the German page number, "85fn," for example (here citing the longest of the marginal remarks). Heidegger's 157 marginal remarks, ranging from a single word to a lengthy paragraph or a long list, are to be found on 101 pages of the German text of 437 pages.

Heidegger's footnotes to the original edition (1927) also appear as footnotes in this translation and are numbered within each chapter; these footnotes are accordingly identified in the Lexicon by German page number followed by the designation "n." and the number assigned to the note, "190 n. 4," for example.

I have, on occasion, italicized particularly important page numbers within the individual entry and, on rare occasion, parenthesized pivotal page numbers where the entry is clearly being alluded to without actually being named. Subentries are by and large ordered, not alphabetically, but more or less *seriatim* or "genealogically," in the order in which the idea is initially broached in the sequence of the text. For some of the entries, I have borrowed subentries extensively from the abundant Index by Macquarrie and Robinson. For others, I have relied heavily on the computer-generated list of page numbers presented by Bast/Delfosse, without however trying to be exhaustive with the highest-frequency terms, especially in their combining forms so prolific in the German language.

Since the Lexicon is not designed to serve as an index for this work, I have avoided the temptation to cite extensive "exergues" that would serve to define terms in greater depth, as is the case in Feick's "Index" and in Albert Hofstadter's lengthy Lexicon to his translation of Heidegger's *The Basic Problems of Phenomenology*. In lieu of this, I have relied both on the briefest of subentries and on the "*See also*" cross-referencing at the end of especially some of the less familiar entries, in order to suggest the relational context from which the terms spring.

I have omitted a number of minor entries provided in the English index by Macquarrie and Robinson, amplifying however its focus on ordinary expressions and phrases like "giving to understand" and "having to be," which play crucial albeit often incipient roles in Heidegger's argument, and added and/or developed some neglected terms which have come to be recognized as important to the developmental infrastructure of the published Divisions of *Being and Time*. Finally, in a few rare instances, I have made a halting beginning toward illustrating how some common idioms ("in the Light of," "with Regard to") also function in a methodological way for Heidegger in the development of his opus magnum. A shift in emphasis in entries chosen was in part also dictated by the incorporation of Heidegger's later marginal remarks into this English translation. But Feick's index in particular is devoted to relating the terms of *Being and Time* to those of Heidegger's later works, thereby turning it, as Heidegger himself remarks in praise of Feick's index, into a "Way–not a Work."

In sum, the following important conventions should be noted by users of this Lexicon:

Usage of the *German* pagination of the Niemeyer edition of *Sein und Zeit*, which is to be found in the margins of the translation

n. 1 = reference to Heidegger's original Notes, which appear in this translation as numbered footnotes

fn = reference to Heidegger's later marginal remarks, which are to be found as footnotes which are noted by symbols (*, †, ‡) on the appropriate pages within the body of the translation

—Theodore Kisiel

Lexicon of English Expressions

Free, freedom (*frei, Freiheit*): for
authenticity or inauthenticity,
188, 268fn, 191, 195, 232, 344;
for Being-guilty, 288; for the call
of conscience, 287; for care, 122;
for death, 264, 266, 384–385;
for freedom of choice, 188; for
oneself, 122; for possibilities, 191,
193, 199, 264, 285, 312, 344; for
one's ownmost potentiality-for-
Being, 144, 191; for repetition,
385. *See also* Liberate
Free (*freigeben*, verb), 83–86, 85fn,
104, 110–111, 118, 120–123, 129,
141, 144, 227, 264, 297–298, 310,
313, 343, 363
Free-floating (*freischwebend*), 9, 19, 28,
36, 123, 144, 156, 272, 276, 279,
298, 309, 325, 339, 388, 424
Fulfillment (*Erfüllung*) of intentional
project of meaning, 31, 52, 76,
97, 151, 153, 191–192, 195, 320 n.
19, 326, 343. *See also* Directing-
oneself-toward; Meaning;
Significance; the Upon-which
Function (*Funktion*), 80, 106, 157–158,
160–161, 182, 190, 202, 289–293,
310, 332, 340, 349, 368, 379, 415,
436; functional concepts, 88.
Fundamental analysis
(*Fundamentalanalyse*): of Dasein,
39, 41, 131, 181, 184, 213, 230–235;
equated with analytic of Dasein,
213; ontological f. a. of "life"
(Dilthey), 210; as preparatory, 231;
of being, 360. *See also* Existental
analytic
Fundamental ontology
(*Fundamentalontologie*), 13–14, 19,
37, 38fn, 131, 143fn, 154, 182–183,
194, 196, 200–202, 213, 232, 235fn
to n. 6, 268, 301, 310, 314, 316,
377, 403, 405–406, 436–437; related
to genealogy of being, (11). *See
also* Ontology; Phenomenology
Future, futural (*Zukunft, zukünftig*),
20, 141, 325–330, 336–348, 350,
360, 365, 378, 381, 385–387, 391,

395–397, 410, 423–427, 431.
See also Being-ahead-of-itself;
Coming toward; For-the-sake-of-
which

Genealogy (*Genealogie*) of being, 11
Generation (*Generation*), 20, 28, 377,
385, 385 n. 8. *See also* Destiny;
Hand down; Heritage, Historicity;
Tradition
Genesis (*Genesis*): ontological, 68,
357, 362, 392; existential, 171,
358; of reference, 68; of science,
171 (Aristotle), 358; of theoretical
behavior, 357–358, 360–362; of
historiography as science out of
historicity, 392; of vulgar concept
of time, 420. *See also* Origin;
Primordiality; Temporalizing
Genuine (*echt, genuin*), 95–96, 127,
165, 168–169 (understanding),
173–174, 177–178, 234, 237, 239,
279, 303, 317, 333, 372, 374, 421, *et
passim*; versus authentic, 142,
146; inauthentic as, 146, 148,
326. *See also* Authenticity;
Inauthentic
Genus (*Gattung*), 3, 14, 37fn, 38,
42, 77, 128, 433 (Hegel); generic,
399, 403; being is not a, 3, 14,
38; Dasein is not a, 42, 128;
relation, as formal, is not a, 77.
See also Jeweiligkeit; Mineness;
Transcendentals; Universal
Geographical (*geographisch*), 70, 80,
103
Geometry (*Geometrie*), 68, 112
Give to understand (*zu verstehen
geben*, 148, 267, 269–271, 279–
280, 287, 296. *See also* Call of
conscience; Givenness
Given(ness), (*Gegeben[heit]; Gebung*),
36, 68, 115–116, 129, 191, 237,
265, 271, 279–280, 284, 309, 401,
et passim; (self-)giving, 115–116,
216; giving meaning, 324–325;
"giving" of surroundings (*Gebe
der Umgebung*), 58fn. *See also* Call

Horizon: Four major ambits of the
term "Horizon" *(continued)*
2. phenomenological:
disclosed by Husserl, 51
n. 11
3. ontological: 116, 194, 289
(for interpretation), 293
(appropriate for analysis
of conscience), 320 n.
19 (of inappropriate
ontology of objective
presence), 377, 387, 422
(of idea of objective
presence)
4. phenomenal: 167, 271,
334
5. of interpretation: 168,
223, 289
6. conceptual: inappropriate
"categorial," 322; of a
clarified being of the
beings unlike Dasein, 333
See also Foundation; Ground;
Light ("in the light of");
Schema; Understanding of
being
Horizonal *(horizonal)*: leeway, 355;
unity, 360, 366, 396; schema(ta),
365; structure of having-been,
365; unity of schemata, 365;
constitution of the ecstatic unity,
365, 408, 414, 419, 422, 426;
phenomenon of the Moment, 427
"How," the *(das Wie)*, 27, 34–35,
218–219, 224, 348, 370
"How one is" *("Wie einem ist")*, 138,
188, 340. *See also* Attunement;
Facticity; It is concerned in its
being about . . . ; It "knows"
where it itself stands; Thrownness
Human sciences, humanistic
disciplines *(Geisteswissenschaften)*,
10, 38, 46, 129, 376, 397–399

I, the *(das Ich)*, 41fn (in each case
"I"), 87fn (egoistic), 109 + fn
(egotism), 114–119, 316–323, 318fn,

319fn, 332, *et passim;* as subject,
22, 179, 317, 322; as the "who"
of Dasein, 114–115, 129, 267, 313,
317, 322; abstract, 401 (Yorck);
isolated, 116, 118, 179, 298; pure,
229; worldless, 316; I-here, 119,
132; I-hood *(Ichheit)*, 116, 318,
323; I-Thing *(Ichding)*, 107, 119;
the "I am": *(das "ich bin")*, 24,
54, 129, 211, 278, 297, 317, 321,
381; I am-as-having-been: *ich bin
gewesen*, 326, 328, 339; I-am-in-
a-world, 211, 321; I take action,
319; I think, 24, 319–321, 427; the
givenness of, 46, 115–116, 129,
265; the not-I, 116, 433; saying
"I," 318–319, 318fn, 321–323;
Descartes on, 46, 95, 98; Hegel
on, 433–434; Humbolt on,
119; Kant on, 109, 318–321,
320 n. 19; the "I" and the Self,
129–130, 317–323, 348. *See also
cogito sum*; Mineness; Self;
Subject
Ideal *(Ideal, ideal)* 153 (scientific
strictness, 156 (validity),
216–217 (content of judgment),
229 (subject versus factic
"subject," Dasein), 266 (of
existence), 280 (potentiality-
of-being individualized in this
particular Dasein), 285 (versus
lack of nullity); of existence
(Existenzideal), 266, 300; factical
existentiell ontic ideal, 266,
300, 310. *See also* Existentiell,
versus existential; Ontic, versus
ontological
Idealism *(Idealismus)*, 34, 39fn, 183,
203–204, 206–208, 320 n. 20
Idle talk *(Gerede)*, 134–135, 165, 167–
170 (§ 35), 173–175, 177, 180, 222,
252–253, 255, 271, 277, 296, 346.
See also Ambiguity; Averageness;
Publicness
"If-then," the *(das "wenn-so")*, 359.
See also Schema

355. *See also* Present; Temporality, inauthentic

Making room (*Einräumen*, an existential), 111, 299, 368–369. *See also* Freeing; Letting; Spatiality of Dasein

Man, human being (*Mensch*), 51, 54, 60, 97, 120, 134fn, 187, 179, 198, 203, 246, 246fn (human life), 371, 379, 382, 396, 400–401, 425; being of, 12 fn, 25, 45, 48–49, 57fn, 87fn, 165fn; his being toward God, 10, 190 n. 4; his good, 199; his substance, 117, 212, 314; his transcendence, 49; as rational animal, 48–49, 165, 183, 197; as the being which talks, 165; as unity of body, soul, spirit, 48, 117, 198; as made in God's image, 48–49; as the 'subject' of events, 379; as an 'atom' in world-history, 382; and Dasein, 25, 46, 87fn, 134fn, 182; and the world, 57, 105, 152; and the surrounding world, 57; and the *lumen naturale*, 133; Aristotle on, 171; Calvin on, 49; Dilthey on, 398; Hyginus on, 198–199; Seneca on, 199; Zwingli on, 49

Manifestness (*Offenbarkeit*), 85fn (of beings), 134 (of burden of being), 371. *See also* Openness

Material (*Material*), 68 (materiality), 366 (formed), 394 (historical); used in a craft, 70, 73, 117. *See also* Matter; Reality; Nature; Substance

Mathematics (*Mathematik*), 9, 63, 65, 88, 95–96, 153, 362, 402; mathematical physics, 9–10, 96 + fn ("the mathematical as such"), 362

Matter (*Materie*), 10, 68, 91, 97, 362; material nature (thing), 47, 98–99, 238. *See also* Corporeal thing; Extension; Motion; Natural thing; Natural science

"Matters," esp. in "the world matters to us" (*angehen, Angänglichkeit*), 94fn, 106, 121–122, 137–139, 141, 170; *sich-angehen-lassen*, 141 (lets us be concerned), 170 (lets itself be affected). *See also* Affection; Attunement; Letting

Meaning (*Sinn*), 1, 37fn (of difference), 137, 151–153, 156, 161, 324; meaningful (*sinnvoll*), 151; meaningless (*sinnlos*), 151; unmeaning (*unsinnig*), 152; give meaning (*Sinn geben*), 324–325; have meaning (*Sinn haben*), 151, 154, 221, 324, 348, 361. *See also* Significance; Signification; the Upon-which

Meaning of being (*Seinssinn, Sinn von Sein*), 1–6, 7fn, 11, 14–15, 17–21, 24, 26–27, 37, 39, 55, 86, 93 (*Bedeutung*), 115, 123, 145, 152, 183, 194, 196, 200–201, 209, 211–213, 226, 228, 230–231, 234–235, 286, 303–304, 310, 314, 316, 323–330, 332–333, 346, 357, 370–374, 392, 397, 406, 419–420, 435–437. *See also* Question of being

Measurement (*Mass, messen*), 68, 262, 358, 417 n. 4; of space, 102–103, 105–106, 110–111, 369; of time, 71, 404, 413–415, 417–419, 417 n. 4. *See also* Calculating; Estimating

Medical (*medizinisch*), 241, 247

Medieval ontology, 3, 25, 40, 93–94

Metaphysics (*Metaphysik*), 2, 21–22, 38fn, 39, 56, 59, 85fn, 231fn, 248, 293, 401–402 (Yorck), 433 n. 30; in the marginal remarks (=fn), 38 fn, 318fn

Method, methodology (*Methode, Methodologie, Methodik*), 2, 10, 27–39 (§ 7), 49, 66–67, 131, 139, 156, 160, 182, 185, 190, 202, 205 n. 15 (Dilthey), 208, 214, 230, 248, 255 n. 13, 280, 301–305 (§ 61), 303fn, 309, 310–316 (§ 63), 324,

New, the *(continued)*
   (curiosity), 24–25, 172–174, 271,
   346, 348, 391. *See also* Ancient
   ontology
Nihilism *(Nihilismus)*, 176fn
No longer *(nicht-mehr)*, 373, 430–431
   (Hegel), *et passim*; being-able-to-
   be-there, 250; no-longer-being-
   there, 236; no longer Dasein,
   237–238, 240, 242, 330; no-longer-
   now *(nicht-mehr-jetzt)*, 327, 421,
   424; now-no-longer *(jetz nicht
   mehr)*, 380, 406, 421; no longer
   objectively present, 374, 378, 380;
   Being-no-longer-in-the-world, 176,
   238, 240
Nobody, no one *(Niemand)*, 128, 177,
   253, 255, 268, 278, 299, 425. *See
   also* Everyone; the They
Nonbeing *(Nichtsein)* and being,
   170, 243, 431, 434 (Hegel);
   being-toward-death as being of
   nonbeing, 234fn. *See also* Being
   and becoming; Transition
Not, the *(das Nicht, Nichtheit)*, 29,
   283–286; ontological origin of, 286
Not-yet *(noch-nicht)*, 145, 242–246,
   250, 259, 317, 325, 347, 373, 380,
   393–394, 427, 430–431 (Hegel),
   et passim; the "not-yet-now" *(das
   Noch-nicht-jetzt)*, 327, 421, 424,
   427; the "not-not-yet" *(das Jetzt-
   noch-nicht)*, 406, 409, 421; Being-
   not-yet *(Noch-nicht-sein)*, 237, 246;
   not yet objectively present, 144,
   237, 243, 374
Nothing(ness), the *(das Nichts)*,
   7fh, 43, 128, 177, 186–188, 266,
   273, 276–277, 308, 343, 352, 431;
   indifference not, 43; no one not,
   128, 177; and nowhere, 186, 188;
   of anxiety, 186–188, 308; of
   possible impossibility of one's
   existence, 266; of the world,
   187, 276–277, 343; self thrown
   into, 7fn, 277; being and, 431
   (Hegel)
Now *(jetzt)*, 325, 388 n. 3, 373, 378,
   406–411, 414, 416–418, 421–427,

430–432, 432 n. 30; now-here
   (Jetzt-hier), 421, 430; now-point
   (Jetzt-Punkt), 430; now that
   (jetzt, da . . .), 406–408, 410–411,
   414, 422; now-time *(Jetzt-Zeit)*,
   421, 423, 426; just now *(soeben)*,
   407, 424; multiplicity of "nows"
   (Jetztmannigfaltigkeit), 417; saying
   "now" *(Jetzt-sagen)*, 406, 408, 416,
   418, 421; succession of "nows"
   (Jetztfolge), 329, 373, 422–426, 431–
   432; stream of "nows" *(Jetztfluss)*,
   410, 436
Nowhere *(nirgends)*, 174, 177, 186–188,
   347. *See also* Curiosity; anxiety
Nullity *(Nichtigkeit*, sometimes
   "nothingness"), 23, 206, 219;
   being the ground of a, 283–
   287, 305–306, 308, 330, 348;
   "nothingness" of inauthentic
   everydayness, 178
Numbed, benumbed *(benommen)*,
   76, 271 (by everyday ambiguity),
   344 (by uncanniness); "taken in"
   by everyday world, 61, 113, 176.
   See also Absorption; Falling prey;
   Flight; Lostness
Number *(Zahl)*, 215, 412, 417–418,
   432 n. 30 (Bergson). *See also*
   Calculating

Object *(Objekt)*: and judgment, 156,
   216; and subject, 59–60, 156, 176,
   179, 192, 204, 208, 216, 219, 366,
   388; and world, 60, 179, 203, 366;
   of historiography, etc., 10, 375–
   376, 392, 397, 401
Object *(Gegenstand)*: of a statement,
   157; of taking care, 238; of
   historiography, 152, 375, 380,
   392–395; of judgment, 214,
   224, 273; of knowing, 60, 215
   (Kant), 218; of mathematics, 9;
   of phenomenology, 34–36; of
   a science, 9–10, 238, 361; to be
   disclosed, 232, 303
Objectivation *(Objektivierung)*, 48, 82,
   363, 375–376, 378, 381, 419–420.
   See also Thematization

Place (*Platz, platzieren*), 97, 102–104, 104fn, 107–108, 111–112, 361–362, 368, 413, 416. *See also* Dwelling; Location; Orientation; Position

Plunge (*Sturz*), 178. *See also* Alienation; Eddying; Falling prey; Temptation; Tranquillization

Poetry (*Dichtung*), 16, 162, 249, 260. *See also* Art; Literature

Point (*Punkt*), 105, 107, 119, 179, 362, 374, 407, 429–430, 432, *et passim*. *See also* Now; Place; Space

Point out (*Augzeigung*): first function of a statement, 154–158, 160, 218, 227–228. *See also* Apophantic "as"; Statement

Political (*politisch*), 16, 193, 400 (Yorck); occurrence-with, of a community and people (384). *See also* Destiny; Heroes; Ideal; ontic; Rhetoric

Position:
  (*Stelle*): spatial, 102–104, 107, 109–110 (order), 112 (multiplicity), 119, 362, 368–369, 420 (of a pointer); *et passim*.
  (*Stand*): of the sun, 71, 416; social class or status (standing), 239, 274; of Dasein, 253, 322 (self having gained a *stand*), 388; p. or state of "status" of a science (esp. our present problematic), 10, 55, (ontological analytic), 156 (question of being), 159, 166, 301, 323.
  (*Lage*), also translated as location: 110, 193, 226, 249, 299–300, 359, 368, 371. *See also* Circumstances; Dwelling; Location; Orientation; Place; Situation; Space (and time)

Positive (*positive*): versus privative, negative, critical, etc., 19, 52, 75, 141, 260, 279, 286, 378, *et passim*; sciences, etc. 9–11, 50–52, 58, 324, 398, *et passim*; call of conscience as, 279, 294, 300; forgetting as, 339. *See also* Deficient; Lack; Negating; Privation

Possibility (*Möglichkeit*): as an existential, it is the most original and positive ontological determination of Dasein, 143–144; higher than actuality, 38, 262, 299; of Dasein, 7, 12–13, 19–20, 42–43, 50, 62, 104, 125–126, 144, 148, 170, 173, 177–178, 181, 187–188, 191, 193–195, 199, 211, 236, 239–240, 244, 250, 260, 264, 266, 270, 273, 284, 288, 295, 325, 384, 394, 396; of impossibility of existence, 250, 262, 265–266, 306; impossible and possible, 342; death as, 248, 250–266, 302–303, 307, 309, 391; being toward, 262; extreme 122, 182; existentiell, 267–270 (§ 54), 336–337; factical, 264, 299, 383–384; ontic and ontological, 312; category of logical, 143; as project of understanding, 145–148, 151, 194, 260, 270, 274, 284–285, 295, 302, 306, 312, 324, 336, 339, 383, 387, 394, 397. *See also* Condition of possibility; Future; Meaning; Potentiality of being; Project

Possible, the (*das Mögliche*), 143, 261–262, 299; Dasein as being-, 42–43, 143–145, 188, 248–249, 259; historical repetition of the silent power of the, 385, 394–395. *See also* Future; Potentiality of being; Power

Potentiality of being (*Seinkönnen, lit.* can-be, potential-to-be), 86–87, 122, *143–148*, 153, 173, 186–188, 191–195, 221, 231–236, 250–255, 257, 262–270, 277–280, 287–289, 298–302, 305–313, 334, *336–339*, 341–344, 359, 363–365, 382–383, 414, *et passim*; authentic, 233, 235, 267–301 (II.II), 302, 313, 317, 322, 343; chosen, 288, 298, 394; existentiell, 260, 280, 313, 385; factical, 145, 187, 268, 280, 298, 306, 325, 341–342; ontic, 260; ownmost, 163, 181, 188, 191, 221, 228, 250–255, 259, 262–263, 265, 267, 273, 276–279, 287–288, 296,

299, 306–308, 317, 325, 336–337, 339, 348; projected potentiality-of-being, 336–337, 365; whole potentiality-of-being, 264, 266, 303, 317; Dasein as potentiality-of-being, 143, 145, 191, 231, 250, 252, 264–265, 277, 284, 287, 312–313, 337; Dasein as delivered over to its potentiality-of-being, 383; as something for the sake of which, 86, 191, 193–194, 334, 336, 359, 412, 414; of Others, 264; potentiality-for-Being-guilty, 289, 306–307; potentiality-of-being-in-the-world, 144, 179, 187, 191–192, 228, 252, 296, 412; in the 'truth', 363; potentiality-of-being-one's-Self, 175, 184, 267–269, 273–275, 294, 298, 307, 316, 322–323; Being-a-whole, 233–235, 237, 266, 301–333 (II.III), 345, 372. *See also* Future; Meaning; Possibility; Project

Power:

(*Kraft*), 91, 127 (of mystery), 220 (of the most elementary words); of the possible, possibilities of Dasein, 173, 394–395; light of clearedness not an innate ontic, 350–351; as ontic "force," 360, 361.

(*Macht*): of the they's chatter, 126, 174; of the call of conscience, 275, 278, 291, 296, 310, 403; of forgetting, 345; of time is the Concept (spirit), 435 (Hegel); gaining p. (*mächtig werden*) over ground, existence, fate, destiny, 284, 310, 384–385 (powerless superpower); *Mächtigkeit* of temporality, 331 (constitutive), 334 (as ontological origin), 344 (through anxiety), 369 (and space). *See also* Clearing; Condition of possibility; Origin; Possibility

Practical (*praktisch*), 57, 59, 69, *193*, 294, 300, 315–316, 319, 320 n. 19,

357–358, 364, 402 (Yorck). *See also* Action; Praxis; Theory

Pragmatic (*pragmatisch*), 68; *pragmata* ("things"), 68, 214. *See also* Circumspection

Praxis (*Praxis*), 68, 193, 357–358, 402 (Yorck). *See also* Theory

Pre-(*vor-*): pre-ontological, 12–13, 15–17, 44, 65, 68, 72, 86, 130, 182–184, 196–200 (§ 42), 197 n. 6, 201, 222, 225, 289, 312, 315, 356; pre-phenomenological, 51, 59, 63, 72, 99, 219, 318; pre-philosophical, 19, 165, 219; pre-predicative, 149, 359; pre-scientific, 9, 393

Predicate, predication (*Prädikat[ion]*), 94, 99, 154–155, 157, 215, 281, 318, 359. *See also* Communication; Pointing out; Statement

Predilection (*Hang*), 182, 194–196, 345. *See also* Care; Urge; Willing; Wishing

Prefiguration, prescription (*Vorzeichnung*), 39, 41, 45, 52, 101, 114, 127, 129, 232, 237, 247, 249–250, 252, 255, 275, 301, 313, 322, 334, 363, 393, 426. *See also* Horizon; Schema

Presence (*Anwesenheit*), 25–26, 71, 415–418, 423 (constant); *das Anwesende*, 326 ("what presences"), 417 ("what is present"); being present (*anwesend sein*), 346 ("bodily," *leibhaftig*), 359, 369, 389, 417 ("having presence"), 423 (constantly); in the later marginal remarks (=fn): 105fn (constant), 153fn, 320fn; also "presencing," 39fn (*Anwesen*), 235fn (*Anwesenheit*). *See also* Objective presence

Presencing: 39fn (*Anwesen*), 85fn (*wesen lassen*), 235fn (*Anwesenheit*); *es west*, 87fn, 165fn, 252fn. *See also* Essence as presencing

Present (*Gegenwart*), 25–26, 326, 328–239, 337–340, 342, 344–348, 350, 355, 360, 363, 365, 369, 378–381,

18, 333, 432 n. 30; Descartes on, 89, Hegel on, 429–430, 432 n. 30; Kant on space, 89

Spatiality (*Räumlichkeit*): of Dasein, 56, 89, 101–113 (§§ 22–24), 119–120, 132, 141, 299, 335, 367–369 (§ 70); of the "there," 132, 299; of being-in-the-world, 79, 101, 104–110, (§ 23), 141, 299; of being-in, 54, 105–106, 186; of the world, 101 + fn, 110–112, 369; of the 'world', 112; of the surrounding world, 60, 66, 89, 112; of beings encountered in the surrounding world, 101; of things at hand, 102–104 (§ 22), 110, 112, 418; of extended things, 112–113, 368; spatializing (*verräumlichen*), 108, 112, 418

Spirit, spiritual (*Geist, geistig*), 22, 26, 47 n. 2 (Husserl), 48, 56, 89, 117, 152, 198, 320 n. 19 (Hegel), 368, 379, 395, 397, 401, (Yorck), 404, 405–406 (Hegel), 427, 428–436 (§ 82)

Stand (*Standgewonnenhaben*), gained through constancy of resoluteness, 322. *See also* Constancy; Existence (*ek-sistence*); Position; Self-constancy

Standpoint (*Standpunkt*), 19, 21–22, 61, 208, 402; phenomenology versus, 27, 152

Statement, proposition (*Aussage, Satz*); as apophantical discourse, § 7B; derivative from the hermeneutical "as," §§ 13, 33, 69b; as traditional place of truth, § 44a, b; interim stages between hermeneutical and theoretical, 157fn, 158; transcendental [example], 31, § 82; ontological 76, 82; phenomenological [example], 120; existential-ontological [examples], 153, 207, 332; ontic-suprasensuous, 318fn. *See also* Apophantic "as"; Diairesis; Judgment; Predicate

Staying, lingering (*Verweilen*), 61, 120, 138, 172 (not-staying of curiosity), 222, 238, 356–347, 358. *See also* Dwelling

Steadfastness (*Standfestigkeit*), 322; *See also* Constancy; Persistence; Resoluteness; Situation; Stand

Steadiness (*Stätigkeit*), 390–391, 398. *See also* Constancy; Persistence; Resoluteness

Stream (*Strom, Fluss*): of experiences, 194, 344, 388; of "nows," 410, 422, 426; of time, 426, 432. *See also* Succession

Stretch (*Strecke*), 23, 106, 285, 374, 417–418. *See also* Connection, between birth and death; Span; Stretching along; Succession

Stretching along (*Erstrecktkheit, Erstreckung*), 371 (temporal), 373–375 (between birth and death), 390–391 (ecstatic), 409–410, 417–418, 423. *See also* Being and movement; Constant; Persistence; Steadiness

Structure, Structural (*Struktur, structural*): of the question of being, 5–8 (§ 2), 14; of existence, 12, 44, *et passim*; of being-in-the-world, 41, 53, 58, 64, 65, *et passim*; of experience, 46; of Dasein, 54, 56, *et passim*; as-, 149, 151, 154, 158, 359–360; of care, 196, 259, 317, 323, 328, 331–332, 346, 350; end-structure, 244, 246; fore-structure, 151–153; of temporalization-structure, 332; of world, 366, 414; of truth, 216, 223, 226; structural moments, 5, 24, 41, 53, 63, 101, 117, 130–131, 140, 162, 176, 181, 209, 230, 232; totality (*Strukturganzheit*), 131, 182, 13, 209, 234, 334; whole (*Strukturganze*), 65, 131, 180–184 (§ 39), 191–192, 231–233, 236, 252, 316–317, 323–325, 350. *See also* Constitution; Whole

Subject, subjectivity (*Subjekt,
Subjektivität*), 14, 22, 24, 46–47,
59–62, 109–111, 113–114, 123–126,
128–130, 154–156, 204, 227, 229–
230, 316–322, 366, 388, 419, 427,
et passim; subject-object relation,
59, 216, 388; versus Dasein, 60,
229; versus self, 303, 322; versus
objectivity, 395, 405, 411, 419;
isolated, 118, 179, 188, 204, 206,
321; worldless, 110–111, 116, 192,
206, 366; of others, 119, 121, 123,
126, 128, 384; objectively present,
119, 121, 123, 128, 131–132, 176,
320; 'idea', 229; absolute, 228, 318;
knowing, 47, 69; of everydayness,
114, 128; 'factual' and factical,
229; 'theoretical', 316; 'historical',
382; 'logical', 319 (Kant); versus
predicate, 154–155, 318–319; the *a
priori* and the, 229, (*Cf.* 110); truth
and the, 227, 229; time and, 62,
109–110, 164; subjectivity of the,
24; 'subjectivity' of the world, 65,
366; 'subjectivity' of time, 419,
427; 'subjectivity' of world-time,
419; Kant on the, 24, 109, 204,
319–321

Submission to (*Angewiesenheit auf*
= dependence/reliance on [the
world]), 87, 137, 139, 161, 297,
348, 383, 412. *See also* Abandon;
Surrender; Thrownness

Subsist, subsistence (*Bestehen,
Bestand*), 7, 153, 216, 284, 288,
303, 333, 348, 351, 420, 430; in a
marginal remark: the there is to
"perdure" (*bestehen*) being as
such, 42fn. *See also* Constancy;
Persistence; Self-subsistence;
Steadfastness; Substance

Substance, substantiality (*Substanz,
Substantialität*), 22, 46–47, 63, 68,
87–90, 92–96, 98, 100, 114, 117,
201, 212, 303, 314, 317–323, 320 n.
19, 398, 425; as basic attribute of
being, 201; ontology of the, 319,
320 n. 19; and subject, 2, 317, 321,
332; and self, 114, 303, 317, 320 n.

10, 323, 332; and the "I" 317–318,
320, 322; and person, 47, 320 n.
19; and spirit, 117; and Dasein's
subsistence, 303; soul-substance,
26, 114, 318; man's substance,
117b, 212, 314; and world, 90,
94, 96; corporeal substance, 90,
92; *and* the objectively present,
114, 318; and reality, 212; and
function, 88; Descartes on, 90,
92–96, 98, 100; Kant on, 318–323;
Scheler on, *etc.*, 47. *See also*
Reality; Subject

Succession (*Folge, Abfolge*): also
Nacheinander, 242 (debt paid in
sequence), 291, 350, 374, 422
(sequence of nows), 426, 430; s.
of experiences, 291 (in "serial
connection"), 293, 355, 373–374,
387–388, 390; of ecstasies, 350;
of days, 371; of processes, 379;
of resolutions, 387; of nows, 329,
373, 409, 422–426, 431–432. *See
also* Connection

Suicide (*Selbstmord*), 229

Sum (*Summe*), 125–127, 187, 210,
242–244, 244 n. 3, 370, 374. *See
also* Totality; Whole

Summons (*Anruf, Aufruf*, of
conscience), 269–275, 277–280,
287–290, 292, 294–297, 299–300,
307, 3313, 317. *See also* Call of
conscience

Sun (*Sonne*), 71, 103, 412–413, 415–
416, 432 n. 30 (Hegel). *See also*
Brightness; Light

Surrender (*sich ausliefern*) to, 128
(they), 139 ("world"), 144
(thrownness), 199 (world taken
care of), 299 (interpretedness of
they), 412 (changes of day and
night). *See also* Abandonment;
Delivered over; Thrownness

Surrounding world (*Umwelt*), 57–58,
58fn ("surroundings"), 65–66,
70–72, 75, 79–80, 82–83, 89, 101,
104–107, 112–133, 117, 126, 136,
158, 172, 209, 239, 300, 334, 342,
349, 352, 354, 356, 359, 361–362,

344, 348; ground, 284, 287, 306, 325; individualization, 280, 343; possibility, 144–145; potentiality of being, 188, 339; and facticity, 135, 179, 221, 276, 284, 328, 348, 410, 414, 436; and everydayness, 167; and disclosedness, 221, 276, 215; and falling, 179, 415, 424; and turbulence, 179, and abandonment, 347, 365, 406, 413; and mood, attunement, *etc.*, 135–136, 139, 144, 179, 181, 251, 270, 276, 328, 340, 365; and fear, 342; and anxiety, 187, 191, 251, 343–344; and predilection and urge, 196; and care, 383, 406, 412; and project, 145, 138, 195, 199, 223, 285, 335, 406; and conscience, 291; and being-towards-death, 344, 348, 374; and time-reckoning, 412–413; taking over one's, 383, 385; coming back behind one's thrownness, 284, 383; throwing against, (*entgegenwerfen*), 363. *See also* Abandonment; Deliver over; Facticity; Submission to; Surrender

Time (*Zeit*): as the horizon for the understanding or interpretation of being, 1, 17, 39, 41, 235, 437; the ordinary way to understanding or interpreting time, 18, 235, 304, 326, 328–329, 338 n. 3, 404–437 (II.VI; *esp.* § 81); traditional conception of, 18, 24, 235, 349, 428, 432 n. 30; everyday experience of, 333, 405, 420; primordial *and* derivative, 329–332, 405, 426, 436; *and* care, 235, 327, 424; *and* the they, 425; *and* idle talk, 174; *and* space, (*see entry under 'space'*) *and* history, 379, 404–405; *and* spirit, 406–406, 428–436 (§ 82), 427; allowing time, (*Zeit-lassen*), 404, 409–410, 414; assigning time, (*Zeit angeben, Zeit-angabe*), 408–410, 413, 418, 422; counting time, 421–422; dating time, 408–409, 412, 413,

415, 417, 422; expressing time, (*Zeit aussprechen*), 406–408, 410–411, 421–422; giving the time, (*Zeit geben*), 412–413, 420, 422, 432 n. 30; having time, (*Zeit haben*), 404, 409–410, 418–419, 422, 425; interpreting time, 407–414; leveling down time, 329, 405, 422, 424–426, 431–432, 432 n. 30, 435; losing time, 404, 410, 418, 425; measuring, 71, 413–419, 417 n. 4, 422; reading off (*Zeit ablesen*), tell time, 70, 415–417; taking time, taking one's time, (*sich Zeit nehmen*), 404–405, 410–413, 416, 418, 421, 424–425; using time (*Zeit brauchen*), 235, 333, 409; taking care of time, 353, 406–420 (§ 79, 80), 422; time-reckoning, 235, 333, 411–412, 414–418 n. 5; reckoning with time, 235, 333, 371, 404, 411–413, 422; taking time into one's reckoning, 371, 404–405, 411, 413; 'time goes on', 330, 425; 'time passes away', 330, 425–426; time *as* finite, endless, in-finite, 330–331, 424–426; *as* continuous, 423; *as* irreversible, 426; stream of time, 426, 432; course of time, 400, 422, 432 n. 30; point of time, (*Zeitpunkt*), 374, 407; 'past', 380; 'fugitive', 425; 'qualitative', 333; 'psychical', 349; 'subjective' *or* 'objective', 326, 405, 411, 419, 427; "immanent" *or* "transcendent," 326, 419; "time for . . . ," 412, 414; in time (*in der Zeit*), 18–19, 204, 330, 338, 338 n. 3, 340–341, 349, 355, 367, 373–374, 376, 379, 381–382, 404–405, 412, 419–420, 425, 428, 435; timeless, 18, 156, 382; now-time, (*Jetzt-Zeit*), 421, 423, 426; world-time (*Weltzeit*), 405–406, 414, 417, 417 n. 4, 417 n. 5, 419–423, 426–428, 428–436 (§ 82); Aristotle on, 26, 40, 42, 427, 427 n. 14, 432 n. 30; Augustine on, 427, 427 n. 15; Bergson on, 18, 26, 333, 432 n. 30; Hegel on,

Supplemental Lexicon of the Later Marginal Remarks (=fn)

Listed below are some of the German terms in the marginal remarks that could not easily be accommodated into the above Lexicon of the most important terms in the vocabulary of *Being and Time* (1927). These constitute terms proper to the later Heidegger, notably from the period of the *Beiträge zur Philosophie* (1936–38) and reflecting its major divisions in terms like resonance (*Anklang*), play (*Zuspiel*), and leap (*Sprung*). *See also* above, for some of the connective links between the two works, in terms like Belonging; Clearing; Constancy; Ek-sistence: Essence as presencing; Event; Presence; Leaping; Relation; Truth; Understanding of being. Along with the usage in the later marginal remarks, we include here, where relevant, instances of pages of minor usage of the same and related terms in *Being and Time* itself.

Umkehr, return into (reversal), 39fn, 426 (and irreversibility of time)

Wahrend, perduring, 7fn; *das "während,"* the "during," 409, 413. *See also* Preservation

das Zu-sagende, what calls for saying, 165fn; *das Zugesagte*, what is said to us, 227fn; to say directly (*auf den Kopf zusagen*), to one's face, to "categorize," 44. *See also* Call of conscience; Gives to understand
Zu-spiel: *See Bei-spiel*

Latin Expressions

a potiori fit demoninatio (The name originates from the more potent), 329
a priori (from the former), 4, 11, 41, 44–45, 50, n. 10, 53, 58, 65, 85, 85fn, 101, 110–111, 115, 141, 149–150, 152, 165, 183, 193, 199, 206, 229, 321, 362
adaequatio intellectus et rei (adequation of intellect and thing), 214–217
anima (soul), 14
animal rationale (rational animal), 48, 165

bonum (good), 286, 345

capax mutationum (capable of changes), 91, 96
circulus vitiosus (vicious circle), 152
cogitare, cogitationes (thinking, thoughts), 46, 49, 211, 433
cogito me cotigare rem (I think myself thinking the thing), 433
cogito sum (I think, I am), 24, 40, 46, 89, 211

colo (I cultivate, inhabit), 54
color (color), 91
commercium (commerce, intercourse), 62, 132, 176
communis opinio (common opinion), 403 (*Yorck*)
compositum (composite), 244 n. 3
concupiscentia (concupiscence), 171
conscientia (consciousness), 433
convenientia (agreement), 132, 214–215

contritio (contrition), 190 n. 4
correspondentia (correspondence), 214
cura (care), 183, 197–199

definitio (definition), 4
diligo (I take care of, esteem highly, cherish), 54
durities (hardness), 91

ego (I), 43, 46 211
ego cogito (I think), 22, 89
ens (being), 24; *ens creatum* (created being, 24, 92; *ens finitum* (finite being), 49, *ens increatum* (uncreated being), 24; *ens infinitum* (infinite being), 24; *ens perfectissimum* (most perfect being), 92; *ens quod natum est convenire cum omni enti* (being that is born to come together with all beings), 14; *ens realissimum* (most real being), 128; *modus specialis entis* (special mode of being), 14
essentia (essence), 42–43
existentia (existence), 42–43
existit, ad existendum (it exists, for existing), 95
exitus (finish, end), 241
extensio (extension, 89–91, 93–95, 97, 99–101

factum brutum (brute fact), 135
fundamentum inconcussum (unshakable foundation), 24
futurum (future): *bonum futurum* (good future), 345; *malum futurum* (bad future), 141, 341, 345;

Greek Expressions

ἀληθές, ἀλήθεια, ἀληθεύειν; *alēthes, alētheia, alētheuein* (true; truth; being-true), 33, 133fn, 212fn, 213, 219–220, 223 n. 39; ἀ–λήθεια (taken out of hiding), 222

ἅμα, *hama* (at the same time, at once), 423

ἄνθρωπος, *anthrōpos* (human being, man), 48, 171

ἀπό, *apo* (from itself), 32

ἀποφαίνεσθαι, ἀπόφανσις, *apophain-esthai, apophansis* (letting be seen from itself [showing]; what allows this display [speech]), 32–34, 154, 213, 218–219

ἀριθμὸς κινήσεως, *arithmos kinēseōs* (number [numeration, count] of motion), 421, 423, 427, 432 n. 30

ἀρχαί, *archai* (beginnings, origins), 212

γένος, *genos* (genus), 3

γιγαντομαχία περὶ τῆς οὐσίας, *gigantomachia peri tēs ousias* (battle of the giants over being), 2

δηλοῦν, *dēloun* (making manifest), 1, 32

διαγωλή, *diagōgē* (staying, tarrying, "whiling"), 138

διαίρεσις, *diairesis* (division, separation), 159

διανοεῖν, *dianoein* (understanding), 96, 147fn, 226

δόξα, *doxa* (opinion), 223 n. 39

εἰδέναι, *eidenai* (to know, to see), 171

εἶδος, *eidos* (outward appearance, look), 61, 61fn, 68fn, 319

εἰκών, εἰκόνα, *eikona* (image, semblance), 48, 423

εἶναι, *einai* (to be, be-ing), 85fn, 171, 212, 427

ἔκστασις, ἐκστατικόν, *ekstasis, ekstatikon* (ecstasy; ecstatic), 329

ἐπιλανθάνομαι, *epilanthanomai* (escape notice, be forgotten), 219

ἐπιστήμη, *epistēmē* (knowledge, science), 213

ἑρμηνεύειν, ἑρμηνεία, *hermēneuein, hermēneia* (understanding interpreting; interpretation), 37, 158

εὐχή, *euchēœ* (request, petition), 32

ζῷον λόγον ἔχον, *zōon logon echon* (living being having speech, rational animal), 25, 48, 165

θαυμάζειν, *thaumazein* (to wonder), 172

θεός, *theos* (God), 48

θεωρεῖν, *theōrein* (to behold, consider), 213

ἰδέα, *idea* (idea), 68fn, 201fn

ἴδια, *idia* (its own, unique), 33

καθ᾽ αὐτό, *kath' auto* (according to itself), 17fn

καθόλου, *katholou* (on the whole, in general), 3, 17fn, 37fn

κατηγορεῖσθαι, κατηγορία, *katēgoreisthai, katēgoria* (to accuse publicly, denounce, "categorize"; category), 44–45

κοινόν, *koinon* (common, universal), 38fn

κρίνειν λόγῳ, *krinein logō* (to distinguish intelligently, "understandingly"), 222–223

κρύπτεσθαι (φιλεῖ), *kryptesthai (philei)* ([it loves] to hide), 212fn

λανθάνω, *lanthanō* (I conceal), 219

λέγειν, *legein* (discoursing), 25–26, 34, 43

λόγος, *logos* (speech, reason, ground), 25, 28, 31–35, 37, 44–45, 59, 154, 158–160, 165, 219–220, 222, 225–226

λύπη, *lupē* (pain, depression), 342

μάθημα, *mathēma* (learning, the mathematical), 96fn

Proper Names